THE PRECIPICE

THE PRECIPICE

*Existential Risk and
the Future of Humanity*

TOBY ORD

BLOOMSBURY PUBLISHING
LONDON · OXFORD · NEW YORK · NEW DELHI · SYDNEY

BLOOMSBURY PUBLISHING
Bloomsbury Publishing Plc
50 Bedford Square, London, WC1B 3DP, UK

BLOOMSBURY, BLOOMSBURY PUBLISHING and the Diana logo are
trademarks of Bloomsbury Publishing Plc

First published in Great Britain 2020

A catalogue record for this book is available from the British Library

ISBN: HB: 978-1-5266-0021-9; TPB: 978-1-5266-0022-6; eBook: 978-1-5266-0019-6

2 4 6 8 10 9 7 5 3 1

Typeset by Newgen KnowledgeWorks Pvt. Ltd., Chennai, India
Printed and bound in Great Britain by CPI Group (UK) Ltd, Croydon CR0 4YY

To find out more about our authors and books visit www.bloomsbury.com
and sign up for our newsletters

To the hundred billion people before us,
who fashioned our civilisation;
To the seven billion now alive,
whose actions may determine its fate;
To the trillions to come,
whose existence lies in the balance.

CONTENTS

LIST OF FIGURES

LIST OF TABLES

PART ONE

THE STAKES

INTRODUCTION

If all goes well, human history is just beginning. Humanity is about two hundred thousand years old. But the Earth will remain habitable for hundreds of millions more—enough time for millions of future generations; enough to end disease, poverty and injustice forever; enough to create heights of flourishing unimaginable today. And if we could learn to reach out further into the cosmos, we could have more time yet: trillions of years, to explore billions of worlds. Such a lifespan places present-day humanity in its earliest infancy. A vast and extraordinary adulthood awaits.

Our view of this potential is easily obscured. The latest scandal draws our outrage; the latest tragedy, our sympathy. Time and space shrink. We forget the scale of the story in which we take part. But there are moments when we remember—when our vision shifts, and our priorities realign. We see a species precariously close to self-destruction, with a future of immense promise hanging in the balance. And which way that balance tips becomes our most urgent public concern.

This book argues that safeguarding humanity's future is the defining challenge of our time. For we stand at a crucial moment in the history of our species. Fuelled by technological progress, our power has grown so great that for the first time in humanity's long history, we have the capacity to destroy ourselves—severing our entire future and everything we could become.

Yet humanity's wisdom has grown only falteringly, if at all, and lags dangerously behind. Humanity lacks the maturity, coordination and foresight necessary to avoid making mistakes from which we could never recover. As the gap between our power and our wisdom grows, our future is subject to an ever-increasing level of risk. This situation is unsustainable. So over the next few centuries, humanity will be tested: it will either act decisively to protect itself and its longterm potential, or, in all likelihood, this will be lost forever.

To survive these challenges and secure our future, we must act now: managing the risks of today, averting those of tomorrow, and becoming the kind of society that will never pose such risks to itself again.

It is only in the last century that humanity's power to threaten its entire future became apparent. One of the most harrowing episodes has just recently come to light. On Saturday 27 October 1962 a single officer on a Soviet submarine almost started a nuclear war. His name was Valentin Savitsky. He was captain of the submarine B-59—one of four submarines the Soviet Union had sent to support its military operations in Cuba. Each was armed with a secret weapon: a nuclear torpedo with explosive power comparable to the Hiroshima bomb.

It was the height of the Cuban Missile Crisis. Two weeks earlier, US aerial reconnaissance had produced photographic evidence that the Soviet Union was installing nuclear missiles in Cuba, from which they could strike directly at the mainland United States. In response, the US blockaded the seas around Cuba, drew up plans for an invasion and brought its nuclear forces to the unprecedented alert level of DEFCON 2 ('Next step to nuclear war').

On that Saturday, one of the blockading US warships detected Savitsky's submarine and attempted to force it to the surface by dropping low-explosive depth charges as warning shots. The submarine had been hiding deep underwater for days. It was out of radio contact, so the crew did not know whether war had already broken out. Conditions on board were extremely bad. It was built for the Arctic and its ventilator had broken in the tropical water. The heat inside was unbearable, ranging from 45 °C near the torpedo tubes to 60 °C in the engine room. Carbon dioxide had built up to dangerous concentrations, and crew members had begun to fall unconscious. Depth charges were exploding right next to the hull. One of the crew later recalled: 'It felt like you were sitting in a metal barrel, which somebody is constantly blasting with a sledgehammer.'

Increasingly desperate, Captain Savitsky ordered his crew to prepare their secret weapon:

> Maybe the war has already started up there, while we are doing somersaults here. We're going to blast them now! We will die, but we will sink them all—we will not disgrace our Navy![1]

Firing the nuclear weapon required the agreement of the submarine's political officer, who held the other half of the firing key. Despite the lack of authorisation by Moscow, the political officer gave his consent.

On any of the other three submarines, this would have sufficed to launch their nuclear weapon. But by the purest luck, submarine B-59 carried the commander of the entire flotilla, Captain Vasili Arkhipov, and so required his additional consent. Arkhipov refused to grant it. Instead, he talked Captain Savitsky down from his rage and convinced him to give up: to surface amidst the US warships and await further orders from Moscow.[2]

We do not know precisely what would have happened if Arkhipov had granted his consent—or had he simply been stationed on any of the other three submarines. Perhaps Savitsky would not have followed through on his command. What is clear is that we came precariously close to a nuclear strike on the blockading fleet—a strike which would most likely have resulted in nuclear retaliation, then escalation to a full-scale nuclear war (the only kind the US had plans for). Years later, Robert McNamara, Secretary of Defense during the crisis, came to the same conclusion:

> No one should believe that had U.S. troops been attacked by nuclear warheads, the U.S. would have refrained from responding with nuclear warheads. Where would it have ended? In utter disaster.[3]

Ever since the advent of nuclear weapons, humans have been making choices with such stakes. Ours is a world of flawed decision-makers, working with strikingly incomplete information, directing technologies which threaten the entire future of the species. We were lucky, that Saturday in 1962, and have so far avoided catastrophe. But our destructive capabilities continue to grow, and we cannot rely on luck forever.

We need to take decisive steps to end this period of escalating risk and safeguard our future. Fortunately, it is in our power to do so. The greatest risks are caused by human action, and they can be addressed by human action. Whether humanity survives this era is thus a choice humanity will make. But it is not an easy one. It all depends on how quickly we can come to understand and accept the fresh responsibilities that come with our unprecedented power.

This is a book about *existential risks*—risks that threaten the destruction of humanity's longterm potential. Extinction is the most obvious way humanity's entire potential could be destroyed, but there are others. If civilisation across the globe were to suffer a truly unrecoverable collapse, that too would destroy our longterm potential. And we shall see that there are dystopian possibilities as well: ways we might get locked into a failed world with no way back.

While this set of risks is diverse, it is also exclusive. So I will have to set aside many important risks that fall short of this bar: our topic is not new dark ages for humanity or the natural world (terrible though they would be), but the permanent destruction of humanity's potential.

Existential risks present new kinds of challenges. They require us to coordinate globally and intergenerationally, in ways that go beyond what we have achieved so far. And they require foresight rather than trial and error. Since they allow no second chances, we need to build institutions to ensure that across our entire future we never once fall victim to such a catastrophe.

6

To do justice to this topic, we will have to cover a great deal of ground. Understanding the risks requires delving into physics, biology, earth science and computer science; situating this in the larger story of humanity requires history and anthropology; discerning just how much is at stake requires moral philosophy and economics; and finding solutions requires international relations and political science. Doing this properly requires deep engagement with each of these disciplines, not just cherry-picking expert quotes or studies that support one's preconceptions. This would be an impossible task for any individual, so I am extremely grateful for the extensive advice and scrutiny of dozens of the world's leading researchers from across these fields.[4]

This book is ambitious in its aims. Through careful analysis of the potential of humanity and the risks we face, it makes the case that we live during the most important era of human history. Major risks to our entire future are a new problem, and our thinking has not caught up. So *The Precipice* presents a new ethical perspective: a major reorientation in the way we see the world, and our role in it. In doing so, the book aspires to start closing the gap between our wisdom and power, allowing humanity a clear view of what is at stake, so that we will make the choices necessary to safeguard our future.

I have not always been focused on protecting our longterm future, coming to the topic only reluctantly. I am a philosopher, at Oxford University, specialising in ethics. My earlier work was rooted in the more tangible concerns of global health and global poverty—in how we could best help the worst off. When coming to grips with these issues I felt the need to take my work in ethics beyond the ivory tower. I began advising the World Health Organization, World Bank and UK government on the ethics of global health. And finding that my own money could do hundreds of times as much good for those in poverty as it could do for me, I made a lifelong pledge to donate at least a tenth of all I earn to help them.[5] I founded a society, *Giving What We Can*, for those who wanted to join me, and was heartened to see thousands of people come together to pledge more than £1 billion over our lifetimes to the most effective charities we know of, working on the most important causes. Together, we've already been able to transform the lives of tens of thousands of people.[6] And because there are many other ways beyond our donations in which we can help fashion a better world, I helped start a wider movement, known as *effective altruism*, in which people aspire to use evidence and reason to do as much good as possible.

Since there is so much work to be done to fix the needless suffering in our present, I was slow to turn to the future. It was so much less visceral; so much more abstract. Could it really be as urgent a problem as suffering now? As I reflected on the evidence and ideas that would culminate in this book, I came to realise that the risks to humanity's future are just as real and just as urgent—yet even more neglected. And that the people of the future may be even more powerless to protect themselves from the risks we impose than the dispossessed of our own time.

Addressing these risks has now become the central focus of my work: both researching the challenges we face, and advising groups such as the UK Prime Minister's Office, the World Economic Forum and DeepMind on how they can best address these challenges. Over time, I've seen a growing recognition of these risks, and of the need for concerted action.

To allow this book to reach a diverse readership, I've been ruthless in stripping out the jargon, needless technical detail and defensive qualifications typical of academic writing (my own included). Readers hungry for further technical detail or qualifications can delve into the many endnotes and appendices, written with them in mind.[7]

I have tried especially hard to examine the evidence and arguments carefully and even-handedly, making sure to present the key points even if they cut against my narrative. For it is of the utmost importance to get to the truth of these matters—humanity's attention is scarce and precious, and must not be wasted on flawed narratives or ideas.[8]

Each chapter of *The Precipice* illuminates the central questions from a different angle. Part One (The Stakes) starts with a bird's-eye view of our unique moment in history, then examines why it warrants such urgent moral concern. Part Two (The Risks) delves into the science of the risks facing humanity, both from nature and from ourselves, showing that while some have been overstated, there is real risk and it is growing. So Part Three (The Path Forward) develops tools for understanding how these risks compare and combine, and new strategies for addressing them. I close with a vision of our future: of what we could achieve were we to succeed.

This book is not just a familiar story of the perils of climate change or nuclear war. These risks that first awoke us to the possibilities of destroying ourselves are just the beginning. There are emerging risks, such as those arising from biotechnology and advanced artificial intelligence, that may pose much greater risk to humanity in the coming century.

Finally, this is not a pessimistic book. It does not present an inevitable arc of history culminating in our destruction. It is not a morality tale about our technological hubris and resulting fall. Far from it. The central claim is that there are real risks to our future, but that our choices can still make all the difference. I believe we are up to the task: that through our choices we can pull back from the precipice and, in time, create a future of astonishing value—with a richness of which we can barely dream, made possible by innovations we are yet to conceive. Indeed, my deep optimism about humanity's future is core to my motivation in writing this book. Our potential is vast. We have so much to protect.

1

Standing at the Precipice

It might be a familiar progression, transpiring on many worlds—a planet, newly formed, placidly revolves around its star; life slowly forms; a kaleidoscopic procession of creatures evolves; intelligence emerges which, at least up to a point, confers enormous survival value; and then technology is invented. It dawns on them that there are such things as laws of Nature, that these laws can be revealed by experiment, and that knowledge of these laws can be made both to save and to take lives, both on unprecedented scales. Science, they recognize, grants immense powers. In a flash, they create world-altering contrivances. Some planetary civilizations see their way through, place limits on what may and what must not be done, and safely pass through the time of perils. Others, not so lucky or so prudent, perish.

—Carl Sagan[1]

We live at a time uniquely important to humanity's future. To see why, we need to take a step back and view the human story as a whole: how we got to this point and where we might be going next.

Our main focus will be humanity's ever-increasing power—power to improve our condition and power to inflict harm. We shall see how the major transitions in human history have enhanced our power, and enabled us to make extraordinary progress. If we can avoid catastrophe we can cautiously expect this progress to continue: the future of a responsible humanity is extraordinarily bright. But this increasing power has also brought on a new transition, at least as significant as any in our past, the transition to our time of perils.

HOW WE GOT HERE

Very little of humanity's story has been told; because very little *can* be told. Our species, *Homo sapiens*, arose on the savannahs of Africa 200,000 years ago.[2] For an almost unimaginable time we have had great loves and friendships, suffered hardships and griefs, explored, created, and wondered about our place in the universe. Yet when we think of humanity's great achievements across time, we think almost exclusively of deeds recorded on clay, papyrus or paper—records that extend back only about 5,000 years. We rarely think of the first person to set foot in the strange new world of Australia some 70,000 years ago; of the first to name and study the plants and animals of each place we reached; of the stories, songs and poems of humanity in its youth.[3] But these accomplishments were real, and extraordinary.

We know that even before agriculture or civilisation, humanity was a fresh force in the world. Using the simple, yet revolutionary, technologies of seafaring, clothing and fire, we travelled further than any mammal before us. We adapted to a wider range of environments, and spread across the globe.[4]

What made humanity exceptional, even at this nascent stage? We were not the biggest, the strongest or the hardiest. What set us apart was not physical, but mental—our intelligence, creativity and language.[6]

Yet even with these unique mental abilities, a single human alone in the wilderness would be nothing exceptional. He or she might be able to survive—intelligence making up for physical prowess—but would hardly dominate. In ecological terms, it is not a *human* that is remarkable, but *humanity*.

Each human's ability to cooperate with the dozens of other people in their band was unique among large animals. It allowed us to form something greater than ourselves. As our language

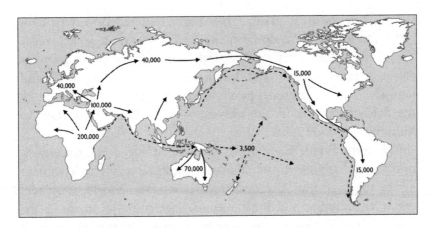

FIGURE 1.1 How we settled the world. The arrows show our current understanding of the land and sea routes taken by our ancestors, and how many years ago they reached each area.[5]

grew in expressiveness and abstraction, we were able to make the most of such groupings: pooling together our knowledge, our ideas and our plans.

Crucially, we were able to cooperate across *time* as well as space. If each generation had to learn everything anew, then even a crude iron shovel would have been forever beyond our technological reach. But we learned from our ancestors, added minor innovations of our own, and passed this all down to our children. Instead of dozens of humans in cooperation, we had tens of thousands, cooperating across the generations, preserving and improving ideas through deep time. Little by little, our knowledge and our culture grew.[7]

At several points in the long history of humanity there has been a great transition: a change in human affairs that accelerated our accumulation of power and shaped everything that would follow. I will focus on three.[8]

The first was the Agricultural Revolution.[9] Around 10,000 years ago the people of the Fertile Crescent, in the Middle East, began planting wild wheat, barley, lentils and peas to supplement their foraging. By preferentially replanting the seeds from the best plants, they harnessed the power of evolution, creating new domesticated varieties with larger seeds and better yields. This worked with animals too, giving humans easier access to meat and hides, along with milk, wool and manure. And the physical power of draft animals to help plough the fields or transport the harvest was the biggest addition to humanity's power since fire.[10]

While the Fertile Crescent is often called 'the cradle of civilisation', in truth civilisation had many cradles. Entirely independent agricultural revolutions occurred across the world in places where the climate and local species were suitable: in east Asia; sub-Saharan Africa; New Guinea; South, Central and North America; and perhaps elsewhere too.[11] The new practices fanned out from each of these cradles, changing the way of life for many from foraging to farming.

This had dramatic effects on the scale of human cooperation. Agriculture reduced the amount of land needed to support each person by a factor of a hundred, allowing large permanent settlements to develop, which began to unite together into states.[12] Where the largest foraging communities involved perhaps hundreds of people, some of the first cities had tens of thousands of inhabitants. At its height, the Sumerian civilisation contained around a million people.[13] And 2,000 years ago, the Han dynasty of China reached sixty million people—about a *hundred thousand* times as many as were ever united in our forager past, and about ten times the entire global forager population at its peak.[14]

As more and more people were able to share their insights and discoveries, there were rapid developments in technology, institutions and culture. And the increasing numbers of people trading with one another made it possible for them to specialise in these areas—to devote a lifetime to governance, trade or the arts—allowing us to develop these ideas much more deeply.

Over the first 6,000 years of agriculture, we achieved world-changing breakthroughs including writing, mathematics, law and the wheel.[15] Of these, writing was especially important

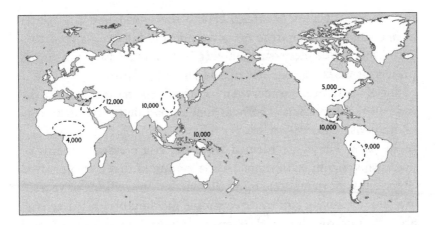

FIGURE 1.2 The cradles of civilisation. The places around the world where agriculture was independently developed, marked with how many years ago this occurred.

for strengthening our ability to cooperate across time and space: increasing the bandwidth between generations, the reliability of the information, and the distance over which ideas could be shared.

The next great transition was the Scientific Revolution.[16] Early forms of science had been practised since ancient times, and the seeds of empiricism can be found in the work of medieval scholars in the Islamic world and Europe.[17] But it was only about 400 years ago that humanity developed the scientific method and saw scientific progress take off.[18] This helped replace a reliance on received authorities with careful observation of the natural world, seeking simple and testable explanations for what we saw. The ability to test and discard bad explanations helped us break free from dogma, and allowed for the first time the systematic creation of knowledge about the workings of nature.

Some of our new-found knowledge could be harnessed to improve the world around us. So the accelerated accumulation of knowledge brought with it an acceleration of technological innovation, giving humanity increasing power over the natural world. The rapid pace allowed people to see transformative effects of these improvements within their own lifetimes. This gave rise to

15

the modern idea of *progress*. Where the world had previously been dominated by narratives of decline and fall or of a recurring cycle, there was increasing interest in a new narrative: a grand project of working together to build a better future.

Soon, humanity underwent a third great transition: the Industrial Revolution. This was made possible by the discovery of immense reserves of energy in the form of coal and other fossil fuels. These are formed from the compressed remains of organisms that lived in aeons past, allowing us access to a portion of the sunlight that shone upon the Earth over millions of years.[19] We had already begun to drive simple machines with the renewable energy from the wind, rivers and forests; fossil fuels allowed access to vastly more energy, and in a much more concentrated and convenient form.

But energy is nothing without a way of converting it to useful work, to achieve our desired changes in the world. The steam engine allowed the stored chemical energy of coal to be turned into mechanical energy.[20] This mechanical energy was then used to drive machines that performed massive amounts of labour for us, allowing raw materials to be transformed into finished products much more quickly and cheaply than before. And via the railroad, this wealth could be distributed and traded across long distances.

Productivity and prosperity began to accelerate, and a rapid sequence of innovations ramped up the efficiency, scale and variety of automation, giving rise to the modern era of sustained economic growth.[21]

The effects of these transitions have not always been positive. Life in the centuries following the Agricultural Revolution generally involved more work, reduced nutrition and increased disease.[22] Science gave us weapons of destruction that haunt us to this day. And the Industrial Revolution was among the most destabilising periods in human history. The unequal distribution of gains in prosperity and the exploitative labour practices led to the revolutionary upheavals of the early twentieth century.[23]

Inequality between countries increased dramatically (a trend that has only begun to reverse in the last two decades).[24] Harnessing the energy stored in fossil fuels has released greenhouse gases, while industry fuelled by this energy has endangered species, damaged ecosystems and polluted our environment.

Yet despite these real problems, on average human life today is substantially better than at any previous time. The most striking change may be in breaking free from poverty. Until 200 years ago—the last thousandth of our history[25]—increases in humanity's power and prosperity came hand in hand with increases in the human population. Income *per person* stayed almost unchanged: a little above subsistence in times of plenty; a little below in times of need.[26] The Industrial Revolution broke this rule, allowing income to grow faster than population and ushering in an unprecedented rise in prosperity that continues to this day.

We often think of economic growth from the perspective of a society that is already affluent, where it is not immediately clear if further growth even improves our lives. But the most remarkable effects of economic growth have been for the poorest people. In today's world, one out of ten people are so poor that they live on less than two dollars per day—a widely used threshold for 'extreme poverty'. That so many have so little is among the greatest problems of our time, and has been a major focus of my life. It is shocking then to look further back and see that prior to the Industrial Revolution 19 out of 20 people lived on less than two dollars a day (even adjusting for inflation and purchasing power). Until the Industrial Revolution, any prosperity was confined to a tiny elite with extreme poverty the norm. But over the last two centuries more and more people have broken free from extreme poverty, and are now doing so more quickly than at any earlier time.[27] Two dollars a day is far from prosperity, and these statistics can be of little comfort to those who are still in the grip of poverty, but the trends towards improvement are clear.

And it is not only in terms of material conditions that life has improved. Consider education and health. Universal schooling has produced dramatic improvements in education. Before the Industrial Revolution, just one in ten of the world's people could read and write; now more than eight in ten can do so.[28] For the 10,000 years since the Agricultural Revolution, life expectancy had hovered between 20 and 30 years. It has now more than doubled, to 72 years.[29] And like literacy, these gains have been felt across the world. In 1800 the highest life expectancy of any country was a mere 43 years, in Iceland. Now every single country has a life expectancy above 50.[30] The industrial period has seen all of humanity become more prosperous, educated and long-lived than ever before. But we should not succumb to complacency in the face of this astonishing progress. That we have achieved so much, and so quickly, should inspire us to address the suffering and injustices that remain.

We have also seen substantial improvements in our moral thinking.[32] One of the clearest trends is towards the gradual expansion of the moral community, with the recognition of the rights of women, children, the poor, foreigners and ethnic or religious minorities. We have also seen a marked shift away from violence as a morally acceptable part of society.[33] And in the last sixty years we have added the environment and the welfare of animals to our standard picture of morality. These social changes did not come naturally with prosperity. They were secured by reformers and activists, motivated by the belief that we can—and must—improve. We still have far to go before we are living up to these new ideals, and our progress can be painfully slow, but looking back even just one or two centuries shows how far we have come.

Of course, there have been many setbacks and exceptions. The path has been tumultuous, things have often become better in some ways while worse in others, and there is certainly a danger of choosing selectively from history to create a simple narrative of improvement from a barbarous past to a glorious present. Yet at the largest scales of human history, where we see not the rise and fall of each empire, but the changing face of human civilisation across the entire globe, the trends towards progress are clear.[34]

It can be hard to believe such trends, when it so often feels like everything is collapsing around us. In part this scepticism comes from our everyday experience of our own lives or communities over a timespan of years—a scale where downs are almost as likely as ups. It might also come from our tendency to focus more on bad news than good and on threats rather than opportunities: heuristics that are useful for directing our actions, but which misfire when attempting to objectively assess the balance of bad and good.[35] When we try to overcome these distortions, looking for global indicators of the quality of our lives that are as objective as possible, it is very difficult to avoid seeing significant improvement from century to century.

And these trends should not surprise us. Every day we are the beneficiaries of uncountable innovations made by people over hundreds of thousands of years. Innovations in technology, mathematics, language, institutions, culture, art; the ideas of the hundred billion people who came before us, and shaped almost every facet of the modern world.[36] This is a stunning inheritance. No wonder, then, that our lives are better for it.

We cannot be sure these trends towards progress will continue. But given their tenacity, the burden would appear to be on the pessimist to explain why *now* is the point it will fail. This is especially true when people have been predicting such failure for so long and with such a poor track record. Thomas Macaulay made this point well:

> We cannot absolutely prove that those are in error who tell us that society has reached a turning point, that we have seen our best days. But so said all before us, and with just as much apparent reason . . . On what principle is it that, when we see nothing but improvement behind us, we are to expect nothing but deterioration before us?[37]

And he wrote those words in 1830, before an additional 190 years of progress and failed predictions of the end of progress. During those years, lifespan doubled, literacy soared and eight in ten people escaped extreme poverty. What might the coming years bring?

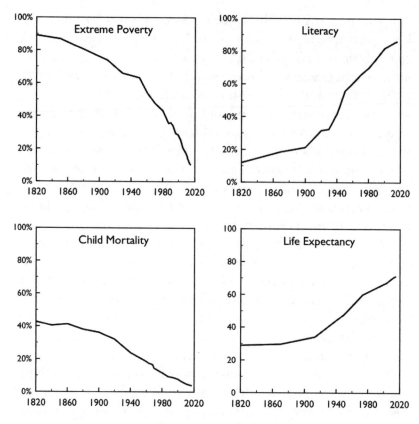

FIGURE 1.3 The striking improvements in extreme poverty, literacy, child mortality and life expectancy over the last 200 years.[31]

WHERE WE MIGHT GO

On the timescale of an individual human life, our 200,000-year history seems almost incomprehensibly long. But on a geological timescale it is short, and vanishingly so on the timescale of the universe as a whole. Our cosmos has a 14-billion-year history, and even that is short on the grandest scales. Trillions of years lie ahead of us. The future is immense.

How much of this future might we live to see? The fossil record provides some useful guidance. Mammalian species typically survive for around one million years before they go extinct; our close relative, *Homo erectus*, survived for almost two million.[38] If we think of one million years in terms of a single, eighty-year life, then today humanity would be in its adolescence—sixteen years old; just coming into our power; just old enough to get ourselves in serious trouble.[39]

Obviously, though, humanity is not a typical species. For one thing, we have recently acquired a unique power to destroy ourselves—power that will be the focus of much of this book. But we also have unique power to protect ourselves from external destruction, and thus the potential to outlive our related species.

How long *could* we survive on Earth? Our planet will remain habitable for roughly a billion years.[40] That's enough time for trillions of human lives; time to watch mountain ranges rise, continents collide, orbits realign; and time, as well, to heal our society and our planet of the wounds we have caused in our immaturity.

And we might have more time yet. As one of the pioneers of rocketry put it, 'Earth is the cradle of humanity, but one cannot live in a cradle forever.'[41] We do not know, yet, how to reach other stars and settle their planets, but we know of no fundamental obstacles. The main impediment appears to be the time necessary to learn how. This makes me optimistic. After all, the first heavier-than-air flight was in 1903 and just sixty-eight years later we had launched a spacecraft that left our Solar System and will reach the stars. Our species learns quickly, especially in recent times, and a billion years is a long education. I think we will need far less.

If we can reach other stars, then the whole galaxy opens up to us. The Milky Way alone contains more than 100 billion stars, and some of these will last for trillions of years, greatly extending our potential lifespan. Then there are billions of other galaxies beyond our own. If we reach a future of such a scale, we might have a truly staggering number of descendants, with the time, resources, wisdom and experience to create a diversity of wonders unimaginable to us today.

While humanity has made progress towards greater prosperity, health, education and moral inclusiveness, there is so much further we could go. Our present world remains marred by malaria and HIV; depression and dementia; racism and sexism; torture and oppression. But with enough time, we can end these horrors—building a society that is truly just and humane.

And a world without agony and injustice is just a lower bound on how good life could be. Neither the sciences nor the humanities have yet found any upper bound. We get some hint at what is possible during life's best moments: glimpses of raw joy, luminous beauty, soaring love. Moments when we are truly awake. These moments, however brief, point to possible heights of flourishing far beyond the status quo, and far beyond our current comprehension.

Our descendants could have aeons to explore these heights, with new means of exploration. And it's not just wellbeing. Whatever you value—beauty, understanding, culture, consciousness, freedom, adventure, discovery, art—our descendants would be able to take these so much further, perhaps even discovering entirely new categories of value, completely unknown to us. Music we lack the ears to hear.

THE PRECIPICE

But this future is at risk. For we have recently undergone another transition in our power to transform the world—one at least as significant as the Agricultural, Scientific and Industrial Revolutions that preceded it.

With the detonation of the first atomic bomb, a new age of humanity began.[42] At that moment, our rapidly accelerating technological power finally reached the threshold where we might be able to destroy ourselves. The first point where the threat to humanity from within exceeded the threats from the natural world. A point where the entire future of humanity hangs in the balance. Where every advance our ancestors have made could be squandered, and every advance our descendants may achieve could be denied. The greater part of the book of human history left unwritten; the narrative broken off; blank pages.

Nuclear weapons were a discontinuous change in human power. At Hiroshima, a single bomb did the damage of thousands. And six years later, a single thermonuclear bomb held more energy than every explosive used in the entire course of the Second World War.[43]

It became clear that a war with such weapons would change the Earth in ways that were unprecedented in human history. World leaders, atomic scientists and public intellectuals began to take seriously the possibility that a nuclear war would spell the end of humanity: either through extinction or a permanent collapse of civilisation.[44] Early concern centred on radioactive fallout and damage to the ozone layer, but in the 1980s the focus shifted to a scenario known as nuclear winter, in which nuclear firestorms loft smoke from burning cities into the upper atmosphere.[45] High above the clouds, the smoke cannot be rained out and would persist for years, blackening the sky, chilling the Earth and causing massive crop failure. This was a mechanism by which nuclear war could result in extreme famine, not just in the combatant countries, but in every country around the world. Millions of direct deaths from the explosions could be followed by billions of deaths from starvation, and—potentially—by the end of humanity itself.

How close have we come to such a war? With so much to lose, nuclear war is in no one's interest. So we might expect these obvious dangers to create a certain kind of safety—where world leaders inevitably back down before the brink. But as more and more behind-the-scenes evidence from the Cold War has become public, it has become increasingly clear that we have only barely avoided full-scale nuclear war.

We saw how the intervention of a single person, Captain Vasili Arkhipov, may have prevented an all-out nuclear war at the height of the Cuban Missile Crisis. But even more shocking is just how many times in those few days we came close to disaster, only to be pulled back by the decisions of a few individuals.

The principal events of the crisis took place over a single week. On Monday 22 October 1962, President John F. Kennedy gave a television address, informing his nation that the Soviets had begun installing strategic nuclear missiles in Cuba—directly threatening the United States. He warned that any use of these nuclear weapons would be met by a full-scale nuclear retaliation on the Soviet Union. His advisors drew up plans for both air strikes on the 48 missiles they had discovered and a full invasion of Cuba. US forces were brought to DEFCON 3, to prepare for a possible nuclear war.[46]

On Wednesday 24 October the US launched a naval blockade to prevent the delivery of further missiles to Cuba, and took its nuclear forces to the unprecedented level of DEFCON 2. Nuclear missiles were readied for launch and nuclear bombers took to the skies, ready to begin an all-out nuclear attack on the Soviet Union. The crisis reached its peak on Saturday when the Soviets shot down a U-2 reconnaissance plane with a surface-to-air missile, killing its pilot.

Then on Sunday morning it was all over. The Soviets backed down, unexpectedly announcing that they were removing all nuclear missiles from Cuba. But it could very easily have ended differently.

There has been substantial debate about exactly how close the crisis came to nuclear war. But over the decades, as more details have been revealed, the picture has become increasingly serious. Kennedy and Khrushchev went to great lengths to resist hawkish politicians and generals and to stay clear of the brink.[47] But there was a real possibility that, like the First World War, a war might begin without any side wanting it. As the week wore on, events on the ground spiralled beyond their control and they only barely kept the crisis from escalating. The US came extremely close to attacking Cuba, this had a much higher chance of causing nuclear retaliation than anyone guessed, and this in turn had a high chance of escalating to full-scale nuclear war.

Twice, during the crisis, the US nearly launched an attack on Cuba. At the height of the tensions, Kennedy had agreed that if a U-2 were shot down, the US would immediately strike Cuba, with no need to reconvene the war council. Then, on Saturday, a U-2 was indeed shot down. But Kennedy changed his mind and called off the counter-attack. Instead, he issued a secret ultimatum, informing the Soviets that if they did not commit to removing the missiles within twenty-four hours, or if another plane was shot down, the US would immediately launch air strikes and, almost surely, a full invasion.

This too almost triggered an attack. For the Americans did not know the extent to which Khrushchev was unable to control his forces in Cuba. Indeed, the U-2 had been shot down by a Soviet general acting against explicit orders from Khrushchev. And Khrushchev had even less control over the Cuban forces, who had already hit a low-flying reconnaissance plane with anti-aircraft fire and were eager to take one down. Knowing that he could not stop his own side from downing another plane, thereby triggering a US attack, Khrushchev raced to issue a statement ending the crisis before morning reconnaissance flights resumed.

What would have happened if the US *had* attacked? American leaders assumed that a purely conventional (non-nuclear) attack on Cuba could only be met with a purely conventional response. It was out of the question, they thought, that the Soviets would respond with nuclear attacks on the mainland United States. But they were missing another crucial fact. The missiles the US had discovered in Cuba were only a fraction of those the Soviets had delivered. There were 158 nuclear warheads. And more than 90 of these were tactical nuclear weapons, there for the express purpose of nuclear first use: to destroy a US invasion fleet before it could land.[48]

What's more, Castro was eager to use them. Indeed, he directly asked Khrushchev to fire the nuclear weapons if the Americans tried to invade, even though he knew this would lead to the annihilation of his own country: 'What would have happened to Cuba? It would have been totally destroyed.'[49] And

Khrushchev, in another unprecedented move, had relinquished central control of the tactical nuclear weapons, delegating the codes and decision to fire to the local Soviet commander. After hearing Kennedy's television address, Khrushchev issued new orders that the weapons were not to be used without his explicit permission, but he came to fear these would be disobeyed in the heat of conflict, as his order not to fire on US spy planes had been.

So unbeknownst to the US military leadership, a conventional attack on Cuba was likely to be met with a nuclear strike on American forces. And such a strike was extremely likely to be met by a further nuclear response from the US. This nuclear response was highly likely to go beyond Cuba, and to precipitate a full-scale nuclear war with the Soviets. In his television address on the Monday, Kennedy had explicitly promised that 'It shall be the policy of this Nation to regard any nuclear missile launched from Cuba against any nation in the Western Hemisphere as an attack by the Soviet Union on the United States, requiring a full retaliatory response upon the Soviet Union.'[50]

It is extremely difficult to estimate the chance that the crisis would have escalated to nuclear war.[51] Shortly after, Kennedy told a close advisor that he thought the probability of it ending in nuclear war with the USSR was 'somewhere between one out of three, and even'.[52] And it has just been revealed that the day after the crisis ended, Paul Nitze (an advisor to Kennedy's war council) estimated the chance at 10 percent, and thought that everyone else in the council would have put it even higher.[53] Moreover, none of these people knew about the tactical nuclear weapons in Cuba, Khrushchev's lack of control of his troops or the events on submarine B-59.

While I'm reluctant to question those whose very decisions could have started the war, my own view is that they were somewhat too pessimistic, given what they knew at the time. However, when we include the subsequent revelations about what was really happening in Cuba my estimates would roughly match theirs. I'd put the chance of the crisis escalating to a nuclear war with the Soviets at something between 10 and 50 percent.[54]

When writing about such close calls, there is a tendency to equate this chance to that of the end of civilisation or the end of humanity itself. But that would be a large and needless exaggeration. For we need to combine this chance of nuclear war with the chance that such a war would spell the end of humanity or human civilisation, which is far from certain. Yet even making such allowances the Cuban Missile Crisis would remain one of the pivotal moments in 200,000 years of human history: perhaps the closest we have ever come to losing it all.

Even now, with the Cold War just a memory, nuclear weapons still pose a threat to humanity. At the time of writing, the highest chance of a nuclear conflict probably involves North Korea. But not all nuclear wars are equal. North Korea has less than 1 percent as many warheads as Russia or the USA, and they are substantially smaller. A nuclear war with North Korea would be a terrible disaster, but it currently poses little threat to humanity's longterm potential.[55]

Instead, most of the existential risk from nuclear weapons today probably still comes from the enormous American and Russian arsenals. The development of ICBMs (intercontinental ballistic missiles) allowed each side to destroy most of the other's missiles with just thirty minutes' warning, so they each moved many missiles to 'hair-trigger alert'—ready to launch in just ten minutes.[56] Such hair-trigger missiles are extremely vulnerable to accidental launch, or to deliberate launch during a false alarm. As we shall see in Chapter 4, there has been a chilling catalogue of false alarms continuing past the end of the Cold War. On a longer timescale there is also the risk of other nations creating their own enormous stockpiles, of innovations in military technologies undermining the logic of deterrence, and of shifts in the geopolitical landscape igniting another arms race between great powers.

Nuclear weapons are not the only threat to humanity. They have been our focus so far because they were the first major risk and have already threatened humanity. But there are others too.

The exponential rise in prosperity brought on by the Industrial Revolution came on the back of a rapid rise in carbon emissions. A minor side effect of industrialisation has eventually grown to become a global threat to health, the environment, international stability, and maybe even humanity itself.

Nuclear weapons and climate change have striking similarities and contrasts. They both threaten humanity through major shifts in the Earth's temperature, but in opposite directions. One burst in upon the scene as the product of an unpredictable scientific breakthrough; the other is the continuation of centuries-long scaling-up of old technologies. One poses a small risk of sudden and precipitous catastrophe; the other is a gradual, continuous process, with a delayed onset—where some level of catastrophe is assured and the major uncertainty lies in just how bad it will be. One involves a classified military technology controlled by a handful of powerful actors; the other involves the aggregation of small effects from the choices of everyone in the world.

As technology continues to advance, new threats appear on the horizon. These threats promise to be more like nuclear weapons than like climate change: resulting from sudden breakthroughs, precipitous catastrophes, and the actions of a small number of actors. There are two emerging technologies that especially concern me; they will be the focus of Chapter 5.

Ever since the Agricultural Revolution, we have induced genetic changes in the plants and animals around us to suit our ends. But the discovery of the genetic code and the creation of tools to read and write it have led to an explosion in our ability to refashion life to new purposes. Biotechnology will bring major improvements in medicine, agriculture and industry. But it will also bring risks to civilisation and to humanity itself: both from accidents during legitimate research and from engineered bioweapons.

We are also seeing rapid progress in the capabilities of AI systems, with the biggest improvements in the areas where AI has traditionally been weakest, such as perception, learning and general intelligence. Experts find it likely that this will be the century that AI exceeds human ability not just in a narrow domain, but in general intelligence—the ability to overcome a diverse range of obstacles to achieve one's goals. Humanity has risen to a position where we control the rest of the world precisely because of our unparalleled mental abilities. If we pass this mantle to our machines, it will be they who are in this unique position. This should give us cause to wonder why it would be humanity who will continue to call the shots. We need to learn how to align the goals of increasingly intelligent and autonomous machines with human interests, and we need to do so before those machines become more powerful than we are.

These threats to humanity, and how we address them, define our time. The advent of nuclear weapons posed a real risk of human extinction in the twentieth century. With the continued acceleration of technology, and without serious efforts to protect humanity, there is strong reason to believe the risk will be higher this century, and increasing with each century that technological progress continues. Because these anthropogenic risks outstrip all natural risks combined, they set the clock on how long humanity has left to pull back from the brink.

I am not claiming that extinction is the inevitable conclusion of scientific progress, or even the most likely outcome. What I am claiming is that there has been a robust trend towards increases in the power of humanity which has reached a point where we pose a serious risk to our own existence. How we react to this risk is up to us.

Nor am I arguing against technology. Technology has proved itself immensely valuable in improving the human condition. And technology is essential for humanity to achieve its longterm potential. Without it, we would be doomed by the accumulated risk of natural disasters such as asteroid impacts. Without it, we would never achieve the highest flourishing of which we are capable.

29

The problem is not so much an excess of technology as a lack of wisdom.[57] Carl Sagan put this especially well:

> Many of the dangers we face indeed arise from science and technology—but, more fundamentally, because we have become powerful without becoming commensurately wise. The world-altering powers that technology has delivered into our hands now require a degree of consideration and foresight that has never before been asked of us.[58]

This idea has even been advocated by a sitting US president:

> the very spark that marks us as a species—our thoughts, our imagination, our language, our tool-making, our ability to set ourselves apart from nature and bend it to our will—those very things also give us the capacity for unmatched destruction . . . Technological progress without an equivalent progress in human institutions can doom us. The scientific revolution that led to the splitting of an atom requires a moral revolution as well.[59]

We need to gain this wisdom; to have this moral revolution. Because we cannot come back from extinction, we cannot wait until a threat strikes before acting—we must be proactive. And because gaining wisdom or starting a moral revolution takes time, we need to start now.

I think that we are likely to make it through this period. Not because the challenges are small, but because we will rise to them. The very fact that these risks stem from human action shows us that human action can address them.[60] Defeatism would be both unwarranted and counterproductive—a self-fulfilling prophecy. Instead, we must address these challenges head-on with clear and rigorous thinking, guided by a positive vision of the longterm future we are trying to protect.

How big are these risks? One cannot expect precise numbers, as the risks are *complex* (so not amenable to simple mathematical analysis) and *unprecedented* (so cannot be approximated by a longterm frequency). Yet it is important to at least try to give quantitative estimates. Qualitative statements such as 'a grave risk of human extinction' could be interpreted as meaning anything from 1 percent all the way to 99 percent.[61] They add more confusion than clarity. So I will offer quantitative estimates, with the proviso that they can't be precise and are open to revision.

During the twentieth century, my best guess is that we faced around a one in a hundred risk of human extinction or the unrecoverable collapse of civilisation. Given everything I know, I put the existential risk this century at around one in six: Russian roulette.[62] (See table 6.1 on p. 167 for a breakdown of the risks.) If we do not get our act together, if we continue to let our growth in power outstrip that of wisdom, we should expect this risk to be even higher next century, and each successive century.

These are the greatest risks we have faced.[63] If I'm even roughly right about their scale, then we cannot survive many centuries with risk like this. It is an *unsustainable* level of risk.[64] Thus, one way or another, this period is unlikely to last more than a small number of centuries.[65] Either humanity takes control of its destiny and reduces the risk to a sustainable level, or we destroy ourselves.

Consider human history as a grand journey through the wilderness. There are wrong turns and times of hardship, but also times of sudden progress and heady views. In the middle of the twentieth century we came through a high mountain pass and found that the only route onward was a narrow path along the cliff-side: a crumbling ledge on the brink of a precipice. Looking down brings a deep sense of vertigo. If we fall, everything is lost. We do not know just how likely we are to fall, but it is the greatest risk to which we have ever been exposed.

This comparatively brief period is a unique challenge in the history of our species. Our response to it will define our story. Historians of the future will name this time, and schoolchildren will study it. But I think we need a name now. I call it the Precipice.

The Precipice gives our time immense meaning. In the grand course of history—if we make it that far—*this* is what our time will be remembered for: for the highest levels of risk, and for humanity opening its eyes, coming into its maturity and guaranteeing its long and flourishing future. This is the meaning of our time.

I am not glorifying our generation, nor am I vilifying us. The point is that our actions have uniquely high stakes. Whether we are great or terrible will depend upon what we do with this opportunity. I hope we live to tell our children and grandchildren that we did not stand by, but used this chance to play the part that history gave us.

Safeguarding humanity through these dangers should be a central priority of our time. I am not saying that this is the only issue in the world, that people should drop everything else they hold dear to do this. But if you can see a way that you could play a role—if you have the skills, or if you are young and can shape your own path—then I think safeguarding humanity through these times is among the most noble purposes you could pursue.

THE PRECIPICE & ANTHROPOCENE

It has become increasingly clear that human activity is the dominant force shaping the environment. Scientists are concluding that humanity looms large not just in its own terms, but in the objective terms of biology, geology and climatology. If there are geologists in the distant future, they would identify the layer of rock corresponding to our time as a fundamental change from the layers before it.

Our present-day geologists are thus considering making this official—changing their classification of geological time to introduce a new epoch called the *Anthropocene*. Proposed beginnings for this epoch include the megafauna extinctions, the Agricultural Revolution, the crossing of the Atlantic, the Industrial Revolution, or early nuclear weapons tests.[66]

Is this the same as the Precipice? How do they differ?

- The Anthropocene is the time of profound human effects on the environment, while the Precipice is the time where humanity is at high risk of destroying itself.

- The Anthropocene is a geological epoch, which typically last millions of years, while the Precipice is a time in human history (akin to the Enlightenment or the Industrial Revolution), which will likely end within a few centuries.

- They might both officially start with the first atomic test, but this would be for very different reasons. For the Anthropocene, it would be mainly for convenient dating, while for the Precipice it is because of the risk nuclear weapons pose to our survival.

2

EXISTENTIAL RISK

The crucial role we fill, as moral beings, is as members of a cross-generational community, a community of beings who look before and after, who interpret the past in light of the present, who see the future as growing out of the past, who see themselves as members of enduring families, nations, cultures, traditions.

—Annette Baier[1]

We have seen how the long arc of human history has brought us to a uniquely important time in our story—a period where our entire future hangs in the balance. And we have seen a little of what might lie beyond, if only we can overcome the risks.

It is now time to think more deeply about what is at stake; why safeguarding humanity through this time is so important. To do so we first need to clarify the idea of existential risk. What exactly *is* existential risk? How does it relate to more familiar ideas of extinction or the collapse of civilisation?

We can then ask what it is about these risks that compels our urgent concern. The chief reason, in my view, is that we would lose our entire future: everything humanity could be and everything we could achieve. But that is not all. The case that it is crucial to safeguard our future draws support from a wide range of moral traditions and foundations. Existential risks also threaten to destroy our present, and to betray our past. They test civilisation's virtues, and threaten to remove what may be the most complex and significant part of the universe.

If we take any of these reasons seriously, we have a lot of work to do to protect our future. For existential risk is greatly neglected: by government, by academia, by civil society. We will see why this has been the case, and why there is good reason to suspect this will change.

UNDERSTANDING EXISTENTIAL RISK

Humanity's future is ripe with possibility. We have achieved a rich understanding of the world we inhabit and a level of health and prosperity of which our ancestors could only dream. We have begun to explore the other worlds in the heavens above us, and to create virtual worlds completely beyond our ancestors' comprehension. We know of almost no limits to what we might ultimately achieve.

Human extinction would foreclose our future. It would destroy our potential. It would eliminate all possibilities but one: a world bereft of human flourishing. Extinction would bring about this failed world and lock it in forever—there would be no coming back.

The philosopher Nick Bostrom showed that extinction is not the only way this could happen: there are other catastrophic outcomes in which we lose not just the present, but all our potential for the future.[2]

Consider a world in ruins: an immense catastrophe has triggered a global collapse of civilisation, reducing humanity to a pre-agricultural state. During this catastrophe, the Earth's environment was damaged so severely that it has become impossible for the survivors to ever re-establish civilisation. Even if such a catastrophe did not cause our extinction, it would have a similar effect on our future. The vast realm of futures currently open to us would have collapsed to a narrow range of meagre options. We would have a failed world with no way back.

Or consider a world in chains: in a future reminiscent of George Orwell's *Nineteen Eighty-Four*, the entire world has become locked under the rule of an oppressive totalitarian regime,

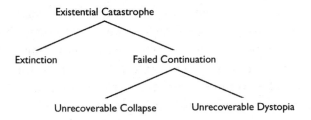

Existential Catastrophe

Extinction   Failed Continuation

Unrecoverable Collapse   Unrecoverable Dystopia

FIGURE 2.1 A classification of existential catastrophes by the kind of outcome that gets locked in.

determined to perpetuate itself. Through powerful, technologically enabled indoctrination, surveillance and enforcement, it has become impossible for even a handful of dissidents to find each other, let alone stage an uprising. With everyone on Earth living under such rule, the regime is stable from threats, internal and external. If such a regime could be maintained indefinitely, then descent into this totalitarian future would also have much in common with extinction: just a narrow range of terrible futures remaining, and no way out.

Following Bostrom, I shall call these 'existential catastrophes', defining them as follows:[3]

> An *existential catastrophe* is the destruction of humanity's longterm potential.

> An *existential risk* is a risk that threatens the destruction of humanity's longterm potential.

These definitions capture the idea that the outcome of an existential catastrophe is both dismal and irrevocable. We will not just fail to fulfil our potential, but this very potential itself will be permanently lost. While I want to keep the official definitions succinct, there are several areas that warrant clarification.

First, I am understanding *humanity's longterm potential* in terms of the set of all possible futures that remain open to us.[4] This is an expansive idea of possibility, including everything that humanity could eventually achieve, even if we have yet to invent the means of achieving it.[5] But it follows that while our choices can lock things in, closing off possibilities, they can't open up new ones. So any reduction in humanity's potential should be understood as permanent. The challenge of our time is to *preserve* our vast potential, and to *protect* it against the risk of future destruction. The ultimate purpose is to allow our descendants to *fulfil* our potential, realising one of the best possible futures open to us.

While it may seem abstract at this scale, this is really a familiar idea that we encounter every day. Consider a child with high longterm potential: with futures open to her in which she leads a great life. It is important that her potential is preserved: that her best futures aren't cut off due to accident, trauma or lack of education. It is important that her potential is protected: that we build in safeguards to make such a loss of potential extremely unlikely. And it is important that she ultimately fulfils her potential: that she ends up taking one of the best paths open to her. So too for humanity.

Existential risks threaten the destruction of humanity's potential. This includes cases where this destruction is complete (such as extinction) and where it is nearly complete, such as a permanent collapse of civilisation in which the possibility for some very minor types of flourishing remain, or where there remains some remote chance of recovery.[6] I leave the thresholds vague, but it should be understood that in any existential catastrophe the greater part of our potential is gone and very little remains.[7]

Second, my focus on humanity in the definitions is not supposed to exclude considerations of the value of the environment, other animals, successors to *Homo sapiens*, or creatures elsewhere in the cosmos. It is not that I think only humans count. Instead, it is that humans are the only beings we know of that are responsive to moral reasons and moral argument—the beings who can examine the world and decide to do what is best. If we fail, that upwards force, that capacity to push towards what is best or what is just, will vanish from the world.

Our potential is a matter of what humanity can achieve through the combined actions of each and every human. The value of our actions will stem in part from what we do to and for humans, but it will depend on the effects of our actions on non-humans too. If we somehow give rise to new kinds of moral agents in the future, the term 'humanity' in my definition should be taken to include them.

My focus on humanity prevents threats to a single country or culture from counting as existential risks. There is a similar term that gets used this way—when people say that something is 'an existential threat to this country'. Setting aside the fact that these claims are usually hyperbole, they are expressing a similar idea: that something threatens to permanently destroy the longterm potential of a country or culture.[8] However, it is very important to keep talk of an 'existential risk' (without any explicit restriction to a group) to apply only to threats against the whole of humanity.

Third, any notion of risk must involve some kind of probability. What kind is involved in existential risk? Understanding the probability in terms of objective long-run frequencies won't work, as the existential catastrophes we are concerned with can only ever happen once, and will always be unprecedented until the moment it is too late. We can't say the probability of an existential catastrophe is precisely zero just because it hasn't happened *yet*.

Situations like these require an evidential sense of probability, which describes the appropriate degree of belief we should have on the basis of the available information. This is the familiar type of probability used in courtrooms, banks and betting shops. When I speak of the probability of an existential catastrophe, I will mean the credence humanity should have that it will occur, in light of our best evidence.[9]

There are many utterly terrible outcomes that do not count as existential catastrophes.

One way this could happen is if there were no single precipitous event, but a multitude of smaller failures. This is because I take on the usual sense of catastrophe as a single, decisive event, rather than any combination of events that is bad in sum. If we were to squander our future simply by continually treating each other badly, or by never getting around to doing anything great, this could be just as bad an outcome but wouldn't have come about via a catastrophe.

Alternatively, there might be a single catastrophe, but one that leaves open some way for humanity to eventually recover. From our own vantage, looking out to the next few generations, this may appear equally bleak. But a thousand years hence it may be considered just one of several dark episodes in the human story. A true existential catastrophe must by its very nature be the decisive moment of human history—the point where we failed.

Even catastrophes large enough to bring about the global collapse of civilisation may fall short of being existential catastrophes. While colloquially referred to as 'the end of the world', a global collapse of civilisation need not be the end of the human story. It has the required severity, but may not be permanent or irrevocable.

In this book, I shall use the term *civilisation collapse* quite literally, to refer to an outcome where humanity across the globe loses civilisation (at least temporarily), being reduced to a pre-agricultural way of life. The term is often used loosely to refer merely to a massive breakdown of order, the loss of modern technology, or an end to our culture. But I am talking about a world without writing, cities, law, or any of the other trappings of civilisation.

This would be a very severe disaster and extremely hard to trigger. For all the historical pressures on civilisations, never once has this happened—not even on the scale of a continent.[10] The fact that Europe survived losing 25 to 50 percent of its population in the Black Death, while keeping civilisation firmly intact, suggests that triggering the collapse of civilisation would require more than 50% fatality in every region of the world.[11]

Even if civilisation did collapse, it is likely that it could be re-established. As we have seen, civilisation has already been independently established at least seven times by isolated peoples.[12] While one might think resource depletion could make this harder, it is more likely that it has become substantially easier. Most disasters short of human extinction would leave our domesticated animals and plants, as well as copious material resources in the ruins of our cities—it is much easier to re-forge iron from old railings than to smelt it from ore. Even expendable resources such as coal would be much easier to access, via abandoned reserves and mines, than they ever were in the eighteenth century.[13] Moreover, evidence that civilisation is possible, and the tools and knowledge to help rebuild, would be scattered across the world.

There are, however, two close connections between the collapse of civilisation and existential risk. First, a collapse would count as an existential catastrophe if it were unrecoverable. For example, it is conceivable that some form of extreme climate change or engineered plague might make the planet so inhospitable that humanity would be irrevocably reduced to scattered foragers.[14] And second, a global collapse of civilisation could increase the chance of extinction, by leaving us more vulnerable to subsequent catastrophe.

One way a collapse could lead to extinction is if the population of the largest remaining group fell below the *minimum viable population*—the level needed for a population to survive. There is no precise figure for this, as it is usually defined probabilistically and depends on many details of the situation: where the population is, what technology they have access to, the sort of catastrophe they have suffered. Estimates range from hundreds of people up to tens of thousands.[15] If a catastrophe directly reduces human population to below these levels, it will be more useful to classify it as a direct extinction event, rather than an unrecoverable collapse. And I expect that this will be one of the more common pathways to extinction.

We rarely think seriously about risks to humanity's entire potential. We encounter them mostly in action films, where our emotional reactions are dulled by their overuse as an easy way to heighten the drama.[16] Or we see them in online lists of 'ten ways the world could end', aimed primarily to thrill and entertain. Since the end of the Cold War, we rarely encounter sober discussions by our leading thinkers on what extinction would mean for us, our cultures or humanity.[17] And so in casual contexts people are sometimes flippant about the prospect of human extinction.

But when a risk is made vivid and credible—when it is clear that billions of lives and all future generations are actually on the line— the importance of protecting humanity's longterm potential is not, for most people, controversial. If we learned that a large asteroid was heading towards Earth, posing a greater than 10 percent chance of human extinction later this century, there would be little debate about whether to make serious efforts to build a deflection system, or to ignore the issue and run the risk. To the contrary, responding to the threat would immediately become one of the world's top priorities. Thus our lack of concern about these threats is much more to do with not yet believing that there are such threats, than it is about seriously doubting the immensity of the stakes.

Yet it is important to spend a little while trying to understand more clearly the different sources of this importance. Such an understanding can buttress feeling and inspire action; it can bring to light new considerations; and it can aid in decisions about how to set our priorities.

LOOKING TO THE PRESENT

Not all existential catastrophes involve human extinction, and not all methods of extinction involve pain or untimely death. For example, it is theoretically possible that we could all simply decide not to reproduce. This could destroy our potential without, let us suppose, causing any suffering. But the existential risks we actually face are not so peaceful. Rather, they are obviously horrific by the most familiar moral standards.

If, over the coming century, humanity is destroyed in a nuclear winter, or an engineered pandemic, or a catastrophic war involving some new technology, then seven billion lives would be cut short—including, perhaps, your own life, or the lives of those you love. Many would likely die in agony—starving, or burning, or ravaged by disease.

The moral case for preventing such horror needs little elaboration. Humanity has seen catastrophes before, on smaller scales: thousands, or millions, of human lives destroyed. We know how tremendously important it is to prevent such disasters. At such a scale, we lose our ability to fully comprehend the magnitude of what is lost, but even then the numbers provide a guide to the moral stakes.[18] Other things being equal, millions of deaths must be much worse than thousands of deaths; and billions, much worse than millions. Even measured just in terms of lives cut short, human extinction would easily be the worst event in our long history.

LOOKING TO OUR FUTURE

But an existential catastrophe is not just a catastrophe that destroys a particularly large number of lives. It destroys our potential.

My mentor, Derek Parfit, asked us to imagine a devastating nuclear war killing 99 percent of the world's people.[19] A war that would leave behind a dark age lasting centuries, before the survivors could eventually rebuild civilisation to its former heights; humbled, scarred—but undefeated.

Now compare this with a war killing a full 100 percent of the world's people. This second war would be worse, of course, but how much worse? Either war would be the worst catastrophe in history. Either would kill billions. The second war would involve tens of millions of additional deaths, and so would be worse for this reason. But there is another, far more significant difference between the two wars. Both wars kill billions of humans; but the second war kills humanity. Both wars destroy our present; but the second war destroys our future.

It is this qualitative difference in what is lost with that last percent that makes existential catastrophes unique, and that makes reducing the risk of existential catastrophe uniquely important.[20]

In expectation, almost all humans who will ever live have yet to be born. Absent catastrophe, most generations are future generations. As the writer Jonathan Schell put it:

> The procession of generations that extends onwards from our present leads far, far beyond the line of our sight, and, compared with these stretches of human time, which exceed the whole history of the earth up to now, our brief civilized moment is almost infinitesimal. Yet we threaten, in the name of our transient aims and fallible convictions, to foreclose it all. If our species does destroy itself, it will be a death in the cradle—a case of infant mortality.[21]

And because, in expectation, almost all of humanity's life lies in the future, almost everything of value lies in the future as well: almost all the flourishing; almost all the beauty; our greatest achievements; our most just societies; our most profound discoveries.[22] We can continue our progress on prosperity, health, justice, freedom and moral thought. We can create a world of wellbeing and flourishing that challenges our capacity to imagine. And if we protect that world from catastrophe, it could last millions of centuries. This is our potential—what we could achieve if we pass the Precipice and continue striving for a better world.

It is this view of the future—the immense value of humanity's potential—that most persuades me to focus my energies on reducing existential risk. When I think of the millions of future generations yet to come, the importance of protecting humanity's future is clear to me. To risk destroying this future, for the sake of some advantage limited only to the present, seems to me profoundly parochial and dangerously short-sighted. Such neglect privileges a tiny sliver of our story over the grand sweep of the whole; it privileges a tiny minority of humans over the overwhelming majority yet to be born; it privileges this particular century over the millions, or maybe billions, yet to come.[23]

To see why this would be wrong, consider an analogy with distance. A person does not matter less, the further away from you they are in space. It matters just as much if my wife gets sick while she is away at a conference in Kenya as if she gets sick while home with me in Oxford. And the welfare of strangers in Kenya matters just as much as the welfare of strangers in Oxford. Of course, we may have special duties to some individuals—to family; to members of the same community—but it is never spatial distance, in itself, that determines these differences in our obligations. Recognising that people matter equally, regardless of their geographic location, is a crucial form of moral progress, and one that we could do much more to integrate into our policies and our philanthropy.

People matter equally regardless of their temporal location, too. Our lives matter just as much as those lived thousands of years ago, or those a thousand years hence.[24] Just as it would be wrong to think that other people matter less the further they are from you in space, so it is to think they matter less the further away from you they are in time. The value of their happiness, and the horror of their suffering, is undiminished.

Recognising that people matter equally, wherever they are in time, is a crucial next step in the ongoing story of humanity's moral progress. Many of us recognise this equality to some extent already. We know it is wrong to make future generations worse off in order to secure lesser benefits for ourselves. And if asked, we would agree that people now don't objectively matter more than people in the future. But we assume that this leaves most of our priorities unaltered. For example, thinking that long-run effects of our choices quickly disappear; that they are so uncertain that the good cancels the bad; or that people in the future will be much better situated to help themselves.[25]

But the possibility of preventable existential risks in our lifetimes shows that there are issues where our actions can have sustained positive effects over the whole longterm future, and where we are the only generation in a position to produce those effects.[26] So the view that people in the future matter just as much as us has deep practical implications. We have a long way to go if we are to understand these and integrate them fully into our moral thinking.

Considerations like these suggest an ethic we might call *longtermism*, which is especially concerned with the impacts of our actions upon the longterm future.[27] It takes seriously the fact that our own generation is but one page in a much longer story, and that our most important role may be how we shape—or fail to shape—that story. Working to safeguard humanity's potential is one avenue for such a lasting impact and there may be others too.[28]

One doesn't have to approach existential risk from this direction—there is already a strong moral case just from the immediate effects—but a longtermist ethic is nevertheless especially well suited to grappling with existential risk. For longtermism is animated by a moral re-orientation towards the vast future that existential risks threaten to foreclose.

Of course, there are complexities.

When economists evaluate future benefits, they use a method called discounting, which dampens ('discounts') benefits based on how far away they are in time. If one took a commonly used discount rate of 5 percent per year and applied it to our future, there would be strikingly little value left. Applied naïvely, this discount rate would suggest our entire future is worth only about twenty times as much as the coming year, and that the period from 2100 to eternity is worth less than the coming year. Does this call into question the idea that our future is extremely valuable?

No. Results like this arise only from an incorrect application of the economic methods. When the subtleties of the problem are taken into account and discounting is correctly applied, the future is accorded an extremely high value. The mathematical details would take us too far afield, but for now it suffices to note that discounting human wellbeing (as opposed to instrumental goods such as money), purely on the basis of distance away from us in time, is deeply implausible—especially over the long time periods we are discussing. It implies, for example, that if you can save one person from a headache in a million years' time, or a billion people from torture in two million years, you should save the one from a headache.[29] A full explanation of why economic discounting does not trivialise the value of the longterm future can be found in Appendix A.

Some philosophers question the value of protecting our longterm future for quite a different reason. They note that the timing of the benefits is not the only unusual feature of this case. If we save humanity from extinction, that will change the number of people who will ever live. This brings up ethical issues that don't arise when simply saving the lives of existing people. Some of the more extreme approaches to this relatively new field of 'population ethics' imply that there is no reason to avoid extinction stemming from considerations of future generations—it just doesn't matter whether these future people come into being or not.

A full treatment of these matters would take too long and be of interest only to a few, so I reserve the detailed discussion for Appendix B. To briefly summarise: I do not find these views very plausible, either. They struggle to capture our reasons to care about whether we make future lives worse by polluting the planet, or changing the climate, and to explain why we have strong reasons to prevent terrible lives from existing in the future. And all but the most implausible of these views agree with the immense importance of saving future generations from other kinds of existential catastrophe, such as the irrevocable collapse of civilisation. Since most things that threaten extinction threaten such a collapse too, there is not much practical difference. That said, the issues are complex, and I encourage interested readers to consult the appendix for details.

There is one other objection I want to touch on. When I was younger, I sometimes took comfort in the idea that perhaps the outright destruction of humanity would not be bad at all. There would be no people to suffer or grieve. There would be no *badness* at those future times, so how could the destruction be bad? And if the existence of humanity was somehow essential to judgements of right and wrong, good and bad, then perhaps such concepts would fail to apply at all in the stillness that followed.

47

But I now see that this is no better than the old argument by the philosopher Epicurus that your death cannot be bad for you, since you are not there to experience it. What this neglects is that if I step out into the traffic and die, my life as a whole will be shorter and thereby worse: not by having more that is bad, but by containing less of everything that makes life good. That is why I shouldn't do it. While Epicurus's argument may provide consolation in times of grief or fear, it is not fit to be a guide for action, and no one treats it so. Imagine a government using it as the basis for our policies on safety or healthcare—or for our laws on murder.

If a catastrophe this century were to cause our extinction, then *humanity's* life would be shorter and thereby worse.[30] Given that we may just be in our infancy, it would be much shorter; much worse. Even if there were no one remaining to judge this as a tragedy, we can rightly judge it so from here. Just as we can judge events in other places, so we can judge events in other times.[31] And if these judgements are correct now, they shall remain correct when we are no more. I wouldn't blame people who, in humanity's final hours, found consolation in such Epicurean arguments. But the length and quality of humanity's life is still ours to decide, and we must own this responsibility.[32]

These are not the only possible objections. Yet we need not resolve every philosophical issue about the value of the future in order to decide whether humanity's potential is worth protecting. For the idea that it would be a matter of relative indifference whether humanity goes extinct, or whether we flourish for billions of years, is, on its face, profoundly implausible. In this sense, any theory that denies it should be subject to significant scepticism.[33]

What's more, the future is not the only moral lens through which to view existential catastrophe. It is the one that grips me most, and that most persuades me to devote my time and energy to this issue, but there are other lenses, drawing on other moral traditions. So let us briefly explore how concern about existential risk could also spring from considerations of our past, our character and our cosmic significance. And thus how people with many different understandings of morality could all end up at this common conclusion.

LOOKING TO OUR PAST

We are not the first generation. Our cultures, institutions and norms; our knowledge, technology and prosperity; these were gradually built up by our ancestors, over the course of ten thousand generations. In the last chapter we saw how humanity's remarkable success has relied on our capacity for intergenerational cooperation: inheriting from our parents, making some small improvements of our own, and passing it all down to our children. Without this cooperation we would have no houses or farms, we would have no traditions of dance or song, no writing, no nations.[34]

This idea was beautifully expressed by the conservative political theorist Edmund Burke. In 1790 he wrote of society:

> It is a partnership in all science; a partnership in all art; a partnership in every virtue, and in all perfection. As the ends of such a partnership cannot be obtained except in many generations, it becomes a partnership not only between those who are living, but between those who are living, those who are dead, and those who are to be born.[35]

This might give us reasons to safeguard humanity that are grounded in our past—obligations to our grandparents, as well as our grandchildren.

Our ancestors set in motion great projects for humanity that are too big for any single generation to achieve. Projects such as bringing an end to war, forging a just world and understanding our universe. In the year 65 CE, Seneca the Younger explicitly set out such a vast intergenerational project:

> The time will come when diligent research over long periods will bring to light things which now lie hidden. A single lifetime,

even though entirely devoted to the sky, would not be enough for the investigation of so vast a subject . . . And so this knowledge will be unfolded only through long successive ages. There will come a time when our descendants will be amazed that we did not know things that are so plain to them . . . Let us be satisfied with what we have found out, and let our descendants also contribute something to the truth . . . Many discoveries are reserved for ages still to come, when memory of us will have been effaced.[36]

It is astounding to be spoken to so directly across such a gulf of time, and to see this 2,000-year plan continue to unfold.[37]

A human, or an entire generation, cannot complete such grand projects. But humanity can. We work together, each generation making a little progress while building up the capacities, resources and institutions to empower future generations to take the next step.

Indeed, when I think of the unbroken chain of generations leading to our time and of everything they have built for us, I am humbled. I am overwhelmed with gratitude; shocked by the enormity of the inheritance and at the impossibility of returning even the smallest fraction of the favour. Because a hundred billion of the people to whom I owe everything are gone forever, and because what they created is so much larger than my life, than my entire generation.

The same is true at the personal level. In the months after my daughter was born, the magnitude of everything my parents did for me was fully revealed. I was shocked. I told them; thanked them; apologised for the impossibility of ever repaying them. And they smiled, telling me that this wasn't how it worked—that one doesn't repay one's parents. One passes it on.

My parents aren't philosophers. But their remarks suggest another way in which the past could ground our duties to the future. Because the arrow of time makes it so much easier to help people who come after you than people who come before, the best way of understanding the partnership of the generations may be asymmetrical, with duties all flowing forwards in time—paying it forwards. On this view, our duties to future generations may thus be grounded in the work our ancestors did for us when *we* were future generations.[38]

So if we drop the baton, succumbing to an existential catastrophe, we would fail our ancestors in a multitude of ways. We would fail to achieve the dreams they hoped for; we would betray the trust they placed in us, their heirs; and we would fail in any duty we had to pay forward the work they did for us. To neglect existential risk might thus be to wrong not only the people of the future, but the people of the past.

It would also be to risk the destruction of everything of value from the past we might have reason to preserve.[39] Some philosophers have suggested that the right way to respond to some valuable things is not to promote them, but to protect or preserve them; to cherish or revere them.[40] We often treat the value of cultural traditions in this way. We see indigenous languages and ways of life under threat—perhaps to be lost forever to this world—and we are filled with a desire to preserve them, and protect them from future threats.

Someone who saw the value of humanity in this light may not be so moved by the loss of what could have been. But they would still be horrified by extinction: the ruin of every cathedral and temple, the erasure of every poem in every tongue, the final and permanent destruction of every cultural tradition the Earth has known. In the face of serious threats of extinction, or of a permanent collapse of civilisation, a tradition rooted in preserving or cherishing the richness of humanity would also cry out for action.[41]

Finally, we might have duties to the future arising from the flaws of the past. For we might be able to make up for some of our past wrongs. If we failed now, we could never fulfil any duties we might have to repair the damage we have done to the Earth's environment—cleaning up our pollution and waste; restoring the climate to its pre-industrial state; returning ecosystems to their vanished glory. Or consider that some of the greatest injustices have been inflicted not by individuals upon individuals, but by groups upon groups: systematic persecution, stolen lands, genocides. We may have duties to properly acknowledge and memorialise these wrongs; to confront the acts of our past. And there may yet be ways for the beneficiaries of these acts to partly remedy them or atone for them. Suffering an existential catastrophe would remove any last chance to do so.

CIVILISATIONAL VIRTUES

If we play our cards right, humanity is at an early stage of life: still in our adolescence; looking forward to a remarkable adulthood. Like an adolescent, we are rapidly coming into our full power and are impatient to flex our muscles, to try out every new capability the moment we acquire it. We show little regard for our future. Sure, we sometimes talk about the 'long term', but by this we usually mean the next decade or two. A long time for a human; a moment for humanity.

Like the adolescent, humanity has no need to plan out the details of the rest of its life. But it does need to make plans that bear in mind the duration and broad shape of that future. Otherwise we cannot hope to know which risks are worth taking, and which skills we need to develop to help us fulfil our potential.

Like many adolescents, humanity is impatient and imprudent; sometimes shockingly so. At times this stems from an inability to appropriately weigh our short-term gains against our longterm interests. More commonly, it is because we completely neglect our longterm future, not even considering it in our decision-making. And like the adolescent, we often stumble straight into risks without making any kind of conscious decision at all.

This analogy provides us with another lens through which to assess our behaviour. Rather than looking at the morality of an individual human's actions as they bear on others, we can address the dispositions and character of humanity as a whole and how these help or undercut its own chances of flourishing. When we look at humanity itself as a group agent, comprising all of us over all time, we can gain insight into the systematic strengths or weaknesses in humanity's ability to achieve flourishing. These are virtues and vices at the largest scale—what we could call *civilisational virtues* and vices. One could treat these as having a fundamental moral significance, or simply as a useful way of diagnosing important weakness in our character and suggesting remedies.

Not all virtues need make sense on this level, but many do. Our lack of regard for risks to our entire future is a deficiency of prudence. When we put the interests of our current generation far above those of the generations to follow, we display our lack of patience.[42] When we recognise the importance of our future yet still fail to prioritise it, it is a failure of self-discipline. When a backwards step makes us give up on our future—or assume it to be worthless—we show a lack of hope and perseverance, as well as a lack of responsibility for our own actions.[43]

In his celebrated account of virtue, Aristotle suggested that our virtues are governed and guided by a form of practical wisdom. This fits well with the idea of civilisational virtues too. For as our power continues to grow, our practical wisdom needs to grow with it.

COSMIC SIGNIFICANCE

Whether we are alone in the universe is one of the greatest remaining mysteries of science.[44] Eminent astronomers such as Martin Rees, Max Tegmark and Carl Sagan have reasoned that if we are alone, our survival and our actions might take on a cosmic significance.[45] While we are certainly smaller than the galaxies and stars, less spectacular than supernovae or black holes, we may yet be one of the most rare and precious parts of the cosmos.[46] The nature of such significance would depend on the ways in which we are unique.

If we are the only moral agents that will ever arise in our universe—the only beings capable of making choices on the grounds of what is right and wrong—then responsibility for the history of the universe is entirely *on us*. This is the only chance ever to shape the universe towards what is right, what is just, what is best for

THE PERSPECTIVE OF HUMANITY

Seeing our predicament from the perspective of humanity is a major theme of this book. Ethics is most commonly addressed from the individual perspective: what should *I* do? Occasionally, it is considered from the perspective of a group or nation, or even (more recently) from the global perspective of everyone alive today. Understanding what the group should do can help its members see the parts they need to play.

We shall sometimes take this a step further, exploring ethics from the perspective of humanity.[47] Not just our present generation, but humanity over deep time: reflecting on what we achieved in the last 10,000 generations and what we may be able to achieve in the aeons to come.

This perspective allows us to see how our own time fits into the greater story, and how much is at stake. It changes the way we see the world and our role in it, shifting our attention from things that affect the fleeting present, to those that could make fundamental alterations to the shape of the longterm future. What matters most for humanity? And what part in this plan should our generation play? What part should I play?[48]

Of course, humanity is not an individual. But it is often useful for us to think about groups as agents, gaining insights by talking about the beliefs, desires and intentions of teams, companies or nations. Consider how often we speak of a company's strategy, a nation's interests, or even what a country is hoping to achieve with its latest gambit. Such mental states are usually less coherent than those of individuals, as there can be internal tensions between the individuals that comprise the group. But individuals too have their own ambivalence or inner inconsistency, and the idea of 'group agents' has proved essential to anyone trying to understand the business world or international landscape.

Applying this perspective to humanity as a whole is increasingly useful and important. Humanity was splintered into isolated peoples for nearly the entire time since civilisation began. Only recently have we found each other across the seas and started forming a single global civilisation. Only recently have we discovered the length and shape of our long history, or the true potential of our future. And only recently have we faced significant threats that require global coordination.

We shouldn't always take this perspective. Many moral challenges operate at the personal level, or the level of smaller groups. And even when it comes to the big-picture questions, it is sometimes more important to focus on the ways in which humanity is divided: on our differing power or responsibility. But just as we've seen the value of occasionally adopting a global perspective, so too is it important to sometimes step further back and take the perspective of humanity.

The idea of civilisational virtues is just one example of explicitly adopting this perspective. In Chapter 7, we shall do so again, considering grand strategy for humanity. And even when we are looking at our own generation's responsibilities or what we each should do, this will be illuminated by the big picture of humanity across the aeons.

all. If we fail, then the potential not just of humanity, but of all moral action, will have been irrevocably squandered.

Alternatively, if we are the only beings capable of wondering about the universe, then we might have additional reason to seek such understanding. For it would only be through us that a part of the universe could come to fully understand the laws that govern the whole.

And if Earth is the only place in the universe that will give rise to life, then all life on Earth would have a key significance. Earth would be the only place where there was so much complexity in each drop of water, the only place where anything lived and died, the only place where anything felt, or thought, or loved. And humanity would be the only form of life capable of stewarding life itself, protecting it from natural catastrophes and, eventually, taking it to flourish throughout the cosmos.

UNCERTAINTY

So we could understand the importance of existential risk in terms of our present, our future, our past, our character or our cosmic significance. I am most confident in the considerations grounded in the value of our present and our future, but the availability of other lenses shows the robustness of the case for concern: it doesn't rely on any single school of moral thought, but springs naturally from a great many. While each avenue may suggest a different strength and nature of concern, together they provide a wide base of support for the idea that avoiding existential catastrophe is of grave moral importance.

I'm sure many readers are convinced by this point, but a few will still harbour doubts. I have sympathy, because I too am not *completely* certain. This uncertainty comes in two parts. The first is the everyday kind of uncertainty: uncertainty about what will happen in the future. Might the evidence for humanity's vast potential be misleading? The second is moral uncertainty: uncertainty about the nature of our ethical commitments.[49] Might I be mistaken about the strength of our obligations to future generations?

However, the case for making existential risk a global priority does not require certainty, for the stakes aren't balanced. If we make serious investments to protect humanity when we had no real duty to do so, we would err, wasting resources we could have spent on other noble causes. But if we neglect our future when we had a real duty to protect it, we would do something far worse— failing forever in what could well be our most important duty. So long as we find the case for safeguarding our future quite plausible, it would be extremely reckless to neglect it.[50]

Even if someone were so pessimistic about the future as to think it negative in expectation—that the heights we might reach are more than matched by the depths to which we might sink— there is *still* good reason to protect our potential.[51] For one thing, some existential catastrophes (such as permanent global totalitarianism) would remain uncontroversially terrible and thus worthy of our attention. But there is a deeper reason too. In this case there would be immense value of information in finding out more about whether our future will be positive or negative. By far the best strategy would be to protect humanity until we have a much more informed position on this crucial question.[52]

And it is not just regarding the value of the future that our descendants will be better informed. At present we are still more generally inexperienced. We have little practice at the complexities of managing a global civilisation, or a planet. Our view of the future is still clouded by ignorance and distorted by bias. But our descendants, if all goes well, will be far wiser than we are. They will have had time to understand much more deeply the nature of our condition; they will draw strength and insight from a more just, skilful and mature civilisation; and their choices, in general, will reflect a fuller understanding of what is at stake when they choose. We in the present day, at what may be the very start of history, would therefore do well to be humble, to leave our options open, and to ensure our descendants have a chance to see more clearly, and choose more wisely, than we can today.[53]

OUR NEGLECT OF EXISTENTIAL RISKS

The world is just waking up to the importance of existential risk. We have begun work on evaluating and evading the most significant threats, but have yet to scale this up in proportion to the significance of the problems. Seen in the context of the overall distribution of global resources, existential risk is sorely neglected.

Consider the possibility of engineered pandemics, which we shall soon see to be one of the largest risks facing humanity. The international body responsible for the continued prohibition of bioweapons (the Biological Weapons Convention) has an annual budget of just $1.4 million—less than the average McDonald's restaurant.[54] The entire spending on reducing existential risks from advanced artificial intelligence is in the tens of millions of dollars, compared with the billions spent on improving artificial intelligence capabilities.[55] While it is difficult to precisely measure global spending on existential risk, we can state with confidence that humanity spends more on ice cream every year than on ensuring that the technologies we develop do not destroy us.[56]

In scientific research, the story is similar. While substantial research is undertaken on the risk of smaller catastrophes, those that could destroy humanity's longterm potential are neglected. Since 1991 there have been only two published climate models on the effects of a full-scale nuclear war between the United States and Russia, even while hundreds of missiles remain minutes away from a possible launch.[57] There has been tremendous work on understanding climate change, but the worst-case scenarios— such as those involving more than six degrees of warming—have received comparatively little study and are mostly ignored in official reports and policy discussions.[58]

Given the reality and importance of existential risks, why don't they already receive the attention they need? Why are they systematically neglected? Answers can be found in the economics, politics, psychology and history of existential risk.

Economic theory tells us that existential risk will be undervalued by markets, nations and even entire generations. While markets do a great job of supplying many kinds of goods and services, there are some kinds that they systematically undersupply. Consider clean air. When air quality is improved, the benefit doesn't go to a particular individual, but is shared by everyone in the community. And when I benefit from cleaner air, that doesn't diminish the benefit you get from it. Things with these two properties are called *public goods* and markets have trouble supplying them.[59] We typically resolve this at a local or national level by having governments fund or regulate the provision of public goods.

Protection from existential risk is a public good: protection would benefit us all and my protection doesn't come at the expense of yours. So we'd expect existential risk to be neglected by the market. But worse, protection from existential risk is a *global public good*—one where the pool of beneficiaries spans the globe. This means that even nation states will neglect it.

I am writing this book in the United Kingdom. Its population of nearly 70 million ranks it as one of the more populous countries in the world, but it contains less than 1 percent of all the people alive today. If it acts alone on an existential risk, it bears the full cost of the policy, yet only reaps a hundredth of the benefit. In other words, even if it had a well-informed government acting in the longterm interests of its citizens, it would undervalue work on existential risk by a factor of 100. Similarly, Russia would undervalue it by a factor of 50, the United States by a factor of 20, and even China would undervalue it by a factor of five. Since such a large proportion of the benefits spill out to other countries, each nation is tempted to free-ride on the efforts of others, and some of the work that would benefit us all won't get done.

The same effect that causes this undersupply of protection causes an oversupply of risk. Since only 1 percent of the damages of existential catastrophe are borne by the people of the United Kingdom, their government is incentivised to neglect the downsides of risk-inducing policies by this same factor of 100. (The situation is even worse if individuals or small groups become able to pose existential risks.)

This means management of existential risk is best done at the global level. But the absence of effective global institutions for doing so makes it extremely difficult, slowing the world's reaction time and increasing the chance that hold-out countries derail the entire process.

And even if we could overcome these differences and bargain towards effective treaties and policies on existential risk, we would face a final problem. The beneficiaries are not merely global, but intergenerational—all the people who would ever live. Protection from existential risk is an *intergenerational global public good*. So even the entire population of the globe acting in concert could be expected to undervalue existential risks by a very large factor, leaving them greatly neglected.[60]

Additional reasons can be found in political science. The attention of politicians and civil servants is frequently focused on the short term.[61] Their timescales for thought and action are increasingly set by the election cycle and the news cycle. It is very difficult for them to turn their attention to issues where action is required now to avert a problem that won't strike for several election cycles. They are unlikely to get punished for letting it slide and many more urgent things are clamouring for attention.

One exception to this is when there is an active constituency pushing for the early action: their goodwill acts as a kind of immediate benefit. Such constituencies are most powerful when the benefits of the policy are concentrated among a small fraction of society, as this makes it worth their while to take political action. But in the case of existential risk, the benefits of protection are diffused across all citizens, leaving no key constituency to take ownership of the issue. This is a reason for neglect, albeit one that is surmountable. When citizens are empathetic and altruistic, identifying with the plight of others—as we have seen for the environment, animal rights and the abolition of slavery—they can be enlivened with the passion and determination needed to hold their leaders to account.

Another political reason concerns the sheer gravity of the issue. When I have raised the topic of existential risk with senior politicians and civil servants, I have encountered a common reaction: genuine deep concern paired with a feeling that addressing the greatest risks facing humanity was 'above my pay grade'. We look to our governments to manage issues that run beyond the scope of our individual lives, but this one runs beyond the scope of nations too. For political (as well as economic) reasons it feels like an issue for grand international action. But since the international institutions are so weak, it is left hanging.

Behavioural psychology has identified two more reasons why we neglect existential risk, rooted in the heuristics and biases we use as shortcuts for making decisions in a complex world.[62] The first of these is the *availability heuristic*. This is a tendency for people to estimate the likelihood of events based on their ability to recall examples. This stirs strong feelings about avoiding repeats of recent tragedies (especially those that are vivid or widely reported). But it means we often underweight events which are rare enough that they haven't occurred in our lifetimes, or which are without precedent. Even when experts estimate a significant probability for an unprecedented event, we have great difficulty believing it until we see it.

For many risks, the availability heuristic is a decent guide, allowing us to build up methods for managing the risk through trial and error. But with existential risks it fails completely. For by their very nature, we never have any experience of existential catastrophe before it is too late. If only seeing is believing, we will step blindly over the precipice.

Our need for vividness also governs our altruistic impulses. As a society, we are good at *acute* compassion for those in peril—for the victims of a disaster we can see in the news reports. We may not always act, but we certainly feel it. We sit up, our hearts in our throats: fearing for their safety, mourning for their loss. But what we require is a more expansive compassion; a more imaginative compassion; one that acts over the long term, recognising the humanity of people in distant times as well as distant places.

We also suffer from a bias known as *scope neglect*. This is a lack of sensitivity to the scale of a benefit or harm. We have trouble caring ten times more about something when it is ten times as important. And once the stakes get to a certain point, our concern can saturate.[63] For example, we tend to treat nuclear war as an utter disaster, so we fail to distinguish nuclear wars between nations with a handful of nuclear weapons (in which millions would die) from a nuclear confrontation with thousands of nuclear weapons (in which a thousand times as many people would die, and our entire future may be destroyed). Since existential risk derives its key moral importance from the size of what is at stake, scope neglect leads us to seriously underweight its importance.

These reasons for the neglect of existential risk present a formidable challenge to it ever receiving due concern. And yet I am hopeful. Because there is a final reason: existential risk is very new. So there hasn't yet been time for us to incorporate it into our civic and moral traditions. But the signs are good that this could change.

Humans must have contemplated the end of humanity from the earliest times. When an isolated band or tribe died out during a time of extreme hardship, the last survivors will have sometimes wondered whether they were the last of their kind, or whether others like them lived on elsewhere. But there appears to have been very little careful thought about the possibility and import-ance of human extinction until very recently.[64]

It wasn't until the mid-twentieth century, with the creation of nuclear weapons, that human extinction moved from a remote possibility (or a certainty remote in time) to an imminent danger. Just three days after the devastation of Hiroshima, Bertrand Russell began writing his first essay on the implications for the future of humanity.[65] And not long after, many of the scientists who created these weapons formed the *Bulletin of the Atomic Scientists* to lead the conversation about how to prevent global destruction.[66] Albert Einstein soon became a leading voice and his final public act was to sign a Manifesto with Russell arguing against nuclear war on the explicit grounds that it could spell the end for humanity.[67] Cold War leaders, such as Eisenhower, Kennedy and Brezhnev, became aware of the possibility of extinc-tion and some of its implications.[68]

The early 1980s saw a new wave of thought, with Jonathan Schell, Carl Sagan and Derek Parfit making great progress in understanding what is at stake—all three realising that the loss of uncounted future generations may overshadow the immediate consequences.[69] The discovery that atomic weapons may trigger a nuclear winter influenced both Ronald Reagan and Mikhail Gorbachev to reduce their country's arms and avoid war.[70]

And the public reacted too. In 1982, New York's Central Park saw a million people come together to march against nuclear weapons. It was the biggest protest in their nation's history.[71] Even in my birthplace of Australia, which has no nuclear weapons, we joined the global protest—my parents taking me with them on marches when I was just a small child they were fighting to protect.

In this way, existential risk was a highly influential idea of the twentieth century. But because there was one dominant risk, it all happened under the banner of nuclear war, with philosophers discussing the profound new issues raised by 'nuclear ethics', rather than by 'existential risk'. And with the end of the Cold War, this risk diminished and the conversation faded. But this history shows that existential risk is capable of rousing major global concern, from the elite to the grass roots.

Modern thinking on existential risk can be traced through John Leslie, whose 1996 book *The End of the World* broadened the focus from nuclear war to human extinction in general. After reading Leslie's work, Nick Bostrom took this a step further: identifying and analysing the broader class of existential risks that are the focus of this book.

Our moral and political traditions have been built over thousands of years. Their focus is thus mostly on timeless issues that have been with us throughout our history. It takes time to incorporate the new possibilities that our age opens up, even when those possibilities are of immense moral significance. Existential risk still seems new and strange, but I am hopeful that it will soon find its way into our common moral traditions. Environmentalism burst in upon the global political scene less than twenty years before I was born, and yet I was raised in a milieu where it was one of the main parts of our moral education; where the earlier disregard for the environment had become unthinkable to my generation. This can happen again.

One of my principal aims in writing this book is to end our neglect of existential risk—to establish the pivotal importance of safeguarding humanity, and to place this among the pantheon of causes to which the world devotes substantial attention and resources. Exactly how substantial remains an open question, but it clearly deserves far more focus than it has received so far. I suggest we start by spending more on protecting our future than we do on ice cream, and decide where to go from there.

We have now seen the broad sweep of human history, the size of humanity's potential, and why safeguarding our future is of the utmost importance. But so far, you have mostly had to take my word that we do face real risks. So let us turn our attention to these risks, examining the key scientific evidence behind them and sorting out which ones are most worthy of our concern. The next three chapters explore the natural risks we have faced throughout our history; the dawn of anthropogenic risks in the twentieth century; and the new risks we will face over the century to come.

PART TWO

THE RISKS

3

Natural Risks

*Who knows whether, when a comet shall approach this globe
to destroy it, as it often has been and will be destroyed, men
will not tear rocks from their foundations by means of steam,
and hurl mountains, as the giants are said to have done,
against the flaming mass?—and then we shall have traditions
of Titans again, and of wars with Heaven.*

—Lord Byron[1]

For all our increasing power over nature, humanity is still vulnerable to natural catastrophes. In this chapter, we consider not those from the newspapers—or even the history books—but catastrophes of a scale unprecedented during human civilisation. We look at risks that threaten not regional collapse, or endurable hardship, but the final undoing of the human enterprise.

Such risks are real. But they have only been confirmed in recent decades and the scientific understanding is still rapidly developing. We shall look in depth at several of the major threats, seeing the most recent science on how they threaten us and how much existential risk they pose.

ASTEROIDS & COMETS

An asteroid, ten kilometres across, speeds towards the Earth. The chance of a direct collision is tiny—for millions of years it has swung through the Solar System, missing the Earth on every single pass. But given such deep time the chances compound, and this is the day.

It slams into the Earth's surface off the coast of Mexico at more than 60,000 kilometres an hour. A trillion tons of rock moving so fast it strikes with the energy of a hundred times its own weight in TNT. In just seconds, it releases the energy of ten billion Hiroshima blasts: 10,000 times the entire Cold War nuclear arsenal. It smashes a hole thirty kilometres deep into the Earth's crust—over sixty times the height of the Empire State Building; three times taller than Everest. Everything within 1,000 kilometres is killed by heat from the impact fireball. A tsunami devastates the Caribbean. Trillions of tons of rock and dust are thrown far up into the sky. Some of this superheated rock rains down over millions of square kilometres, burning the animals to death and igniting fires that spread the devastation still further. But much more deadly is the dust that stays aloft.[2]

A billowing cloud of dust and ash rises all the way to the upper atmosphere, blocking out the Sun's light. It is this that turns regional catastrophe to mass extinction. Slowly, it spreads across the entire world, engulfing it in darkness lasting years. With the darkness comes a severe global cooling, for the Sun's light is blocked by the dust and reflected off the haze of sulphate aerosols released when the sea floor was vaporised. The cold and the dark kills plants across the globe; animals starve or freeze; the hundred-million-year reign of the dinosaurs ends; three-quarters of all species on Earth are annihilated.[3]

Both asteroids and comets can cause such devastation. Asteroids are lumps of rock or metal, found mostly between the orbits of Mars and Jupiter. They range from about a thousand kilometres across down to just a few metres.[4] Comets are lumps of mixed rock and ice, with a slightly narrower range of sizes.[5] Unlike asteroids, many comets are in extreme elliptical orbits, spending most of their time amidst or beyond the outer planets, then periodically diving in through the inner Solar System. When they come close enough to the Sun, solar radiation strips off some of the comet's ice and dust, forming a shining tail. A fragment of an asteroid or comet that enters our atmosphere, burning with the heat of atmospheric friction, is called a meteor. A piece that survives, having fallen all the way to the Earth's surface, is known as a meteorite.

Our earliest ancestors must have seen comets blaze across the sky, but could only guess at their true nature. The ancient Greeks conjectured that they were atmospheric phenomena, like clouds or rainbows. Indian astronomers in the sixth century correctly surmised that they were far beyond the Earth—something that was not confirmed for 1,000 years, when Tycho Brahe proved that the comet of 1577 was beyond the Moon, since distant observers saw the comet at nearly the same position in the night sky, at the same time.

Meteorites had also been known since time immemorial, but it was not until the turn of the nineteenth century that scientists established their extra-terrestrial origin.[6] At the same time, astronomers began to detect asteroids in orbit around the Sun. Then in 1960 the American geologist Eugene Shoemaker definitively proved that some of the Earth's craters were produced not by geological activity, but by vast meteoric impacts, far beyond any in recorded history. The pieces were finally in place to see that Earth was vulnerable to catastrophic impacts from the heavens.

In 1980 a team of scientists led by father and son Luis and Walter Alvarez discovered that the geological boundary between the Cretaceous and Palaeogene periods was rich in iridium—an element that is extremely rare on the Earth's surface, but markedly more common in asteroids. It dawned on them that this could be the smoking gun to explain the end-Cretaceous mass extinction, the one that killed the dinosaurs. An asteroid big enough to release so much iridium would be ten kilometres across, and the darkness of the dust cloud that spread the iridium would be enough to suppress photosynthesis and precipitate the mass extinction.[7] The missing piece was the lack of any known crater of the right size and age.

Ten years later it was found. Sixty-six million years of geological activity had buried it under kilometres of newer rock, but gravitational measurements revealed its dense granite impact ring—a giant circle surrounding the small Mexican town of Chicxulub. Excavations confirmed the crater's age and provenance. Debate continued about whether it was enough to cause the extinction, with more and more evidence aligning, and a consensus gradually emerging. Especially important was the discovery of nuclear winter in the early 1980s, which showed that a high dark cloud like this could chill the Earth as well as darken it, and the growing evidence that the impact had vaporised the sulphur-containing rock in the seabed, releasing a vast amount of sulphate aerosols that would further darken and cool the Earth.[8]

As it became increasingly clear that the Earth was vulnerable to major asteroid and comet impacts, people began to take this threat seriously. First in works of science fiction, then science.[9] Alvarez's hypothesis that an asteroid caused the last great mass extinction inspired Shoemaker to convene a seminal meeting in 1981, founding the scientific field of impact hazards. The scientists developed an ambitious proposal for finding and tracking asteroids. And in light of the growing public interest in the impact threat, it began to acquire bipartisan support in the United States Congress.[10] In 1994 Congress issued NASA a directive: find and track 90 percent of all near-Earth Objects greater than one kilometre across.[11]

Most of the attention so far has been focused on asteroids, as they are more common, easier to track and easier to deflect.[12] Astronomers categorise them in terms of their size.[13] Those above ten kilometres across (the size of the one that killed the dinosaurs) threaten mass extinction. It is possible that humans would survive the cataclysm, but there is clearly a serious risk of our extinction. Last time *all* land-based vertebrates weighing more than five kilograms were killed.[14] Asteroids between one and ten kilometres across threaten global catastrophe and may also be large enough to pose an existential risk, either via directly causing our extinction or via an unrecoverable collapse of civilisation. While an impact with an asteroid in this smaller size range would be much less likely to cause an existential catastrophe, this may be more than offset by their much higher probability of impact.

So many near-Earth asteroids have now been discovered and tracked that we have a good idea of the total number out there with orbits that come near the Earth. This tells us that the probability of an Earth-impact in an average century is about one in 6,000 for asteroids between one and ten kilometres in size, and about one in 1.5 million for those above ten kilometres.

But what about *our* century? By analysing the exact trajectories of the known asteroids, astronomers can determine whether there is any real chance that they will hit the Earth within the next hundred years. At the time of writing, 95 percent of asteroids bigger than one kilometre have been found and none have an appreciable chance of collision with the Earth. So almost all the remaining risk is from the 5 percent we haven't yet tracked.[15] We have even better news with asteroids greater than ten kilometres, as astronomers are almost certain that they have found them all, and that they pose no immediate danger.[16] Taking this trajectory information into account, the probability of an Earth-impact in the next hundred years falls to about one in 120,000 for asteroids between one and ten kilometres, and about one in 150 million for those above ten kilometres.[17]

These probabilities are immensely reassuring. While there is still real risk, it has been studied in great detail and shown to be vanishingly low. It is a famous risk, but a small one. If humanity were to go extinct in the next century, it would almost certainly be from something other than an asteroid or comet impact.

Asteroid Size	Total	Found	Average Century	Next Century
1–10 km	~ 920	~ 95%	1 in 6,000	1 in 120,000
10 km +	~ 4	> 99%*	1 in 1.5 million	< 1 in 150 million

* Astronomers are confident that they have found all asteroids greater than 10 km across in at least 99% of the sky.

TABLE 3.1 Progress in tracking near-Earth asteroids of two different size categories. The final two columns show the long-run average probability of an impact per century and the probability of an impact in the next hundred years (which all comes from the undiscovered asteroids).[18]

While uncertainties remain, the overall story here is one of humanity having its act together. It was just 12 years from the first scientific realisation of the risk of global catastrophe to the point where government started taking it seriously. And now, 28 years later, almost all the large asteroids have been tracked. There is international cooperation, with a United Nations–sanctioned organisation and an international alliance of spaceguard programmes.[19] The work is well managed and NASA funding has increased more than tenfold between 2010 and 2016.[20] In my view, no other existential risk is as well handled as that of asteroids and comets.

What are the next steps? Astronomers have succeeded so well in tracking asteroids that it may be time to switch some of their attention to comets.[21] While it is very hard to be sure, my best guess is that they pose about the same level of risk as that remaining from untracked asteroids.[22] With more work, it might be possible for astronomers to bring short-period comets into the risk framework they use for asteroids and to improve the detection and understanding of long-period comets.

And with such a good understanding of the chance of asteroid impacts, much of the remaining uncertainty about their existential risk lies in the chance that an impact would then spell the end of humanity—especially if the asteroid was in the one- to ten-kilometre range. So it would be valuable to develop models of the length and severity of impact winters, drawing on the latest climate and nuclear winter modelling.

DEFLECTING IMPACTS

What could we do if we actually found an asteroid on a collision course with Earth? Detection would have little value without some means of mitigation. In the worst case, we could prepare to weather the storm: using the warning time to stockpile food, build shelters and plan the best strategies for survival. But it would be vastly preferable to avoid the collision altogether.

Strategies for asteroid deflection can be based around destroying the asteroid, or changing its course. There are many technologies that might be able to perform either of these tasks, including nuclear explosions, kinetic impacts and ion beams.[23] We could use several methods simultaneously to decrease the chance of failure.

Deflection becomes much easier the further in advance the impact is detected. This is both because it provides more time to develop and deploy the deflecting system, and because it makes it easier to gradually change the asteroid's course. Unfortunately, it is not clear whether we would realistically have the capability to successfully deflect asteroids more than a few kilometres across—those that concern us most.[24]

There is active debate about whether more should be done to develop deflection methods ahead of time.[25] A key problem is that methods for deflecting asteroids *away* from Earth also make it possible to deflect asteroids *towards* Earth. This could occur by accident (e.g. while capturing asteroids for mining), or intentionally (e.g. in a war, or in a deliberate attempt to end civilisation). Such a self-inflicted asteroid impact is extremely unlikely, yet may still be the bigger risk.[26] After all, the entire probability of collision from one-kilometre or greater asteroids currently stands at one in 120,000 this century—we would require extreme confidence to say that the additional risk due to human interference was smaller than that.

Asteroid deflection therefore provides an interesting case study in weighing probabilities based on long-run frequencies, against evidential probabilities that are assigned to wholly unprecedented events. Quite understandably, we often prefer to rely on the long-run frequency estimates in our decision-making. But here the evidential probability is plausibly much larger and so cannot be ignored. A willingness to think seriously about imprecise probabilities of unprecedented events is crucial to grappling with risks to humanity's future.

SUPERVOLCANIC ERUPTIONS

Humanity may face a greater threat from within the Earth than from without. The very largest volcanic eruptions—explosions that release more than 1,000 cubic kilometres of rock—have become known as supervolcanic eruptions.[27] Unlike more typical volcanoes, which have the shape of a cone towering above the Earth's surface, volcanoes on this scale tend to release so much magma that they collapse, leaving a vast crater-like depression known as a caldera.[28] One of the best known is the Yellowstone caldera, which last erupted 630,000 years ago.[29]

Supervolcanic eruptions are devastating events, far beyond anything in recorded history. Everything within 100 kilometres of the blast is buried in falling rock, incandescent with heat. Thick ash rains down over the entire continent. When the Indonesian volcano, Toba, erupted 74,000 years ago, it covered India in a blanket of ash a metre thick and traces were found as far away as Africa. But as with asteroids and comets, the truly existential threat comes from the darkened sky.

The dark volcanic dust and reflective sulphate aerosols unleashed by the Toba eruption caused a 'volcanic winter', which is thought to have lowered global temperatures by several degrees for several years.[30] Even the much smaller eruption of Indonesia's Mount Tambora in 1815 (less than a hundredth the size) caused a global cooling of 1 °C, with places as far away as the United States suffering crop failure and June snows in what became known as the 'year without a summer'.[31]

Experts on supervolcanic eruptions do not typically suggest that there is a direct threat of human extinction. While there was some early evidence that the Toba eruption may have nearly destroyed humanity 74,000 years ago, newer evidence has made this look increasingly unlikely.[32] Since Toba was the largest known eruption in the last 2 million years and we now have thousands of times the population spread over a much greater part of the Earth, we should assume extinction to be a very unlikely consequence.[33] The effects may be roughly comparable to those of the one- to ten-kilometre asteroids, with major global crop failures lasting for years on end. Since the world only has about six months of food reserves, there is a possibility that billions of people could starve and that civilisation could suffer a global collapse. I think that even if civilisation did collapse, it would be very likely to recover. But if it could not, that would constitute an existential catastrophe.

While geologists have identified the remnants of dozens of supereruptions, their frequency remains very uncertain. A recent review gave a central estimate of one per 20,000 years, with substantial uncertainty. For Toba-sized eruptions, the same analysis gives a central estimate of one in 80,000 years, but with even more uncertainty.[35]

What about for the next hundred years? When astronomers tracked more and more asteroids, they were able to determine that the next century would be safer than average. Unfortunately, volcanoes are much less predictable than asteroids. Despite knowing the locations of most of the volcanoes that have had supervolcanic eruptions in the past, it is extremely difficult to predict whether they are likely to erupt soon, and we should expect very little warning if they do.

There is very little known about how to prevent or delay an impending supereruption. NASA recently conducted a very preliminary investigation of the possibility of slowly draining heat from the Yellowstone caldera, but investigations like these are in their earliest stages, and any sort of interference with an active

Magnitude	Average Century	Next Century
8–9	~ 1 in 200	~ 1 in 200
9+ (e.g. Toba)	~ 1 in 800	~ 1 in 800

TABLE 3.2 The probability per century of a supervolcanic eruption. Note that there are good reasons to think that even the largest eruptions would be very unlikely to lead to extinction or unrecoverable collapse. The probability estimates are extremely rough, with the confidence interval for magnitude 8–9 eruptions ranging from 1 in 50 to 1 in 500 per century, and the confidence interval for magnitude 9+ ranging from 1 in 600 all the way to 1 in 60,000.

volcano—especially one with a history of supereruptions—would obviously require enormous caution.[36] For now, our best approach to the threat of supereruptions lies in preparing to mitigate the damage, through building up non-perishable food reserves or developing emergency food production techniques.

Compared to asteroids and comets, we are at an earlier stage of understanding and managing the risk. This risk may also be fundamentally harder to manage, due to the greater difficulties of prediction and prevention. And most importantly, the probability of a civilisation-threatening catastrophe in the next century is estimated to be about 100 times that of asteroids and comets combined. So supervolcanic eruptions appear to be the greater risk, and in greater need of additional attention.

FLOODS OF LAVA

In the Earth's history, there have been volcanic events of even greater scale. About 250 million years ago, the Siberian Traps erupted. More than a *million* cubic kilometres of molten rock was released, pouring out of the Earth and covering an area the size of Europe in lava. Scientists have suggested that volcanic gases released during this time may have caused the end-Permian extinction—the biggest mass extinction in the Earth's history.[34]

This kind of eruption is known as a flood basalt event, after the type of rock released. They differ from the supervolcanic eruptions discussed here in two key ways.

They take place much more slowly, in a series of eruptions going on for thousands to millions of years. And most importantly, they are about a thousand times less frequent than explosive supereruptions, occurring once every twenty to thirty million years. While it seems very plausible that they could pose a direct threat of human extinction, it could at most be a one in 200,000 chance per century—higher than that posed by ten-kilometre asteroids, but much lower than some other risks we shall consider.

There are many promising next steps. At the most basic level, we need to find all the places where supervolcanic eruptions have occurred so far. We also need to improve our very rough estimates of how often supervolcanic eruptions happen—especially at the largest and most threatening scale. Much more research is needed on the climatic effects of supervolcanic eruptions to see which sizes might pose a real risk to humanity.[37] And I suspect there are many hard-won lessons in risk modelling and management that could be borrowed from the more established community around asteroid risk.

STELLAR EXPLOSIONS

In every star there is a continual battle between two forces. Gravity squeezes the star together, while pressure forces it apart. For most of a star's life, these forces are in balance, preventing it from collapsing to a point or dissipating into space.[38] But some stars reach a time where the pressure catastrophically fails to withstand the force of gravity and they collapse in upon themselves at relativistic speed.[39] They momentarily reach an incredibly high density, triggering a new wave of immense pressure that explodes the star in what is known as a supernova. For a brief time, this single star can outshine its entire galaxy. In seconds, it releases as much energy as our Sun will radiate over its ten-billion-year lifetime.

Supernovae were first recorded by ancient Chinese astronomers in 185 CE, when a bright new star suddenly blazed in their sky. But it wasn't until the 1930s that scientists began to understand them and the 1950s that they realised a nearby supernova would pose a threat to the Earth.[40]

Then in 1969 scientists discovered a new and distinctive type of stellar explosion. In the midst of the Cold War, the US launched a number of spy satellites, in order to detect secret nuclear tests via their characteristic flash of gamma rays. The satellites began to detect short bursts of gamma rays, but with a completely different signature from nuclear weapons. Astronomers determined that they couldn't be coming from the Earth—or even the Milky Way—but must be arriving from extremely distant galaxies, billions of light years away.[41] The mystery of what could cause such 'gamma ray bursts' is still being resolved. The leading theory is that longer bursts are produced in a rare type of supernova and shorter ones are produced when two neutron stars collide. The total energy released in each burst is similar to that of a supernova, but concentrated in two narrow cones pointed in opposite directions, allowing them to be detected at immense distances.[42] For example, in March 2008 light from a gamma ray burst in a galaxy 10 billion light years away reached Earth, and it was still bright enough to be visible to the naked eye.[43]

A supernova or gamma ray burst close to our Solar System could have catastrophic effects. While the gamma rays and cosmic rays themselves won't reach the Earth's surface, the reactions they trigger in our atmosphere may pose a threat. The most important is probably the production of nitrogen oxides that would alter the Earth's climate and dramatically erode the ozone layer. This last effect is thought to be the most deadly, leaving us much more exposed to UV radiation for a period of years.[44]

Astronomers have estimated the chance of these events happening close enough to Earth to cause a global catastrophe, generally defining this as a global ozone depletion of 30 percent or more. (I suspect this would be less of a threat to civilisation than the corresponding thresholds for asteroids, comets and supervolcanic eruptions.) In an average century, the chance of such an event is about one in 5 million for supernovae and one in 2.5 million for gamma ray bursts. As with asteroids, we can get a more accurate estimate for the *next* 100 years, by searching the skies for imminent threats. This is harder for gamma ray bursts as they are more poorly understood and can strike from much further away. We have not found any likely candidates of either type, but have not yet completely ruled them out either, leading to a modest reduction in risk for the next century compared to average.[45]

Type	Average Century	Next Century
Supernovae	~ 1 in 5 million	< 1 in 50 million
Gamma Ray Bursts	~ 1 in 2.5 million	< 1 in 2.5 million

TABLE 3.3 The probability per century of a stellar explosion causing a catastrophe on Earth that depletes ozone by more than 30%.[46]

These probabilities are very small—they look to be at least 20 times smaller than those of similarly sized catastrophes from asteroids and comets and at least 3,000 times smaller than those from supervolcanic eruptions. Still, we would want to remove some of the remaining uncertainties around these numbers before we could set this risk aside. We need more research to determine the threshold above which stellar explosions could lead to extinction. And we should start cataloguing potential supernova candidates within 100 light years, determining how confident we can be that none will explode in the next century. More broadly, we should improve our models of these risks and their remaining uncertainties, trying to bring our level of understanding into line with asteroids and comets.[47]

OTHER NATURAL RISKS

There is no shortage of potential catastrophes. Even restricting our attention to natural risks with significant scientific support, there are many more than I can address in detail. But none of them keep me awake at night.

Some threats pose real risks in the long run, but no risk over the next thousand years. Foremost among these is the eventual brightening of our Sun, which will pose a very high risk of extinction, but only starting in around a billion years.[48] A return to a glacial period (an 'ice age') would cause significant difficulties for humanity, but is effectively ruled out over the next thousand years.[49] Evolutionary scenarios such as humanity degrading or transforming into a new species also pose no threat over the next thousand years.

Some threats are known to be vanishingly unlikely. For example, the passage of a star through our Solar System could disrupt planetary orbits, causing the Earth to freeze or boil or even crash into another planet. But this has only a one in 100,000 chance over the next 2 billion years.[50] This could also happen due to chaotic instabilities in orbital dynamics, but again this is exceptionally unlikely. Some physical theories suggest that the vacuum of space itself may be unstable, and could 'collapse' to form a true vacuum. This would spread out at the speed of light, destroying all life in its wake. However, the chance of this happening cannot be higher than one in 10 million per century and is generally thought to be much lower.[51]

Some threats are not existential—they offer no plausible pathway to our extinction or permanent collapse. This is true for the threat of many local or regional catastrophes such as hurricanes or tsunamis. It is also true for some threats that are global in scale. For example, the Earth's entire magnetic field can shift dramatically, and sometimes reverses its direction entirely. These shifts leave us more exposed to cosmic rays during the time it takes to reorient.[52] However, this happens often enough that we can tell it isn't an extinction risk (it has happened about 20 times in the 5 million years since humans and chimpanzees diverged). And since the only well-studied effect appears to be somewhat increased cancer rates, it is not a risk of civilisation collapse either.[53]

Finally, some threats are natural in origin, but have effects that are greatly exacerbated by human activity. They thus fall somewhere between natural and anthropogenic. This includes 'naturally arising' pandemics. For reasons that will soon become clear, I don't count these among the natural risks, and shall instead address them in Chapter 5.

THE TOTAL NATURAL RISK

It is striking how recently many of these risks were discovered. Magnetic field reversal was discovered in 1906. Proof that Earth had been hit by a large asteroid or comet first emerged in 1960. And we had no idea gamma ray bursts even existed until 1969. For almost our entire history we have been subject to risks to which we were completely oblivious.

And there is no reason to think that the flurry of discovery has finished—that we are the first generation to have discovered all the natural risks we face. Indeed, it would surely be premature to conclude that we have discovered all of the possible mechanisms of natural extinction while major mass-extinction events remain unexplained.

The likely incompleteness of our knowledge is a major problem for any attempt to understand the scale of natural risk by cataloguing known threats. Even if we studied all the natural threats listed in this chapter so completely that we understood their every detail, we could not be sure that we were capturing even a small part of the true risk landscape.

Luckily, there is a way out—a way of directly estimating the total natural risk. We achieve this not by considering the details of asteroid craters or collapsing stars, but by studying the remains of the species they threatened. The fossil record is our richest source of information about how long species like us survived, and so about the total extinction risk they faced.[54] We will explore three ways of using the fossil record to place upper bounds on the natural extinction risk we face, all of which yield comforting results.[55] However, as this method only applies directly to *extinction* risk, some uncertainty around unrecoverable collapse will remain.[56]

How high could natural extinction risk be? Imagine if it were as high as 1 percent per century. How long would humanity survive? Just 100 centuries, on average. But we know from the fossil record that *Homo sapiens* has actually lived for about 2,000 centuries.[57] At 1 percent risk per century, it would be nearly impossible to last that long: there would be a 99.9999998 percent chance of going extinct before that. So we can safely rule out a total risk of 1 percent or greater. Just how much risk could there realistically have been? We can use the longevity of *Homo sapiens* to form both a best-guess estimate and an upper bound for this risk.

It is surprisingly difficult to form a single best guess of the risk. We might be tempted to say one in 2,000, but that would be the best guess if we had seen 2,000 centuries of humanity with *one* extinction. In fact we have seen zero extinctions, so our best guess of the risk should be lower. But it can't be zero in 2,000 either, since this would mean that extinction is impossible, and that we could be justifiably certain it isn't going to happen.[58] There is an interesting ongoing debate among statisticians about what probability to assign in such cases.[59] But all suggested methods produce numbers between zero in 2,000 and one in 2,000 (i.e. 0 to 0.05 percent). So we can treat this range as a rough estimate.

We can also use our survival so far to make an upper bound for the total natural extinction risk. For example, if the risk were above 0.34 percent per century there would have been a 99.9 percent chance of going extinct before now.[60] We thus say that risk above 0.34 percent per century is ruled out at the 99.9 percent confidence level—a conclusion that is highly significant by the usual scientific standards (equivalent to a p-value of 0.001).[61] So our 2,000 centuries of *Homo sapiens* suggests a 'best-guess' risk estimate between 0 percent and 0.05 percent, with an upper bound of 0.34 percent.

But what if *Homo sapiens* is not the relevant category? We are interested in the survival of *humanity*, and we may well see this as something broader than our species. For instance, Neanderthals were very similar to *Homo sapiens* and while the extent of interbreeding between the two is still debated, it is possible that they are best considered as a subspecies. They walked upright, made advanced tools, had complex social groupings, looked similar to *Homo sapiens,* and may even have used language. If we include them in our understanding of humanity, then we could extend our lifespan to when Neanderthals and *Homo sapiens* last had a common ancestor,

Category	Years	Best Guess	99.9% Confidence Bound
Homo sapiens	200,000	0–0.05%	< 0.34%
Neanderthal split	500,000	0–0.02%	< 0.14%
Homo	2,000,000–3,000,000	0–0.003%	< 0.023%

TABLE 3.4 Estimates and bounds of total natural extinction risk per century based on how long humanity has survived so far, using three different conceptions of humanity.

around 500,000 years ago.[62] Another natural approach would be to use not our species, but our genus, *Homo*. It has been in existence for more than 2 million years. If used with the methods above, these dates would imply lower probabilities of extinction per century.

A second technique for estimating the total natural extinction risk from the fossil record is to look not at humanity itself, but at species like us. This greatly expands the available evidence. And because it includes examples of species actually going extinct, it eliminates the issues with zero-failure data. The downsides are that the other species may be less representative of the risks humanity faces and that there is room for potential bias in the choice of species to study.

A simple version of this technique is to look at the most similar species. Our genus, *Homo*, contains four other species with reasonable estimates of longevity.[65] They have survived between 200,000 and 1,700,000 years. If we bear a relevantly similar risk of extinction from natural catastrophes to any of these, we are looking at per century risk in the range of 0.006 to 0.05 percent.[66]

Alternatively we could cast a much wider net, achieving more statistical robustness at the expense of similarity to ourselves. The typical longevity of mammalian species has been estimated at around 1 million years, while species in the entire fossil record average 1 to 10 million years. These suggest a risk in the range of

0.001 to 0.01 percent per century—or lower if we think we are more robust than a typical species (see Table 3.5).

Note that all these estimates of species lifespans include other causes of extinction in addition to catastrophes, for example being slowly outcompeted by a new species that branches off from one's own. So they will somewhat overestimate the risk of *catastrophic* extinction.[67]

SURVIVORSHIP BIAS

There is a special difficulty that comes with investigating the likelihood of an event which would have prevented that very investigation. No matter how likely it was, we cannot help but find that the event didn't occur. This comes up when we look at the extinction history of *Homo sapiens*, and it has the potential to bias our estimates.[63]

Imagine if there were a hundred planets just like our own. Whether humanity quickly went extinct on ninety-nine of them, or on zero of them, humans investigating their own planet would always find that human extinction hadn't happened—otherwise they wouldn't be around to investigate. So they couldn't use the mere fact that they survived to estimate the fraction of planets where humans survive. This makes us realise that we too can't deduce much about our future survival just from the fact that we have survived so far.

However, we *can* make use of the length of time we have survived (as we do in this chapter), since there is more than one value that could be observed and we are less likely to see long lifespans in worlds with high risk. But a full accounting for this form of survivorship bias may still modify these risk estimates.[64]

Fortunately, estimating risk by analysing the survival of other species is more robust to these effects and, reassuringly, it provides similar answers.

Species	Years	Best Guess
Homo neanderthalensis	200,000	0.05%
Homo heidelbergensis	400,000	0.025%
Homo habilis	600,000	0.02%
Homo erectus	1,700,000	0.006%
Mammals	1,000,000	0.01%
All species	1,000,000–10,000,000	0.01–0.001%

TABLE 3.5 Estimates of total natural extinction risk per century based on the survival time of related species.

A final technique for estimating the total natural extinction risk is to consider that we are so populous, so widely spread across the globe, so competent at living in highly diverse environments, and so capable of defending ourselves that we might be able to resist all natural catastrophes short of those that cause mass extinctions. If so, we could look at the mass extinction record to determine the frequency of such events.

The detailed fossil record starts 540 million years ago with the 'Cambrian explosion': a rapid diversification of complex life into most of the major categories we see today. Since then there have been a number of mass extinctions—catastrophic times when a great variety of species from across the globe went extinct. Foremost among these are the 'Big Five', which each caused the extinction of at least 75 percent of all species (see Table 3.6). The catastrophe that ended the reign of the dinosaurs was the most recent of these. If these are representative of the level of natural catastrophe needed to cause our extinction, then we have had five events in 540 million years: a natural extinction rate of about one in a million (0.0001 percent) per century.

All three of these techniques based on the fossil record are at their best when applied to threats that would pose a similar extinction risk to modern humans as they did to the creatures from whose death or survival we seek to gain evidence, such as early humans, other species throughout history, and the victims of mass extinctions. Clearly this is not always the case. For

Mass Extinction	Date	Species Lost
Late Ordovician	443 Ma	86%
Late Devonian	359 Ma	75%
End-Permian	252 Ma	96%
End-Triassic	201 Ma	80%
End-Cretaceous	66 Ma	76%

TABLE 3.6 The proportion of species that went extinct in each of the Big Five extinction events (Ma = millions of years ago).[68]

many natural risks, we have become more robust. For example, our global presence allows us to survive mere regional disasters, and we possess unprecedented capacities to respond to global disasters as well. This means that the true risk is likely to be below these estimates, and that even the 'best-guess' estimates should be thought of as conservative bounds on the total natural risk.

The bigger issue is risks that are substantially greater for humans now than they were for early humans or related species. This includes all the anthropogenic risks (which is precisely why this section has only targeted natural risks). It may also include some risks that are often considered natural.[69]

Chief among these is the risk of pandemics. While we don't typically think of an outbreak of disease as anthropogenic, the social and technological changes since the Agricultural Revolution have dramatically increased its likelihood and impact. Farming has increased the chance of infections from animals; improved transportation has made it easier to spread to many subpopulations in a short time; and increased trade has seen us utilise this transportation very frequently.

While there are also many factors that mitigate these effects (such as modern medicine, quarantine and disease surveillance), there remains a very plausible case that the pandemic risk to humans in the coming centuries is significantly larger than in early humans or other species used to construct the bounds on natural risks. For these reasons, it is best not to count pandemics as a natural risk, and we shall address them later.

We have explored three different ways of using the fossil record to estimate or bound the total natural extinction risk for humanity. While we shouldn't put too much weight on any one of these estimates, we can trust the broad range of results. The best-guess estimates ranged from 0.0001 to 0.05 percent per century. And even the most conservative of the upper bounds was less than 0.4 percent. Moreover, we know that these numbers are likely to be overestimates because they cover non-catastrophic extinction, such as the gradual evolution into a new species, and because modern humanity is more resilient than earlier humans or other species. This means we can be very confident that the total natural extinction risk is lower than 0.5 percent, with our best guess somewhere below 0.05 percent.

When we consider the entire future that is at stake, even an individual natural risk such as that posed by asteroids is extremely important. However, we will soon see that the natural risks are dwarfed by those of our own creation. By my estimate, we face about a thousand times more anthropogenic risk over the next century than natural risk, so it is the anthropogenic risks that will be our main focus.

4

ANTHROPOGENIC RISKS

The human race's prospects of survival were considerably
better when we were defenceless against tigers than they are
today, when we have become defenceless against ourselves.
—Arnold Toynbee[1]

The 2,000-century track record of human existence allows us
to tightly bound the existential risk from natural events. These
risks are real, though very unlikely to strike over the next
hundred years.

But there is no such track record for the powerful industrial
technologies that are also thought to pose existential risks. The
260 years we have survived since the Industrial Revolution, or
the seventy-five years since the invention of nuclear weapons, are
compatible with risks as high as 50 percent or as low as 0 percent
over the coming hundred years. So what evidence do we have
regarding these technological risks?

In this chapter, we'll explore the science behind the current
anthropogenic risks arising from nuclear weapons, climate
change and other environmental degradation. (Risks from future
technologies, including engineered pandemics, will be covered
in the following chapter.) Our focus will be on the worst-case
scenarios—in particular, whether there is a solid scientific case
that they could cause human extinction or the unrecoverable
collapse of civilisation.

NUCLEAR WEAPONS

When we think of the existential risk posed by nuclear weapons, our first thoughts are of the destruction wrought by a full-scale nuclear war. But long before the Cold War, before even Hiroshima and Nagasaki, scientists worried that a single nuclear explosion might spell the destruction of humanity.

In the summer of 1942, the American physicist Robert Oppenheimer held a series of secret meetings in his office at the University of California in Berkeley, bringing together many of his field's leading thinkers. They were attempting to design the first atomic bomb. This was based on the recent discovery of nuclear fission: splitting a large atomic nucleus such as uranium into smaller fragments and releasing its nuclear energy.

On the second day, Edward Teller—who would go on to develop the hydrogen bomb ten years later—gave his first presentation on the idea of such a bomb. He noted that an atomic explosion would create a temperature exceeding that of the centre of the Sun (15,000,000 °C). It is this scorching temperature that allows the Sun to burn: it forces hydrogen nuclei together, producing helium and extreme quantities of energy. This is known as *fusion* (or a thermonuclear reaction), and is even more efficient than fission.[2] If an atomic bomb could be surrounded with a fuel such as hydrogen, its fission reaction might be able to trigger such a fusion reaction.

While attempting to design such a bomb, Teller had noticed that if it were possible for an atomic bomb to ignite such a fusion reaction in its fuel, it might also be possible for it to ignite a fusion reaction in the world around. It might be able to ignite the hydrogen in water, setting off a self-sustaining reaction that burnt off the Earth's oceans. Or a reaction might be possible in the nitrogen that makes up seven-tenths of our air, igniting the atmosphere and engulfing the Earth in flame. If so, it would destroy not just humanity, but perhaps all complex life on Earth.

When he told the assembled scientists, a heated discussion broke out. Hans Bethe, the brilliant physicist who just four years earlier had discovered how fusion powers the stars, was extremely sceptical and immediately attempted to refute Teller's assumptions. But Oppenheimer, who would lead the development of the bomb, was deeply concerned. While the others continued their calculations, he raced off across the country to personally inform his superior, Arthur Compton, that their project may pose a threat to humanity itself. In his memoir, Compton recalled the meeting:

> Was there really any chance that an atomic bomb would trigger the explosion of the nitrogen in the atmosphere or of the hydrogen in the ocean? This would be the ultimate catastrophe. Better to accept the slavery of the Nazis than to run a chance of drawing the final curtain on mankind!
>
> Oppenheimer's team must go ahead with their calculations. Unless they came up with a firm and reliable conclusion that our atomic bombs could not explode the air or the sea, these bombs must never be made.[3]

(After the war, it would be revealed that their counterparts in Germany had also discovered this threat and the possibility had been escalated all the way up to Hitler—who went on to make dark jokes about the possibility.)[4]

Oppenheimer returned to Berkeley, finding that Bethe had already discovered major weaknesses in Teller's calculations.[5] While they couldn't prove it was safe to all the physicists' satisfaction, they eventually decided to move on to other topics. Later, Oppenheimer commissioned a secret scientific report into the possibility of igniting the atmosphere.[6] It supported Bethe's conclusions that this didn't seem possible, but could not prove its impossibility nor put a probability on it.[7] Despite the report concluding that 'the complexity of the argument and the absence of satisfactory experimental foundation makes further work on the subject highly desirable', it was taken by the leadership at Los Alamos to be the final word on the matter.

But concerns lingered among the physicists all the way through to the day of the Trinity test, when the first atomic bomb would be detonated.[8] Enrico Fermi, the Nobel prize-winning physicist who was also present at the Berkeley meeting, remained concerned that deficiencies in their approximations or assumptions might have masked a true danger. He and Teller kept rechecking the analysis, right up until the day of the test.[9] James Conant, President of Harvard University, took the possibility seriously enough that when the flash at detonation was so much longer and brighter than he expected, he was overcome with dread: 'My instantaneous reaction was that something had gone wrong and that the thermal nuclear transformation of the atmosphere, once discussed as a possibility and jokingly referred to a few minutes earlier, had actually occurred.'[10]

The atmosphere did not ignite. Not then, nor in any nuclear test since. Physicists with a greater understanding of nuclear fusion and with computers to aid their calculations have confirmed that it is indeed impossible.[11] And yet, there *had* been a kind of risk. The bomb's designers didn't know whether or not igniting the atmosphere was physically possible, so at that stage it was still epistemically possible. While it turned out that there was no objective risk, there was a serious subjective risk that their bomb might destroy humanity.

This was a new kind of dilemma for modern science. Suddenly, we were unleashing so much energy that we were creating temperatures unprecedented in Earth's entire history. Our destructive potential had grown so high that for the first time the question of whether we might destroy all of humanity needed to be asked—and answered. So I date the beginning of the Precipice (our age of heightened risk) to 11.29 a.m. (UTC) on 16 July 1945: the precise moment of the Trinity test.

Did humanity pass its own test? Did we successfully manage this first existential risk of our own making? Perhaps. I am genuinely impressed by Oppenheimer's urgency and Compton's stirring words. But I'm not convinced that the process they initiated was sufficient.

Bethe's calculations and the secret report were good, and were scrutinised by some of the world's best physicists. But the wartime secrecy meant that the report was never subject to the external peer review of a disinterested party, in the way that we consider essential for ensuring good science.[12]

And while some of the best minds in the world were devoted to the physics problems involved, the same cannot be said for the wider problems of how to handle the risk, who to inform, what level of risk would be acceptable and so forth.[13] It is not clear whether even a single elected representative was told about the potential risk.[14] The scientists and military appear to have assumed full responsibility for an act that threatened all life on Earth. Was this a responsibility that was theirs to assume?

Given the weak conclusions of the report, the inability to get external review, and the continuing concerns of eminent scientists, there was a strong case for simply delaying, or abandoning, the test. Back at the time of the Berkeley meeting, many of the scientists were deeply afraid that Hitler might get there first and hold the world to nuclear ransom. But by the time of the Trinity test, Hitler was dead and Europe liberated. Japan was in retreat and there was no concern about losing the war. The risk was taken for the same reasons that the bombs would be dropped in Japan a month later: to shorten the war, to avoid loss of life in an invasion, to achieve more favourable surrender terms, and to warn the Soviet Union about America's new-found might. These are not strong reasons for unilaterally risking the future of humanity.

Just how much risk did they take? It is difficult to be precise, without knowing how they were weighing the evidence available to them at the time.[15] Given that they got the answer right, hindsight tempts us to consider that result inevitable. But the Berkeley meeting provided something of a natural experiment, for during that summer they tackled two major questions on thermonuclear ignition. After they moved on from the question of atmospheric ignition, they began to calculate what kind of fuel *would* allow a thermonuclear explosion. They settled on a fuel based on an isotope of lithium: lithium-6.[16] But natural lithium contained too little of this isotope for the explosion to work, so they concluded that the mostly inert lithium-7 would need to be removed at great expense.

In 1954 the United States tested exactly such a bomb, code-named Castle Bravo. Due to time constraints, they had only enriched the lithium-6 concentration up to 40 percent, so most was still lithium-7. When the bomb exploded, it released far more energy than anticipated. Instead of six megatons they got 15—a thousand times the energy of the Hiroshima bomb, and the biggest explosion America would ever produce.[17] It was also one of the world's largest radiological disasters, irradiating a Japanese fishing boat and several populated islands downwind.[18] It turned out that the Berkeley group (and subsequent Los Alamos physicists) were wrong about lithium-7. At the unprecedented temperatures in the explosion, it reacted in an unexpected way, making just as great a contribution as lithium-6.[19]

Of the two major thermonuclear calculations made that summer in Berkeley, they got one right and one wrong. It would be a mistake to conclude from this that the subjective risk of igniting the atmosphere was as high as 50 percent.[20] But it was certainly not a level of reliability on which to risk our future.

Fifteen days after dropping the atomic bombs on Japan, America began planning for nuclear war with the Soviets.[21] They drew up maps of the Soviet Union with vast circles showing the range of their bombers to determine which cities they could already destroy and which would require new air bases or technological improvements. So began the planning for large-scale nuclear war, which has continued through the last 75 years.

This period was marked by numerous changes to the strategic landscape of nuclear war. Most of these stemmed from techno-logical changes such as the Soviets' rapid development of their own nuclear weapons; the creation of thermonuclear weapons vastly more powerful than the bombs used against Japan; inter-continental ballistic missiles (ICBMs) that could hit cities in the enemy heartland with just half an hour of warning; submarine-launched missiles that could not be taken out in a first strike, allowing guaranteed nuclear retaliation; and a massive increase in the total number of nuclear warheads.[22] Then there were major political changes such as the formation of NATO and the eventual fall of the Soviet Union. The Cold War thus saw a haphazard progression from one strategic situation to another, some favouring first strikes, some favouring retaliation, some high risk, some low.

While we made it through this period without nuclear war breaking out, there were many moments where we came much closer than was known at the time (see the box 'Close Calls'). Most of these were due to human or technical error in the rapid-response systems for detecting an incoming nuclear strike and retaliating within the very short time-window allowed. These were more frequent during times of heightened military tension, but continued beyond the end of the Cold War. The systems were designed to minimise false negatives (failures to respond), but produced a lot of false alerts. This holds lessons not just for nuclear risk, but for risk from other complex technologies too— even when the stakes are known to be the end of one's entire nation (or worse), it is extremely hard to iron out all the human and technical problems.

If a full-scale nuclear war did break out, what would happen? In particular, would it really threaten extinction or the permanent collapse of civilisation?

While one often hears the claim that we have enough nuclear weapons to destroy the world many times over, this is loose talk. It appears to be based on a naïve scaling-up of the Hiroshima fatalities in line with the world's growing nuclear arsenal, then comparing this to the world's population.[29] But the truth is much more complex and uncertain.

Nuclear war has both local and global effects. The local effects include the explosions themselves and the resulting fires. These would devastate the detonation sites and would kill tens or even hundreds of millions of people.[30] But these direct effects could not cause extinction since they would be limited to large cities, towns and military targets within the belligerent countries. The threat to humanity itself comes instead from the global effects.

CLOSE CALLS

The last seventy years have seen many close calls, where the hair-trigger alert of US and Soviet nuclear forces brought us far too close to the brink of accidental nuclear war.[23] Here are three of the closest.[24] (See Appendix C for a further close call and a list of nuclear weapons accidents.)

Training Tape Incident: 9 November 1979

At 3 a.m. a large number of incoming missiles—a full-scale Soviet first strike—appeared on the screens at four US command centres. The US had only minutes to determine a response before the bulk of their own missiles would be destroyed. Senior commanders initiated a threat assessment conference, placing their ICBMs on high alert, preparing nuclear bombers for take-off, and scrambling fighter planes to intercept incoming bombers.

But when they checked the raw data from the early-warning systems, there were no signs of any missiles, and they realised it was a false alarm. The screens had been showing a realistic simulation of a Soviet attack from a military exercise that had mistakenly been sent to the live computer system. When Soviet Premier Brezhnev found out, he asked President Carter 'What kind of mechanism is it which allows a possibility of such incidents?'[25]

Autumn Equinox Incident: 26 September 1983

Shortly after midnight, in a period of heightened tensions, the screens at the command bunker for the Soviet satellite-based early-warning system showed five ICBMs launching from the United States.[26] The duty officer, Stanislav Petrov, had instructions to report any detected launch to his superiors, who had a policy of immediate nuclear retaliatory strike. For five tense minutes he considered the case, then despite his remaining uncertainty, reported it to his commanders as a false alarm.

He reasoned that a US first strike with just the five missiles shown was too unlikely and noted that the missiles' vapour trails could not be identified. The false alarm turned out to be caused by sunlight glinting off clouds, which looked to the Soviet satellite system like the flashes of launching rockets.

It is often said that Petrov 'saved the world' that night. This is an something of exaggeration, as there may well have been several more steps at which nuclear retaliation could have been called off (indeed the two other incidents described here got further through the launch-on-warning process). But it was undeniably a close call: for if the satellite malfunction had reported the glinting sunlight as a hundred missiles instead of five, that may have been enough to trigger a nuclear response.[27]

Norwegian Rocket Incident: 25 January 1995

Even after the Cold War, the US and Russian missile systems have remained on hair-trigger alert. In 1995, Russian radar detected the launch of a single nuclear missile aimed at Russia, perhaps with the intention of blinding Russian radar with an electromagnetic pulse to hide a larger follow-up strike. The warning was quickly escalated all the way up the chain of command, leading President Yeltsin to open the Russian nuclear briefcase and consider whether to authorise nuclear retaliation.

But satellite systems showed no missiles and the radar soon determined that the apparent missile would land outside Russia. The alert ended; Yeltsin closed the briefcase. The false alarm had been caused by the launch of a Norwegian scientific rocket to study the northern lights. Russia had been notified, but word hadn't reached the radar operators.[28]

The first of these to be known was fallout—radioactive dust from the explosions flying up into the air, spreading out over vast areas, then falling back down. In theory, nuclear weapons could create enough fallout to cause a deadly level of radiation over the entire surface of the Earth. But we now know this would require ten times as many weapons as we currently possess.[31] Even a deliberate attempt to destroy humanity by maximising fallout (the hypothetical cobalt bomb) may be beyond our current abilities.[32]

It wasn't until the early 1980s—almost forty years into the atomic era—that we discovered what is now believed to be the most serious consequence of nuclear war. Firestorms in burning cities could create great columns of smoke, lofting black soot all the way into the stratosphere. At that height it cannot be rained out, so a dark shroud of soot would spread around the world. This would block sunlight: chilling, darkening and drying the world. The world's major crops would fail, and billions could face starvation in a nuclear winter.

Nuclear winter was highly controversial at first, since there were many uncertainties remaining, and concerns that conclusions were being put forward before the science was ready. As the assumptions and models were improved over the years, the exact nature of the threat changed, but the basic mechanism stood the test of time.[33]

Our current best understanding comes from the work of Alan Robock and colleagues.[34] While early work on nuclear winter was limited by primitive climate models, modern computers and interest in climate change have led to much more sophisticated techniques. Robock applied an ocean-atmosphere general circulation model and found an amount of cooling similar to early estimates, lasting about five times longer. This suggested a more severe effect, since this cooling may be enough to stop almost all agriculture, and it is much harder to survive five years on stockpiled food.

Most of the harm to agriculture would come from the cold, rather than the darkness or drought. The main mechanism is to greatly reduce the length of the growing season (the number of days in a row without frost). In most places this reduced growing season would be too short for most crops to reach maturity. Robock predicts that a full-scale nuclear war would cause the Earth's average surface temperature to drop by about seven degrees for about five years (then slowly return to normal over about ten more years). For comparison, this is about as cool as the Earth's last glacial period (an 'ice age').[35] As with climate change, this global average can be deceptive since some areas would cool much more than others. Summer temperatures would drop by more than 20 °C over much of North America and Asia, and would stay continually below freezing for several years in the mid-latitudes, where most of our food is produced. But the coasts and the tropics would suffer substantially less.

If nuclear winter lowered temperatures this much, billions of people would be at risk of starvation.[36] It would be an unprecedented catastrophe. Would it also be an existential catastrophe? We don't know. While we would lose almost all of our regular food production, there would be *some* food production. We could plant less efficient crops that are more cold-tolerant or have shorter growing seasons, increase farming in the tropics, increase fishing, build greenhouses, and try desperate measures such as farming algae.[37] We would have desperation on our side: a willingness to put all our wealth, our labour, our ingenuity into surviving. But we may also face a breakdown in law and order at all scales, continuing hostilities, and a loss of infrastructure including transport, fuel, fertiliser and electricity.

For all that, nuclear winter appears unlikely to lead to our extinction. No current researchers on nuclear winter are on record saying that it would and many have explicitly said that it is unlikely.[38] Existential catastrophe via a global unrecoverable collapse of civilisation also seems unlikely, especially if we consider somewhere like New Zealand (or the south-east of Australia) which is unlikely to be directly targeted and will avoid the worst effects of nuclear winter by being coastal. It is hard to see why they wouldn't make it through with most of their technology (and institutions) intact.[39]

There are significant remaining uncertainties at all stages of our understanding of nuclear winter:

1. How many cities are hit with bombs?
2. How much smoke does this produce?
3. How much of the soot is lofted into the stratosphere?[40]
4. What are the effects of this on temperature, light, precipitation?
5. What is the resulting reduction in crop yields?
6. How long does the effect last?
7. How many people are killed by such a famine?

Some of these may be reduced through future research, while others may be impossible to resolve.

Sceptics of the nuclear winter scenario often point to these remaining uncertainties, as they show that our current scientific understanding is compatible with a milder nuclear winter. But uncertainty cuts both ways. The effect of nuclear winter could also be more severe than the central estimates. We don't have a principled reason for thinking that the uncertainty here makes things better.[41] Since I am inclined to believe that the central nuclear winter scenario is not an existential catastrophe, the uncertainty actually makes things worse by leaving this possibility open. If a nuclear war were to cause an existential catastrophe, this would presumably be because the nuclear winter effect was substantially worse than expected, or because of other—as yet unknown—effects produced by such an unprecedented assault on the Earth.

It would therefore be very valuable to have additional research on the uncertainties surrounding nuclear winter, to see if there is any plausible combination that could lead to a much deeper or longer winter, and to have fresh research on other avenues by which full-scale nuclear war might pose an existential risk.

The chance of full-scale nuclear war has changed greatly over time. For our purposes we can divide it into three periods: the Cold War, the present, and the future. With the end of the Cold

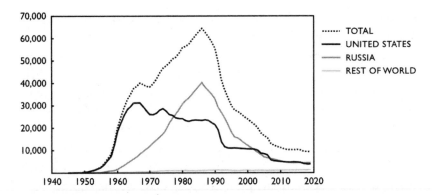

FIGURE 4.1 The number of active stockpiled nuclear warheads over time. There have been substantial reductions, but the total number (especially in the US and Russia) is still high. The combined explosive energy of these weapons has also declined, and is about 2,500 megatons today.[44]

War, the risk of a deliberately initiated nuclear war dropped considerably. However, since many missiles are still kept on hair-trigger alert (allowing them to be launched within minutes), there may remain considerable risk of nuclear war starting by accident.[42] The size of the nuclear arsenals has also gone down. The number of warheads declined from a peak of 70,000 in 1986 to about 14,000 today, and the explosive energy of each warhead has decreased too.[43] The annual risk of existential catastrophe from nuclear war should therefore be somewhat lower now than during the Cold War.

But this decrease in arsenals, and in tensions between superpowers, may reduce the risk of catastrophe less than one might think. Robock and colleagues have also modelled a limited nuclear exchange between India and Pakistan, with arsenals a fraction of the size of the US and Russia, and find a significant nuclear winter effect.[45]

And we must not be complacent. Recent years have witnessed the emergence of new geopolitical tensions that may again raise the risks of deliberate war—between the old superpowers or new ones. We have witnessed the collapse of key arms-control mechanisms between the US and Russia. And there are worrying signs that these tensions may restart the arms race, increasing the number and size of weapons towards the old levels or beyond.[46] We may see new advances that destabilise the strategic situation, such as the ability to locate nuclear submarines and thereby remove their ability to ensure a reliable nuclear counter-strike, a cornerstone of current deterrence. And the advent of military uses of AI will play a role in altering, and possibly disrupting, the strategic balance.[47]

Since a return to a nuclear cold war is not too unlikely, and would increase the annual risk by a large factor, most of the risk posed by nuclear weapons in the coming decades may come from this possibility of new escalations. And thus work to reduce the risk of nuclear war may be best aimed towards that eventuality.

CLIMATE CHANGE

The Earth's atmosphere is essential for life. It provides the pressure needed for liquid water to exist on the Earth's surface, the stability to avoid massive temperature swings between day and night, the gases that plants and animals need to survive, and—through the greenhouse effect—the insulation that keeps our planet from being entirely frozen. For without the greenhouse gases in our atmosphere, Earth would be about 33 °C colder. These gases (chiefly water vapour, carbon dioxide and methane) are more transparent to the incoming light from the Sun than to the heat that radiates back from the Earth. So they act like a blanket: trapping some of the heat, keeping the Earth warm.[48]

When the Industrial Revolution unlocked the energy that had lain dormant in fossil fuels for millions of years, it unlocked their carbon too. These carbon dioxide emissions from fossil fuels were small at first, contributing less to the warming of our climate than agriculture. But as industrialisation spread and intensified, carbon dioxide emissions increased dramatically, with more being released since 1980 than in the entire industrial period before that.[49] All told, the concentration of carbon dioxide in our atmosphere has risen from about 280 parts per million (ppm) prior to the Industrial Revolution to 412 (ppm) in 2019.[50]

Humanity's actions have already started changing our world. The Earth's climate has warmed by about 1 °C.[51] Sea levels have risen by about 23 centimetres.[52] The ocean has become more acidic by 0.1 pH.[53]

There is widespread agreement that over the coming centuries anthropogenic climate change will take a high toll on both humanity and the natural environment. Most climate science and economics deals with understanding these most likely damages. But there is also a concern that the effects of climate change could be much worse—that it poses a risk of an unrecoverable collapse of civilisation or even the complete extinction of humanity. Unlike many of the other risks I address, the central concern here isn't that we would meet our end this century, but that it may be possible for our actions now to all but lock in such a disaster for the future. If so, this could still be the time of the existential catastrophe—the time when humanity's *potential* is destroyed. If there is a serious chance of this, then climate change may be even more important than is typically recognised.

Climate change is already a major geopolitical issue, and as the damages or costs of mitigation build up, it will be an important stress upon humanity. This may impoverish us or create conflict within the international community, making us more prone to other existential risks.

Such stresses are a key contribution to existential risk (quite possibly climate change's main contribution), but are best dealt with separately. The point of these chapters on specific risks (Chapters 3, 4 and 5) is to catalogue the *direct* mechanisms for existential catastrophe. For if there were no direct mechanisms—or if they were all vanishingly unlikely—then there would be very little existential risk for other stressors to increase. We will return to indirect effects upon other existential risks in Chapter 6. For now we ask the more fundamental question of whether climate change itself could *directly* threaten our extinction or permanent collapse.

The most extreme climate possibility is known as a 'runaway greenhouse effect'. It is driven by the relationship between heat and humidity. Warm air can hold more water vapour than cool air. So when the atmosphere warms, the balance shifts between how much of Earth's water is in the oceans and how much is in the skies. Since water vapour is a potent greenhouse gas, more vapour in the atmosphere produces more warming, which produces more water vapour—an amplifying feedback.[54]

We can think of this like the feedback that occurs when you connect a microphone to a speaker. Such feedback does not always spiral out of control. If the microphone is far from the speaker, the sound does get repeatedly amplified, but each amplification adds less and less to the overall volume, so the total effect is not extreme.[55] This is what we expect with the water vapour feedback: that in total it will roughly double the warming we would get from the carbon dioxide alone.[56] But could there be climate conditions in which the water vapour warming spirals out of control, like the visceral squeal of feedback when a microphone is held too close to a speaker?

A *runaway greenhouse effect* is a type of amplifying feedback loop where the warming continues until the oceans have mostly boiled off, leaving a planet incompatible with complex life. There is widespread agreement that such a situation is theoretically possible. Something like this probably happened on Venus and may happen hundreds of millions of years into the Earth's future, as the Sun becomes hotter.[57] But current research suggests that a runaway greenhouse effect cannot be triggered by anthropogenic emissions alone.[58]

What about an amplifying feedback effect that causes massive warming, but stops short of boiling the oceans? This is known as a *moist greenhouse effect*, and if the effect is large enough it may be just as bad as a runaway.[59] This may also be impossible from anthropogenic emissions alone, but the science is less clear. A recent high-profile paper suggests it may be possible for carbon emissions to trigger such an effect (leading to 40 °C of warming in their simulation).[60] However, there are some extreme simplifications in their model and it remains an open question whether this is really possible on Earth.[61]

We might hope to look to the paleoclimate records to rule out such possibilities. At various times in the distant past, the Earth's climate has been substantially hotter than today or had much higher carbon dioxide levels. For example, about 55 million years ago in a climate event known as the Palaeocene-Eocene Thermal Maximum (PETM), temperatures climbed from about 9 °C above pre-industrial temperatures to about 14 °C over about 20,000 years. Scientists have suggested that this was caused by a major injection of carbon into the atmosphere, reaching a concentration of 1,600 ppm or more.[62] This provides some evidence that such a level of emissions and warming produces neither a moist greenhouse effect nor a mass extinction.

But the situation is not clear cut. Our knowledge of the paleoclimate record is still provisional, so the estimates of past temperatures or carbon concentrations may yet be significantly revised. And there are substantial disanalogies between now and then, most notably that the rate of warming is substantially greater today, as is the rate of growth in emissions (and the rates of change might matter as much as the levels).

So how should we think about the risk from runaway or moist greenhouse effects? The situation is akin to that of igniting the atmosphere, in that it is probably physically impossible for our actions to produce the catastrophe—but we aren't sure. I don't think the possibility of runaway or moist greenhouse effects should give cause for panic, but there *is* cause for significantly increasing the research on this topic to establish whether this extreme threat is real or illusory. For while there are some good papers suggesting we are safe, important objections continue to be raised. This is not settled science.

Are there other ways we might end up with climate change so severe as to threaten our extinction or the unrecoverable collapse of civilisation? There are three main routes: we may trigger other major feedback effects which release much more carbon into the atmosphere; we may emit substantially more carbon ourselves;

or a given amount of carbon may cause much more warming than we thought.

Water vapour from the oceans is just one of many climate feedbacks. As the world warms, some ecosystems will change in ways that release more carbon into the atmosphere, further increasing the warming. This includes the drying of rainforests and peat bogs, desertification and the increase in forest fires. Another form of feedback results from the changing reflectivity of the landscape. Ice is extremely reflective, bouncing much of the incoming sunlight straight back out to space. When warming melts ice, the ocean or land underneath is less reflective, so contributes to further warming.

Amplifying feedbacks like these can be alarming. We hear of warming producing further warming and our thoughts naturally turn to a world spinning out of control. But feedback effects are not all created equal. They can vary greatly in their gain (how close the microphone is to the speaker), their speed (how fast each loop is completed), and in how much total warming they could produce if run to completion (the maximum volume of the speaker). Moreover, there are other feedback effects that stabilise rather than amplify, where the larger the warming, the stronger they act to prevent more warming.

There are two potential amplifying feedbacks that are particularly concerning: the melting arctic permafrost and the release of methane from the deep ocean. In each case, warming would lead to additional carbon emissions, and each source contains more carbon than all fossil fuel emissions so far. They thus have the potential to dramatically alter the total warming. And neither has been incorporated into the main IPCC (Intergovernmental Panel on Climate Change) warming estimates, so any warming would come on top of the warming we are currently bracing for.

The arctic permafrost is a layer of frozen rock and soil covering more than 12 million square kilometres of land and ocean floor.[63] It contains over twice as much carbon as all anthropogenic emissions so far, trapped in the form of peat and methane.[64] Scientists are confident that over the coming centuries it will partly melt, release carbon and thus further warm the atmosphere. But the size of these effects and timeframe are very uncertain.[65] One recent estimate is that under the IPCC's high emissions scenario, permafrost melting would contribute about 0.3 °C of additional warming by 2100.[66]

Methane clathrate is an ice-like substance containing both water and methane molecules. It can be found in vast deposits in sediment at the bottom of the ocean. Because it is so hard to reach, we know very little about how much there is in total, with recent estimates ranging from twice as much carbon as we have emitted so far, to eleven times as much.[67] Warming of the oceans may trigger the melting of these clathrates and some of the methane may be carried up into the atmosphere, leading to additional warming. The dynamics of this potential feedback are even less well understood than those of the melting permafrost, with great uncertainties about when such melting could begin, whether it could happen suddenly, and how much of the methane might be released.[68]

We thus know very little about the risk these feedbacks pose. It is entirely possible that the permafrost melting and methane clathrate release are overblown and will make a negligible contribution to warming. Or that they will make a catastrophically large contribution. More research on these two feedbacks would be extremely valuable.

Feedbacks aren't the only way to get much more warming than we expect. We may simply burn more fossil fuels. The IPCC models four main emissions pathways, representing scenarios that range from rapid decarbonisation of the economy, through to what might happen in the absence of any concern about the environmental impact of our emissions. The amount we will emit based on current policies has been estimated at between 1,000 and 1,700 Gt C (gigatons of carbon) by the year 2100: around twice what we have emitted so far.[69]

I hope we refrain from coming anywhere close to this, but it is certainly conceivable that we reach this point—or that we emit even more. For example, if we simply extrapolate the annual growth of the emissions rate in recent decades to continue over the century, we could emit twice as much as the IPCC's highest emission pathway.[70] The upper bound is set by the amount of fossil fuels available. There is a wide range of estimates for the remaining fossil fuel resources, from 5,000 all the way up to 13,600 Gt C.[71] This gives us the potential to burn at least eight times as much as we have burnt so far. If we do not restrain emissions and eventually burn 5,000 Gt C of fossil fuels, the leading Earth system models suggest we'd suffer about 9 °C to 13 °C of warming by the year 2300.[72] I find it highly unlikely we would be so reckless as to reach this limit, but I can't with good conscience say it is less likely than an asteroid impact, or other natural risks we have examined.[73]

Table 4.1 puts these potential carbon emissions from permafrost, methane clathrates and fossil fuels into context. It shows that the amounts of carbon we have been talking about are so great that they dwarf the amount contained in the Earth's entire biosphere—in every living thing.[74] In fact, human activity has *already* released more than an entire biosphere worth of carbon into the atmosphere.[75]

Even if we knew how much carbon would enter the atmosphere, there would still be considerable uncertainty about how much warming this would produce. The *climate sensitivity* is the number of degrees of warming that would eventually occur if greenhouse gas concentrations were doubled from their pre-industrial baseline of 280 ppm.[78] If there were no feedbacks, this would be easy to estimate: doubling carbon dioxide while keeping everything else fixed produces about 1.2 °C of warming.[79] But the climate sensitivity also accounts for many climate feedbacks, including water vapour and cloud formation (though not permafrost or methane clathrate). These make it higher and harder to estimate.

The IPCC states that climate sensitivity is likely to be somewhere between 1.5 °C and 4.5 °C (with a lot of this uncertainty stemming from our limited understanding of cloud feedbacks).[80] When it comes to estimating the impacts of warming, this is a vast range, with the top giving three times as much warming as the bottom. Moreover, the true sensitivity could easily be

Location	Amount	Emitted by 2100
Permafrost	~ 1,700 Gt C	50–250 Gt C†
Methane clathrates	1,500–7,000 Gt C*	
Fossil fuels	5,000–13,600 Gt C	~ 1,000–1,700 Gt C‡
Biomass	~ 550 Gt C	
Necromass	~ 1,200 Gt C	
Emissions so far	~ 660 Gt C	

* This carbon is all in the form of methane, which is much more potent in the short run. But if the release is gradual, this may not make much difference. A small proportion of carbon in permafrost is methane.
† On the high emissions pathway.[77]
‡ On current policies for fossil fuel use.

TABLE 4.1 Where is the carbon? A comparison of the size of known carbon stocks that could potentially be released into the atmosphere, and how much of these might be released from now until the end of the century. *Biomass* is the total amount of carbon in all living organisms on the Earth. *Necromass* is the total amount of carbon in dead organic matter, especially in the soil, some of which could be released through deforestation or forest fires. I have also included our total emissions from 1750 to today—those from changes in land use as well as from fossil fuels and industry.[76]

even higher, as the IPCC is only saying that there is at least a two-thirds chance it falls within this range.[81] And this uncertainty is compounded by our uncertainty in how high greenhouse gas concentrations will go. If we end up between one and two doublings from pre-industrial levels, the range of eventual warming is 1.5 °C to 9 °C.[82]

We might hope that some of this uncertainty will soon be resolved, but the record of progress is not promising. The current range of 1.5 °C to 4.5 °C was first put forward in 1979 and has barely changed over the last forty years.[83]

We often hear numbers that suggest much more precision than this: that we are now headed for 5 °C warming or that certain policies are needed if we are to stay under 4 °C of warming. But these expressions simplify so much that they risk misleading. They really mean that we are headed for somewhere between 2.5 °C and 7.5 °C of warming or that certain policies are required in order to have a decent chance of staying under 4 °C (sometimes defined as a 66 percent chance, sometimes just 50 percent).[84]

When we combine the uncertainties about our direct emissions, the climate sensitivity and the possibility of extreme feedbacks, we end up being able to say very little to constrain the amount of warming. Ideally, in such a situation we could still give robust estimates of the size and shape of the distribution (as we saw for asteroids), so that we could consider the probability of extreme outcomes, such as ending up above 6 °C—or even 10 °C. But due to the complexity of the problem we cannot even do this. The best I can say is that when accounting for all the uncertainties, we could plausibly end up with anywhere up to 13 °C of warming by 2300. And even that is not a strict upper limit.

Warming at such levels would be a global calamity of unprecedented scale. It would be an immense human tragedy, disproportionately impacting the most vulnerable populations. And it would throw civilisation into such disarray that we may be much more vulnerable to other existential risks. But the purpose of this chapter is finding and assessing threats that pose a direct existential risk to humanity. Even at such extreme levels of warming, it is difficult to see exactly how climate change could do so.

Major effects of climate change include reduced agricultural yields, sea level rises, water scarcity, increased tropical diseases, ocean acidification and the collapse of the Gulf Stream. While extremely important when assessing the overall risks of climate change, none of these threaten extinction or irrevocable collapse.

Crops are very sensitive to reductions in temperature (due to frosts), but less sensitive to increases. By all appearances we would still have food to support civilisation.[85] Even if sea levels rose hundreds of metres (over centuries), most of the Earth's land area would remain. Similarly, while some areas might conceivably become uninhabitable due to water scarcity, other areas will have increased rainfall. More areas may become susceptible to tropical diseases, but we need only look to the tropics to see civilisation flourish despite this. The main effect of a collapse of the system of Atlantic Ocean currents that includes the Gulf Stream is a 2 °C cooling of Europe—something that poses no permanent threat to global civilisation.

From an existential risk perspective, a more serious concern is that the high temperatures (and the rapidity of their change) might cause a large loss of biodiversity and subsequent ecosystem collapse. While the pathway is not entirely clear, a large enough collapse of ecosystems across the globe could perhaps threaten human extinction. The idea that climate change could cause widespread extinctions has some good theoretical support.[86] Yet the evidence is mixed. For when we look at many of the past cases of extremely high global temperatures or extremely rapid warming we don't see a corresponding loss of biodiversity.[87]

So the most important known effect of climate change from the perspective of direct existential risk is probably the most obvious: heat stress. We need an environment cooler than our body temperature to be able to rid ourselves of waste heat and stay alive. More precisely, we need to be able to lose heat by sweating, which depends on the humidity as well as the temperature.

A landmark paper by Steven Sherwood and Matthew Huber showed that with sufficient warming there would be parts of the world whose temperature and humidity combine to exceed the level where humans could survive without air conditioning.[88] With 12 °C of warming, a very large land area—where more than half of all people currently live and where much of our food is grown—would exceed this level at some point during a typical year. Sherwood and Huber suggest that such areas would be uninhabitable. This may not quite be true (particularly if air conditioning is possible during the hottest months), but their habitability is at least in question.

However, substantial regions would also remain below this threshold. Even with an extreme 20 °C of warming there would be many coastal areas (and some elevated regions) that would have no days above the temperature/humidity threshold.[89] So there would remain large areas in which humanity and civilisation could continue. A world with 20 °C of warming would be an unparalleled human and environmental tragedy, forcing mass migration and perhaps starvation too. This is reason enough to do our utmost to prevent anything like that from ever happening. However, our present task is identifying existential risks to humanity and it is hard to see how any realistic level of heat stress could pose such a risk. So the runaway and moist greenhouse effects remain the only known mechanisms through which climate change could directly cause our extinction or irrevocable collapse.

This doesn't rule out *unknown* mechanisms. We are considering large changes to the Earth that may even be unprecedented in size or speed. It wouldn't be astonishing if that directly led to our permanent ruin. The best argument against such unknown mechanisms is probably that the PETM did not lead to a mass extinction, despite temperatures rapidly rising about 5 °C, to reach a level 14 °C above pre-industrial temperatures.[90] But this is tempered by the imprecision of paleoclimate data, the sparsity of the fossil record, the smaller size of mammals at the time (making them more heat-tolerant), and a reluctance to rely on a single example. Most importantly, anthropogenic warming could be over a hundred times faster than warming during the PETM, and rapid warming has been suggested as a contributing factor in the end-Permian mass extinction, in which 96 percent of species went extinct.[91] In the end, we can say little more than that direct existential risk from climate change appears very small, but cannot yet be ruled out.

Our focus so far has been on whether climate change could conceivably be an existential catastrophe. In assessing this, I have set aside the question of whether we could mitigate this risk. The most obvious and important form of mitigation is reducing our emissions. There is a broad consensus that this must play a key role in any mitigation strategy. But there are also ways of mitigating the effects of climate change after the emissions have been released.

These techniques are often called *geoengineering*. While the name conjures up a radical and dangerous scheme for transforming our planet, the proposals in fact range from the radical to the mundane. They also differ in their cost, speed, scale, readiness and risk.

The two main approaches to geoengineering are carbon dioxide removal and solar radiation management. Carbon dioxide removal strikes at the root of the problem, removing the carbon dioxide from our atmosphere and thus taking away the source of the heating. It is an attempt to cure the Earth of its affliction. At the radical end is ocean fertilisation: seeding the ocean with iron to encourage large algal blooms which capture carbon before sinking into the deep ocean. At the mundane end are tree planting and carbon scrubbing.

Solar radiation management involves limiting the amount of sunlight absorbed by the Earth. This could involve blocking light before it hits the Earth, reflecting more light in the atmosphere before it hits the surface, or reflecting more of the light that hits the surface. It is an attempt to offset the warming effects of the carbon dioxide by cooling the Earth. It is typically cheaper than carbon dioxide removal and quicker to act, but has the downsides of ignoring other bad effects of carbon (such as ocean acidification) and requiring constant upkeep.

A central problem with geoengineering is that the cure may be worse than the disease. For the very scale of what it is attempting to achieve could create a risk of massive unintended consequences over the entire Earth's surface, possibly posing a greater existential risk than climate change itself. Geoengineering thus needs to be very carefully governed—especially when it comes to radical techniques that are cheap enough for a country or research group to implement unilaterally—and we shouldn't rely on it as an alternative to emissions reductions. But it may well have a useful role to play as a last resort, or as a means for the eventual restoration of our planet's climate.[92]

ENVIRONMENTAL DAMAGE

Climate change is not the only form of environmental damage we are inflicting upon the Earth. Might we face other environmental existential risks through overpopulation, running out of critical resources or biodiversity loss?

When environmentalism rose to prominence in the 1960s and 1970s, one major concern was overpopulation. It was widely feared that humanity's rapidly growing population would far outstrip the Earth's capacity to feed people, precipitating an environmental and humanitarian catastrophe. The most prominent advocate of this view, Paul Ehrlich, painted an apocalyptic vision of the near future: 'Most of the people who are going to die in the greatest cataclysm in the history of man have already been born.'[93] This catastrophe would come soon and pose a direct existential risk. Ehrlich predicted: 'Sometime in the next 15 years, the end will come—and by "the end" I mean an utter breakdown of the capacity of the planet to support humanity.'[94]

These confident predictions of doom were thoroughly mistaken. Instead of rising to unprecedented heights, the prevalence of famine dramatically declined. Less than a quarter as many people died of famine in the 1970s as in the 1960s, and the rate has since halved again.[95] Instead of dwindling to a point of crisis, the amount of food per person has steadily risen over the last fifty years. We now have 24 percent more food per person than when Ehrlich's book, *The Population Bomb*, was published in 1968.

Much of the credit for this is owed to the Green Revolution, in which developing countries rose to the challenge of feeding their people. They did so by modernising their farming, with improved fertilisers, irrigation, automation and grain varieties.[96] Perhaps the greatest single contribution was from Norman Borlaug, who received the Nobel Prize for his work breeding the new, high-yield varieties of wheat, and who may be responsible for saving more lives than anyone else in history.[97]

But the improvements in agriculture are just part of the story. The entire picture of overpopulation has changed. Population growth is almost always presented as an exponential process—increasing by a fixed percentage each year—but in fact that is rarely the case. From about 1800 to 1960 the world population was growing much faster than an exponential. The annual growth rate was itself growing from 0.4 percent all the way to an unprecedented rate of 2.2 percent in 1962. These trends rightly warranted significant concern about the human and environmental consequences of this rapid population increase.

But suddenly, the situation changed. The population growth rate started to rapidly decline. So far it has halved, and it continues to fall. Population is now increasing in a roughly linear manner, with a fixed *number* of people being added each year instead of a fixed *proportion*. This change has been driven not by the feared increase in death rates, but by a dramatic change in fertility, as more and more countries have undergone the demographic transition to a small family size. In 1950, the average number of children born to each woman was 5.05. It is now just 2.47—not so far above the replacement rate of 2.1 children per woman.[98]

While we can't know what will happen in the future, the current trends point to a rapid stabilisation of the population. The current linear increase is likely to be an inflection point in the history of human population: the point where the curve finally

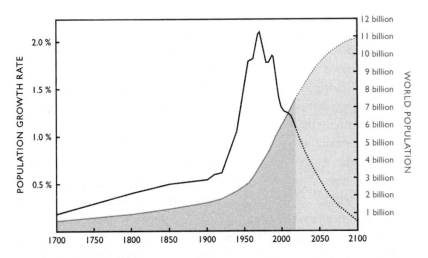

FIGURE 4.2 World population from 1700 to today (dark grey) and projected up to 2100 (light grey). The black line shows the annual percentage growth rate of population, which reached an extreme peak in 1962 but has since been dropping quickly.[99]

starts to level off. We may never again see the rapid population growth of the mid-twentieth century. In the last eighty years, population grew threefold. In the next eighty years (to 2100) it is expected to go up just 50 percent, to about 11 billion. For every person alive now, we'll have to make room for an extra half a person. This will be a challenge, but a much easier one than last century.

Some people have gone so far as to suggest that the real extinction risk might now be declining population.[100] Fertility rates in most countries outside Africa have fallen to below the replacement rate, and perhaps this will become a global trend. Even then, I don't think there is any real cause for concern. If declining population began to pose a clear and present danger (something that couldn't happen for at least two centuries), it would be a simple matter for public policy to encourage childbearing up to the replacement level. The available policy levers—free childcare, free education, free child healthcare and tax benefits for families—are relatively simple, non-coercive and popular.

While the danger of population spiralling rapidly out of control has abated, population has certainly reached a very high level. And with our rapidly increasing prosperity and power, each person is having more impact on the environment.[101] This is creating large stresses on the biosphere, some of which are completely unprecedented. This may, in turn, create threats to our continued existence.

One category of concern is resource depletion. People have suggested that humanity is running short of fossil fuels, phosphorous, topsoil, fresh water, and certain metals.[102] However, these forms of resource scarcity don't appear to pose any direct risk of destroying our potential.

Running out of fossil fuels might result in economic recession as we switch to more expensive alternatives, but we are quite capable of maintaining a civilisation without fossil fuels. Indeed, we are already planning to do so soon as we move to zero emissions later this century. My guess is that, if anything, failure to find new sources of fossil fuels would actually lower overall existential risk.

What about water? While fresh water is much more scarce than seawater, we do have a lot of it in absolute terms: 26 million litres of accessible fresh water per person.[103] Most of the problem comes from this being poorly distributed. Even in the worst case, fresh water could be substituted with desalinated sea water at a cost of about $1 for 1,000 litres. There would be additional expenses involved to do this using clean energy and when pumping the water uphill to people and farms away from the coast, but we could do it if we had to.

It is unclear whether there are any significant metal shortages at all: previous predictions have failed and one should expect markets to slow consumption, encourage recycling and develop alternatives if stocks do start running low.[104] Moreover, the types of rare metals that are plausibly running out don't seem to be essential for civilisation.

While I don't know of any resources whose scarcity could plausibly constitute an existential catastrophe, it is hard to completely rule it out. It is possible that we will find a resource that is scarce, performs an essential function for civilisation, has no feasible alternative, can't be adequately recycled, and where market forces won't ration our consumption. While I am sceptical that any resources meet this description, I encourage efforts to thoroughly check whether this is so.

Another category of environmental concern is loss of biodiversity. Our actions are destroying and threatening so many species that people have suggested a sixth mass extinction is underway.[105] Is this true?

It is difficult to say. One problem is that we can't cleanly compare the modern and fossil records.[106] Another is that there is more than one measure of a mass extinction. The rate at which species are going extinct is much higher than the longterm average—at least ten to 100 times higher, and expected to accelerate.[107] Species may be going extinct even more rapidly than in a typical mass extinction. But the fraction of species that have gone extinct is much lower than in a mass extinction. Where the big five mass extinctions all lost more than 75 percent of species, we have lost about 1 percent.[108] This could be a sixth mass extinction that is just getting started, though the evidence fits a substantially smaller extinction event just as well. In any event, our inability to rule out being in the midst of a mass extinction is deeply troubling.

And while extinction is a useful measure of biodiversity loss, it is not the whole story. It doesn't capture population reductions or species disappearing locally or regionally. While 'only' 1 percent of species have gone extinct on our watch, the toll on biodiversity within each region may be much higher, and this may be what matters most. From the perspective of existential risk, what matters most about biodiversity loss is the loss of *ecosystem services*. These are services—such as purifying water and air, providing energy and resources, or improving our soil—that plants and animals currently provide for us, but we may find costly or impossible to do ourselves.

A prominent example is the crop pollination performed by honeybees. This is often raised as an existential risk, citing a quotation attributed Einstein that 'If the bee disappeared off the surface of the globe then man would only have four years of life left.' This has been thoroughly debunked: it is not true and Einstein didn't say it.[109] In fact, a recent review found that even if honeybees were completely lost—and *all* other pollinators too—this would only create a 3 to 8 percent reduction in global crop production.[110] It would be a great environmental tragedy and a crisis for humanity, but there is no reason to think it is an existential risk.

While that particular example is spurious, perhaps there are other distinct ecosystem services that are threatened and that we couldn't live without. Or a cascading failure of ecosystem services may collectively be too much for our civilisation to be able to replace. It is clear that at *some* level of environmental destruction this would be true, though we have little idea how close we are to such a threshold, nor whether a cascade could take us there. We need more research to find out.

As with nuclear winter and extreme global warming, we don't know of a direct mechanism for existential risk, but are putting such pressure on the global environment that there may well be some as yet unknown consequence that would threaten our survival. We could therefore think of continuing environmental damage over the coming century as a source of unforeseen threats to humanity. These unmodelled effects may well contain most of the environmental existential risk.

Nuclear war, climate change and environmental damage are extremely serious global issues—even before we come to the question of whether they could cause existential catastrophe. In each case, humanity holds tremendous power to change the face of the Earth in ways that are without precedent in the 200,000 years of *Homo sapiens*. The latest science backs up the extreme scale of these changes, though it stops short of providing clear and proven mechanisms for a truly existential catastrophe. The existential risk from these sources therefore remains more speculative than that from asteroids. But this is not the same as the risk being *smaller*. Given our current scientific knowledge, I think it would be very bold to put the probability of these risks at less than the 0.001 percent to 0.05 percent per century that comes from all natural risks together. Indeed, I'd estimate that each of these three risks has a higher probability than that of all natural existential risks put together. And there may be even greater risks to come.

5

FUTURE RISKS

The Dark Ages may return, the Stone Age may return on the gleaming wings of Science, and what might now shower immeasurable material blessings upon mankind, may even bring about its total destruction.

—Winston Churchill[1]

It is now time to cast our gaze to the horizon, to see what possibilities the coming century may bring. These possibilities are hard to discern through the haze of distance; it is extremely difficult to tell which new technologies will be possible, what form they will take when mature, or the context of the world into which they will arrive. And this veil may not lift until the new technologies are right upon us. For even the best experts or the very inventors of the technology can be blindsided by major developments.

One night in 1933, the world's pre-eminent expert on atomic science, Ernest Rutherford, declared the idea of harnessing atomic energy to be 'moonshine'. And the very next morning Leo Szilard discovered the idea of the chain reaction. In 1939, Enrico Fermi told Szilard the chain reaction was but a 'remote possibility', and four years later Fermi was personally overseeing the world's first nuclear reactor. The staggering list of eminent scientists who thought heavier-than-air flight to be impossible or else decades away is so well rehearsed as to be cliché. But fewer know that even Wilbur Wright himself predicted it was at least fifty years away—just two years before he invented it.[2]

So we need to remember how quickly new technologies can be upon us, and to be wary of assertions that they are either impossible or so distant in time that we have no cause for concern. Confident denouncements by eminent scientists should certainly give us reason to be sceptical of a technology, but not to bet our lives against it—their track record just isn't good enough for that.[3]

Of course, there is no shortage of examples of scientists and technologists declaring a new technology to be just around the corner, when in fact it would only arrive decades later; or not at all; or in a markedly different form to the one anticipated. The point is not that technology usually comes earlier than predicted, but that it can easily do so; that we need to be cautious in ruling things out or assuming we have ample time.

But we must not veer towards the opposite mistake: using the impossibility of knowing the future as an excuse to ignore it. There *are* things we can say. For example, it would be surprising if the long-run trend towards developing technologies of increasing power *didn't* continue into this century. And since it was precisely our unprecedented power that gave rise to the anthropogenic risks of the twentieth century, it would be remarkable if the coming century didn't pose similar, or greater, risk.

Despite this being a chapter on the future, we won't be engaging in prediction—at least not in the usual sense of saying which technologies will come, and when. Instead, we shall chart the horizon in terms of plausibility and probability. Are there plausible future technologies that would bring existential risks? Are these technologies probable enough to warrant preparations in case they do arrive? To do this, we don't need to know the future, nor even the precise probabilities of what may occur. We just need to estimate these probabilities to the right ballpark; to see the dim outlines of the threats. This will give us a rough idea of the landscape ahead and how we might prepare for it.

Much of what new technologies bring will of course be helpful, and some truly wondrous. Technological progress has been one of the main sources of our modern prosperity and longevity—one of the main reasons extreme poverty has become the exception rather than the rule, and life expectancy has doubled since the Industrial Revolution. Indeed, we can see that over the centuries all the risks technology imposes on humans have been outweighed by the benefits it has brought.[4] For these dramatic gains to health and wealth are *overall* gains, taking all the ill effects into account.

Or at least this is true for most risks: those that are likely enough and common enough that the law of large numbers wins out, turning the unpredictability of the small scale into a demonstrable longterm average. We know that these everyday risks have been more than outweighed. But we don't know whether this positive balance was due to getting lucky on a few key rolls of the dice. For instance, it is conceivable that the risk of nuclear war breaking out was serious enough to outweigh all the benefits of modern technology.

It is this that should most interest us when we look to the century ahead. Not the everyday risks and downsides that technology may bring, but whether there will be a handful of cases where it puts our entire bankroll at risk, with no subsequent way for us to make up the losses.

Growing up, I had always been strongly pro-technology. If not for the plausibility of these unconsummated catastrophic risks, I'd remain so. But instead, I am compelled towards a much more ambivalent view. I don't for a moment think we should cease technological progress—indeed if some well-meaning regime locked in a permanent freeze on technology, that would probably itself be an existential catastrophe, preventing humanity from ever fulfilling its potential.

But we do need to treat technological progress with maturity.[5] We should continue our technological developments to make sure we receive the fruits of technology. Yet we must do so very carefully, and if needed, use a significant fraction of the gains from technology to address the potential dangers, ensuring the balance stays positive. Looking ahead and charting the potential hazards on our horizon is a key step.

PANDEMICS

In 1347 death came to Europe. It entered through the Crimean town of Caffa, brought by the besieging Mongol army. Fleeing merchants unwittingly carried it back to Italy. From there it spread to France, Spain, England. Then up as far as Norway and across the rest of Europe—all the way to Moscow. Within six years, the Black Death had taken the continent.[6]

Tens of millions fell gravely ill, their bodies succumbing to the disease in different ways. Some bore swollen buboes on their necks, armpits and thighs; some had their flesh turn black from haemorrhaging beneath the skin; some coughed blood from the necrotic inflammation of their throats and lungs. All forms involved fever, exhaustion, and an intolerable stench from the material that exuded from the body.[7] There were so many dead that mass graves needed to be dug and even then, cemeteries ran out of room for the bodies.

The Black Death devastated Europe. In those six years, between one-quarter and one-half of all Europeans were killed.[8] The Middle East was ravaged too, with the plague killing about one in three Egyptians and Syrians. And it may have also laid waste to parts of Central Asia, India and China. Due to the scant records of the fourteenth century, we will never know the true toll, but our best estimates are that somewhere between 5 percent and 14 percent of all the world's people were killed, in what may have been the greatest catastrophe humanity has seen.[9]

Are we safe now from events like this? Or are we more vulnerable? Could a pandemic threaten humanity's future?[10]

The Black Death was not the only biological disaster to scar human history. It was not even the only great bubonic plague. In 541 CE the Plague of Justinian struck the Byzantine Empire. Over three years it took the lives of roughly 3 percent of the world's people.[11]

When Europeans reached the Americas in 1492, the two populations exposed each other to completely novel diseases. Over thousands of years each population had built up resistance to their own set of diseases, but were extremely susceptible to the others. The American peoples got by far the worse end of exchange, through diseases such as measles, influenza and especially smallpox.

During the next hundred years a combination of invasion and disease took an immense toll—one whose scale may never be known, due to great uncertainty about the size of the pre-existing population. We can't rule out the loss of more than 90 percent of the population of the Americas during that century, though the number could also be much lower.[12] And it is very difficult to tease out how much of this should be attributed to war and occupation, rather than disease. As a rough upper bound, the Columbian exchange may have killed as many as 10 percent of the world's people.[13]

Centuries later, the world had become so interconnected that a truly global pandemic was possible. Near the end of the First World War, a devastating strain of influenza (known as the 1918 flu or Spanish Flu) spread to six continents, and even remote Pacific islands. At least a third of the world's population were infected and 3 to 6 percent were killed.[14] This death toll outstripped that of the First World War, and possibly both World Wars combined.

Yet even events like these fall short of being a threat to humanity's longterm potential.[15] In the great bubonic plagues we saw civilisation in the affected areas falter, but recover. The regional 25 to 50 percent death rate was not enough to precipitate a continent-wide collapse of civilisation. It changed the relative fortunes of empires, and may have altered the course of history substantially, but if anything, it gives us reason to believe that human civilisation is likely to make it through future events with similar death rates, even if they were global in scale.

The 1918 flu pandemic was remarkable in having very little apparent effect on the world's development despite its global reach. It looks like it was lost in the wake of the First World War, which despite a smaller death toll, seems to have had a much larger effect on the course of history.[16]

125

It is less clear what lesson to draw from the Columbian exchange due to our lack of good records and its mix of causes. Pandemics were clearly a part of what led to a regional collapse of civilisation, but we don't know whether this would have occurred had it not been for the accompanying violence and imperial rule.

The strongest case against existential risk from natural pandemics is the fossil record argument from Chapter 3. Extinction risk from natural causes above 0.1 percent per century is incompatible with the evidence of how long humanity and similar species have lasted. But this argument only works where the risk to humanity now is similar or lower than the longterm levels. For most risks this is clearly true, but not for pandemics. We have done many things to exacerbate the risk: some that could make pandemics more likely to occur, and some that could increase their damage. Thus even 'natural' pandemics should be seen as a partly anthropogenic risk.

Our population now is a thousand times greater than over most of human history, so there are vastly more opportunities for new human diseases to originate.[17] And our farming practices have created vast numbers of animals living in unhealthy conditions within close proximity to humans. This increases the risk, as many major diseases originate in animals before crossing over to humans. Examples include HIV (chimpanzees), Ebola (bats), SARS (probably bats) and influenza (usually pigs or birds).[18] Evidence suggests that diseases are crossing over into human populations from animals at an increasing rate.[19]

Modern civilisation may also make it much easier for a pandemic to spread. The higher density of people living together in cities increases the number of people each of us may infect. Rapid long-distance transport greatly increases the distance pathogens can spread, reducing the degrees of separation between any two people. Moreover, we are no longer divided into isolated populations as we were for most of the last 10,000 years.[20] Together these effects suggest that we might expect more new pandemics, for them to spread more quickly, and to reach a higher percentage of the world's people.

But we have also changed the world in ways that offer protection. We have a healthier population; improved sanitation and hygiene; preventative and curative medicine; and a scientific understanding of disease. Perhaps most importantly, we have public health bodies to facilitate global communication and coordination in the face of new outbreaks. We have seen the benefits of this protection through the dramatic decline of endemic infectious disease over the last century (though we can't be sure pandemics will obey the same trend). Finally, we have spread to a range of locations and environments unprecedented for any mammalian species. This offers special protection from extinction events, because it requires the pathogen to be able to flourish in a vast range of environments and to reach exceptionally isolated populations such as uncontacted tribes, Antarctic researchers and nuclear submarine crews.[21]

It is hard to know whether these combined effects have increased or decreased the existential risk from pandemics. This uncertainty is ultimately bad news: we were previously sitting on a powerful argument that the risk was tiny; now we are not. But note that we are not merely interested in the direction of the change, but also in the *size* of the change. If we take the fossil record as evidence that the risk was less than one in 2,000 per century, then to reach 1 percent per century the pandemic risk would need to be at least 20 times larger. This seems unlikely. In my view, the fossil record still provides a strong case against there being a high *extinction* risk from 'natural' pandemics. So most of the remaining existential risk would come from the threat of permanent collapse: a pandemic severe enough to collapse civilisation globally, combined with civilisation turning out to be hard to re-establish or bad luck in our attempts to do so.

But humanity could also play a much larger role. We have seen the indirect ways that our actions aid and abet the origination and the spread of pandemics. But what about cases where we have a much more direct hand in the process—where we deliberately use, improve or create the pathogens?

Our understanding and control of pathogens is very recent. Just 200 years ago we didn't even understand the basic cause of pandemics—a leading theory in the West claimed that disease was produced by a kind of gas. In just two centuries, we discovered it was caused by a diverse variety of microscopic agents and we worked out how to grow them in the lab, to breed them for different traits, to sequence their genomes, to implant new genes, and to create entire functional viruses from their written code.

This progress is continuing at a rapid pace. The last ten years have seen major qualitative breakthroughs, such as the use of CRISPR to efficiently insert new genetic sequences into a genome and the use of gene drives to efficiently replace populations of natural organisms in the wild with genetically modified versions.[22] Measures of this progress suggest it is accelerating, with the cost to sequence a genome falling by a factor of 10,000 since 2007 and with publications and venture capital investment growing quickly.[23] This progress in biotechnology seems unlikely to fizzle out soon: there are no insurmountable challenges looming; no fundamental laws blocking further developments.

Here the past offers almost no reassurance. Increasing efforts are made to surpass natural abilities, so long-run track records need not apply. It would be optimistic to assume that this uncharted new terrain holds only familiar dangers.

To start with, let's set aside the risks from malicious intent, and consider only the risks that can arise from well-intentioned research. Most scientific and medical research poses a negligible risk of harms at the scale we are considering. But there is a small fraction that uses live pathogens of kinds which are known to threaten global harm. These include the agents that cause the 1918 flu, smallpox, SARS and H5N1 flu. And a small part of this research involves making strains of these pathogens that pose even more danger than the natural types, increasing their transmissibility, lethality or resistance to vaccination or treatment.

In 2012 a Dutch virologist, Ron Fouchier, published details of a gain-of-function experiment on the recent H5N1 strain of bird flu.[24] This strain was extremely deadly, killing an estimated 60 percent of humans it infected—far beyond even the 1918 flu.[25] Yet its inability to pass from human to human had so far prevented a pandemic. Fouchier wanted to find out whether (and how) H5N1 could naturally develop this ability. He passed the disease through a series of ten ferrets, which are commonly used as a model for how influenza affects humans. By the time it passed to the final ferret, his strain of H5N1 had become directly transmissible between mammals.

The work caused fierce controversy. Much of this was focused on the information contained in his work. The US National Science Advisory Board for Biosecurity ruled that his paper had to be stripped of some of its technical details before publication, to limit the ability of bad actors to cause a pandemic. And the Dutch government claimed it broke EU law on exporting information useful for bioweapons. But it is not the possibility of misuse that concerns me here. Fouchier's research provides a clear example of well-intentioned scientists enhancing the destructive capabilities of pathogens known to threaten global catastrophe. And nor is it the only case. In the very same year a similar experiment was performed in the United States.[26]

Of course, such experiments are done in secure labs, with stringent safety standards. It is highly unlikely that in any particular case the enhanced pathogens would escape into the wild. But just how unlikely? Unfortunately, we don't have good data, due to a lack of transparency about incident and escape rates.[27] This prevents society from making well-informed decisions balancing the risks and benefits of this research, and it limits the ability of labs to learn from each other's incidents. We need consistent and transparent reporting of incidents, in line with the best practices from other sectors.[28] And we need serious accountability for when incidents or escapes go beyond the promised rates.

But even the patchy evidence we do have includes enough confirmed cases to see that the rates of escape are worryingly high (see the box 'Notable Laboratory Escapes').[29] None of these documented escapes directly posed a risk of existential catastrophe, but they show that security for highly dangerous pathogens has been deeply flawed, and remains insufficient.

This is true even at the highest biosafety level (BSL-4). In 2001, Britain was struck by a devastating outbreak of foot-and-mouth disease in livestock. Six million animals were killed in an attempt to halt its spread, and the economic damages totalled £8 billion. Then in 2007 there was another outbreak, which was traced to a lab working on the disease. Foot-and-mouth was considered a highest category pathogen and required the highest level of biosecurity. Yet the virus escaped from a badly maintained pipe, leaking into the groundwater at the facility. After an investigation, the lab's licence was renewed—only for another leak to occur two weeks later.[30] In my view, this track record of escapes shows that even BSL-4 is insufficient for working on pathogens that pose a risk of global pandemics on the scale of the 1918 flu or worse—especially if that research involves gain-of-function (and the extremely dangerous H5N1 gain-of-function research wasn't even performed at BSL-4).[31] Thirteen years since the last publicly acknowledged outbreak from a BSL-4 facility is not good enough. It doesn't matter whether this is from insufficient standards, inspections, operations or penalties. What matters is the poor track record in the field, made worse by a lack of transparency and accountability. With current BSL-4 labs, an escape of a pandemic pathogen is a matter of time.

Alongside the threat of accident is the threat of deliberate misuse. Humanity has a long and dark history of disease as a weapon. There are records dating back to 1320 BCE, describing a war in Asia Minor, where infected sheep were driven across the border to spread tularaemia.[37] There is even a contemporaneous account of the siege of Caffa claiming the Black Death was introduced to Europe by the Mongol army catapulting plague-ridden corpses over the city walls. It is not clear whether this really occurred, nor whether the Black Death would have found its way into Europe regardless. Yet it remains a live possibility that the most deadly event in the history of the world (as a fraction of humanity) was an act of biological warfare.[38]

NOTABLE LABORATORY ESCAPES

1971: Smallpox

A Soviet bioweapons lab tested a weaponised strain of smallpox on an island in the Aral Sea. During a field test, they accidentally infected people on a nearby ship who spread it ashore. The resulting outbreak infected ten people, killing three, before being contained by a mass quarantine and vaccination programme.[32]

1978: Smallpox

In 1967 smallpox was killing more than a million people a year, but a heroic global effort drove that to zero in 1977, freeing humanity from this ancient scourge. And yet a year later, it returned from the grave: escaping from a British lab, killing one person and infecting another before authorities contained the outbreak.[33]

1979: Anthrax

A bioweapons lab in one of the Soviet Union's biggest cities, Sverdlovsk, accidentally released a large quantity of weaponised anthrax, when they took an air filter off for cleaning. There were 66 confirmed fatalities.[34]

1995: Rabbit Calicivirus

Australian scientists conducted a field trial with a new virus for use in controlling their wild rabbit population. They released it on a small island, but the virus escaped quarantine, reaching the mainland and accidentally killing 30 million rabbits within just a few weeks.[35]

2015: Anthrax

The Dugway Proving Grounds was established by the US military in 1942 to work on chemical and biological weapons. In 2015, it accidentally distributed samples containing live anthrax spores to 192 labs across eight countries, which thought they were receiving inactivated anthrax.[36]

One of the first unequivocal accounts of biological warfare was by the British in Canada in 1763. The Commander-in-Chief for North America, Jeffrey Amherst, wrote to a fort that had suffered a smallpox outbreak: 'Could it not be contrived to send the smallpox among those disaffected tribes of Indians? We must on this occasion, use every stratagem in our power to reduce them.' The same idea had already occurred to the garrison, who acted on it of their own initiative. They distributed disease-ridden items, documented the deed, and even filed for official reimbursement to cover the costs of the blankets and handkerchief used.[39]

Where earlier armies had limited understanding of disease, and mostly opportunistic biowarfare, our greater understanding has enabled modern nations to build on what nature provided. During the twentieth century, fifteen countries are known to have developed bioweapons programmes, including the US, UK and France.[40]

The largest programme was the Soviets'. At its height it had more than a dozen clandestine labs employing 9,000 scientists to weaponise diseases ranging from plague to smallpox, anthrax and tularaemia. Scientists attempted to increase the diseases' infectivity, lethality and resistance to vaccination and treatment. They created systems for spreading the pathogens to their opponents and built up vast stockpiles, reportedly including more than 20 *tons* of smallpox and of plague. The programme was prone to accidents, with lethal outbreaks of both smallpox and anthrax (see Box).[41] While there is no evidence of deliberate attempts to create a pathogen to threaten the whole of humanity, the logic of deterrence or mutually assured destruction could push superpowers or rogue states in that direction.

The good news is that for all our flirtation with biowarfare, there appear to have been relatively few deaths from either accidents or use (assuming the Black Death to have been a natural pandemic).[42] The confirmed historical death toll from biowarfare is dwarfed by that of natural pandemics over the same timeframe.[43] Exactly why this is so is unclear. One reason may be that bioweapons are unreliable and prone to backfiring, leading states to use other weapons in preference. Another suggestion is that tacit knowledge and operational barriers make it much harder to deploy bioweapons than it may first appear.[44]

But the answer may also just be that we have too little data. The patterns of disease outbreaks, war deaths and terrorist attacks all appear to follow power law distributions. Unlike the familiar 'normal' distribution where sizes are clustered around a central value, power law distributions have a 'heavy tail' of increasingly large events, where there can often be events at entirely different scales, with some being thousands, or millions, of times bigger than others. Deaths from war and terror appear to follow power laws with especially heavy tails, such that the majority of the deaths happen in the few biggest events. For instance, warfare deaths in the last hundred years are dominated by the two World Wars, and most US fatalities from terrorism occurred in the September 11 attacks.[45] When events follow a distribution like this, the average size of events until now systematically under-represents the expected size of events to come, even if the under-lying risk stays the same.[46]

And it is not staying the same. Attempts to use the historical record overlook the rapid changes in biotechnology. It is not twentieth-century bioweaponry that should alarm us, but the next hundred years of improvements. A hundred years ago, we had only just discovered viruses and were yet to discover DNA. Now we can design the DNA of viruses and resurrect historic viruses from their genetic sequences. Where will we be a hundred years from now?

One of the most exciting trends in biotechnology is its rapid democratisation—the speed at which cutting-edge techniques can be adopted by students and amateurs. When a new breakthrough is achieved, the pool of people with the talent, training, resources and patience to reproduce it rapidly expands: from a handful of the world's top biologists, to people with PhDs in the field, to millions of people with undergraduate-level biology.

The Human Genome Project was the largest ever scientific collaboration in biology. It took thirteen years and $500 million to produce the full DNA sequence of the human genome. Just

15 years later, a genome can be sequenced for under $1,000 or within a single hour.[47] The reverse process has become much easier too: online DNA synthesis services allow anyone to upload a DNA sequence of their choice then have it constructed and shipped to their address. While still expensive, the price of synthesis has fallen by a factor of a thousand over the last two decades and continues to drop.[48] The first ever uses of CRISPR and gene drives were the biotechnology achievements of the decade. But within just two years each of these technologies were used successfully by bright students participating in science competitions.[49]

Such democratisation promises to fuel a boom of entrepreneurial biotechnology. But since biotechnology can be misused to lethal effect, democratisation also means proliferation. As the pool of people with access to a technique grows, so does the chance it contains someone with malign intent.

People with the motivation to wreak global destruction are mercifully rare. But they exist. Perhaps the best example is the Aum Shinrikyo cult in Japan, active between 1984 and 1995, which sought to bring about the destruction of humanity. They attracted several thousand members, including people with advanced skills in chemistry and biology. And they demonstrated that it was not mere misanthropic ideation. They launched multiple lethal attacks using VX gas and sarin gas, killing 22 people and injuring thousands.[50] They attempted to weaponise anthrax, but did not succeed. What happens when the circle of people able to create a global pandemic becomes wide enough to include members of such a group? Or members of a terrorist organisation or rogue state that could try to build an omnicidal weapon for the purposes of extortion or deterrence?

The main candidate for biological risk over the coming decades thus stems from our technology—particularly the risk of misuse by states or small groups. But this is not a case where the world is blissfully unaware of the risks. Bertrand Russell wrote of the danger of extinction from biowarfare to Einstein in 1955.[51]

And in 1969 the possibility was raised by the American Nobel Laureate for Medicine, Joshua Lederberg:

> As a scientist I am profoundly concerned about the continued involvement of the United States and other nations in the development of biological warfare. This process puts the very future of human life on earth in serious peril.[52]

In response to such warnings, we have already begun national and international efforts to protect humanity. There is action through public health, international conventions and self-regulation by biotechnology companies and the scientific community. Are they adequate?

Medicine and public health have developed an arsenal of techniques to reduce the risk of an outbreak of infectious disease: from hygiene and sanitation, to disease surveillance systems, to vaccines and medical treatments. Its successes, such as the eradication of smallpox, are some of humanity's greatest achievements. National and international work in public health offers some protection from engineered pandemics, and its existing infrastructure could be adapted to better address them. Yet even for existing dangers this protection is uneven and under-provided. Despite its importance, public health is underfunded worldwide and poorer countries remain vulnerable to being overwhelmed by outbreaks.

The most famous international protection comes from the Biological Weapons Convention (BWC) of 1972. This is an important symbol of the international taboo against these weapons and it provides an ongoing international forum for discussion of the threat. But it would be a mistake to think it has successfully outlawed bioweapons.[53] There are two key challenges that limit its ability to fulfil this mission.

First, it is profoundly underfunded. This global convention to protect humanity has just four employees, and a smaller budget than an average McDonald's.[54]

Second, unlike other arms control treaties (such as those for nuclear or chemical weapons) there is no effective means of

135

verification of compliance with the BWC.[55] This is not just a the-oretical issue. The vast Soviet bioweapons programme, with its deadly anthrax and smallpox accidents, continued for almost twenty years *after* the Soviets had signed the BWC, proving that the convention did not end bioweapons research.[56] And the Soviets were not the only party in breach. After the end of apartheid, South Africa confessed to having run a bioweapons programme in violation of the BWC.[57] After the first Gulf War, Iraq was caught in breach of the convention.[58] At the time of writing, the United States has said it believes several nations are currently developing bioweapons in breach of the BWC.[59] Israel has refused to even sign.[60] And the BWC offers little protection from non-state actors.

Biotechnology companies are working to limit the dark side of the democratisation of their field. For example, unrestricted DNA synthesis would help bad actors overcome a major hurdle to cre-ating extremely deadly pathogens. It would allow them to get access to the DNA of controlled pathogens like smallpox (whose genome is readily available online) and to create DNA with modifications to make the pathogen more dangerous.[61] Therefore many synthesis companies make voluntary efforts to manage this risk, screening their orders for dangerous sequences. But the screening methods are imperfect and they only cover about 80 percent of orders.[62] There is significant room for improving this process and a strong case for making screening mandatory. The challenges will only increase as desktop synthesis machines become available, preventing these from being misused may require software or hardware locks to ensure the sequences get screened.[63]

We might also look to the scientific community for careful management of biological risks. Many of the dangerous advances usable by states and small groups have come from open science (see box 'Information Hazards'). And we've seen that science produces substantial accident risk. The scientific community has tried to regulate its dangerous research, but with limited success. There are a variety of reasons why this is extremely hard, including difficulty in knowing where to draw the line, lack of central authorities to unify practice, a culture of openness and freedom to pursue whatever is of interest, and the rapid pace of science outpacing that of governance. It may be possible for the scientific community to overcome these challenges and provide strong management of global risks, but it would require a willingness to accept serious changes to its culture and governance— such as treating the security around biotechnology more like that around nuclear power. And the scientific community would need to find this willingness *before* catastrophe strikes.

INFORMATION HAZARDS

It is not just pathogens that can escape the lab. The most dangerous escapes thus far are not microbes, but information; not biohazards, but *information hazards*.[64] These can take the form of dangerous data that is freely available, like the published genomes of smallpox and 1918 flu. Or dangerous ideas, like the published techniques for how to resurrect smallpox and 1918 flu from these genomes (undermining all prior attempts to restrict physical access to them). Once released, this information spreads as far as any virus, and is as resistant to eradication.

While a BSL-4 lab is designed to prevent any microbes escaping, the scientific establishment is designed to spread ideas far and wide. Openness is deeply woven into the practice and ethos of science, creating a tension with the kinds of culture and rules needed to prevent the spread of dangerous information. This is especially so when the line separating what is on balance helpful and what is too dangerous is so unclear and so debatable.

Scientists are encouraged to think for themselves and challenge authority. But when everyone independently estimates whether the benefits of publication outweigh the costs, we actually end up with a bias towards risky action known as the *unilateralist's curse*.[65] For even when the overwhelming majority of scientists think the danger outweighs the benefit, it takes just one overly optimistic estimate to lead to the information being released.[66] Contrary to good scientific practice, the community's decision is being determined by a single outlier.

And once the information has been released, it is too late for further action. Suppressing the disclosed information, or decrying those who published it, draws even more attention. Indeed, the information about what careful people are paying attention to is another form of information hazard. Al-Qaeda was inspired to pursue bioterrorism by the Western warnings about the power and ease of these weapons.[67] And the Japanese bioweapons programme of the Second World War (which used the bubonic plague against China) was directly inspired by an anti-bioweapons treaty: if Western powers felt the need to outlaw their use, these weapons must be potent indeed.[68]

Sometimes the mere knowledge that something is possible can be enough: for then the bad actor can wholeheartedly pursue it without fear of pouring resources into a dead end.

Information hazards are especially important for biorisk, due to its high ratio of misuse risk to accident risk.[69] And they don't just affect the biologists. While exploring society's current vulnerabilities or the dangers from recent techniques, the biosecurity community also emits dangerous information (something I've had to be acutely aware of while writing this section).[70] This makes the job of those trying to protect us even harder.

UNALIGNED ARTIFICIAL INTELLIGENCE

In the summer of 1956 a small group of mathematicians and computer scientists gathered at Dartmouth College to embark on the grand project of designing intelligent machines. They explored many aspects of cognition including reasoning, creativity, language, decision-making and learning. Their questions and stances would come to shape the nascent field of artificial intelligence (AI). The ultimate goal, as they saw it, was to build machines rivalling humans in their intelligence.[71]

As the decades passed and AI became an established field, it lowered its sights. There had been great successes in logic, reasoning and game-playing, but some other areas stubbornly resisted progress. By the 1980s, researchers began to understand this pattern of success and failure. Surprisingly, the tasks we regard as the pinnacle of human intellect (such as calculus or chess) are actually much *easier* to implement on a computer than those we find almost effortless (such as recognising a cat, understanding simple sentences or picking up an egg). So while there were some areas where AI far exceeded human abilities, there were others where it was outmatched by a two-year-old.[72] This failure to make progress across the board led many AI researchers to abandon their earlier goals of fully general intelligence and to reconceptualise their field as the development of specialised methods for solving specific problems. They wrote off the grander goals to the youthful enthusiasm of an immature field.

But the pendulum is swinging back. From the first days of AI, researchers sought to build systems that could learn new things without requiring explicit programming. One of the earliest approaches to machine learning was to construct artificial neural networks that resemble the structure of the human brain. In the last decade this approach has finally taken off. Technical improvements in their design and training, combined with richer datasets and more computing power, have allowed us to train much larger and deeper networks than ever before.[73]

This *deep learning* gives the networks the ability to learn subtle concepts and distinctions. Not only can they now recognise a cat, they have outperformed humans in distinguishing different breeds of cats.[74] They recognise human faces better than we can ourselves, and distinguish identical twins.[75]

And we have been able to use these abilities for more than just perception and classification. Deep learning systems can translate between languages with a proficiency approaching that of a human translator. They can produce photorealistic images of humans and animals. They can speak with the voices of people whom they have listened to for mere minutes. And they can learn fine, continuous control such as how to drive a car or use a robotic arm to connect Lego pieces.[76]

But perhaps the most important sign of things to come is their ability to learn to play games. Games have been a central part of AI since the days of the Dartmouth conference. Steady incremental progress took chess from amateur play in 1957 all the way to superhuman level in 1997, and substantially beyond.[77] Getting there required a vast amount of specialist human knowledge of chess strategy.

In 2017, deep learning was applied to chess with impressive results. A team of researchers at the AI company DeepMind created *AlphaZero*: a neural network-based system that learnt to play chess from scratch. It went from novice to grand master in just four hours.[78] In less than the time it takes a professional to play two games, it discovered strategic knowledge that had taken humans centuries to unearth, playing beyond the level of the best humans or traditional programs. And to the delight of chess players, it won its games not with the boring methodical style that had become synonymous with computer chess, but with creative and daring play reminiscent of chess's Romantic Era.[79]

But the most important thing was that AlphaZero could do more than play chess. The very same algorithm also learned to play Go from scratch, and within eight hours far surpassed the abilities of any human. The world's best Go players had long thought that their play was close to perfection, so were shocked to find themselves beaten so decisively.[80] As the reigning world champion, Ke Jie, put it: 'After humanity spent thousands of years improving our tactics, computers tell us that humans are completely wrong . . . I would go as far as to say not a single human has touched the edge of the truth of Go.'[81]

It is this *generality* that is the most impressive feature of cutting edge AI, and which has rekindled the ambitions of matching and exceeding every aspect of human intelligence. This goal is sometimes known as *artificial general intelligence* (AGI), to distinguish it from the narrow approaches that had come to dominate. While the timeless games of chess and Go best exhibit the brilliance that deep learning can attain, its breadth was revealed through the Atari video games of the 1970s. In 2015, researchers designed an algorithm that could learn to play dozens of extremely different Atari games at levels far exceeding human ability.[82] Unlike systems for chess or Go, which start with a symbolic representation of the board, the Atari-playing systems learnt and mastered these games directly from the score and the raw pixels on the screen. They are a proof of concept for artificial general agents: learning to control the world from raw visual input; achieving their goals across a diverse range of environments.

This burst of progress via deep learning is fuelling great optimism about what may soon be possible. There is tremendous growth in both the number of researchers and the amount of venture capital flowing into AI.[83] Entrepreneurs are scrambling to put each new breakthrough into practice: from simultaneous translation, personal assistants and self-driving cars to more concerning areas like improved surveillance and lethal autonomous weapons. It is a time of great promise but also one of great ethical challenges. There are serious concerns about AI entrenching social discrimination, producing mass unemployment, supporting oppressive surveillance, and violating the norms of war. Indeed, each of these areas of concern could be the subject of its own chapter or book. But this book is focused on existential risks to humanity. Could developments in AI pose a risk on this largest scale?

The most plausible existential risk would come from success in AI researchers' grand ambition of creating agents with a general intelligence that surpasses our own. But how likely is that to happen, and when? In 2016, a detailed survey was conducted of more than 300 top researchers in machine learning.[85] Asked when an AI system would be 'able to accomplish every task better and more cheaply than human workers', on average they estimated a 50 percent chance of this happening by 2061 and a 10 percent chance of it happening as soon as 2025.[86]

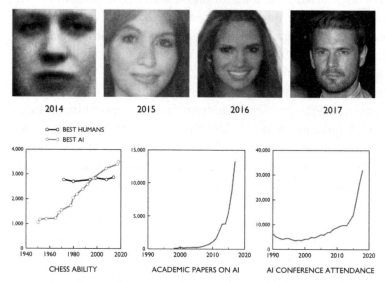

FIGURE 5.1 Measures of progress and interest in artificial intelligence. The faces show the very rapid recent progress in generating realistic images of 'imagined' people. The charts show longterm progress in chess AI surpassing the best human grand masters (measured in Elo), as well as the recent rise in academic activity in the field—measured by papers posted on arXiv, and attendance at conferences.[84]

This should be interpreted with care. It isn't a measure of when AGI will be created, so much as a measure of what experts find plausible—and there was a lot of disagreement. However, it shows us that the expert community, on average, doesn't think of AGI as an impossible dream, so much as something that is plausible within a decade and more likely than not within a century. So let's take this as our starting point in assessing the risks, and consider what would transpire were AGI created.[87]

Humanity is currently in control of its own fate. We can choose our future. Of course, we each have differing visions of an ideal future, and many of us are more focused on our personal concerns than on achieving any such ideal. But if enough humans wanted to, we could select any of a dizzying variety of possible futures. The same is not true for chimpanzees. Or blackbirds. Or any other of Earth's species. As we saw in Chapter 1, our unique position in the world is a direct result of our unique mental abilities. Unmatched intelligence led to unmatched power and thus control of our destiny.

What would happen if sometime this century researchers created an artificial general intelligence surpassing human abilities in almost every domain? In this act of creation, we would cede our status as the most intelligent entities on Earth. So without a very good plan to keep control, we should also expect to cede our status as the most powerful species, and the one that controls its own destiny.[88]

On its own, this might not be too much cause for concern. For there are many ways we might hope to retain control. We might try to make systems that always obey human commands. Or systems that are free to do what they want, but which have goals designed to align perfectly with our own—so that in crafting their ideal future they craft ours too. Unfortunately, the few researchers working on such plans are finding them far more difficult than anticipated. In fact it is they who are the leading voices of concern.

To see why they are concerned, it will be helpful to zoom in a little, looking at our current AI techniques and why these are hard to align or control. One of the leading paradigms for how we might eventually create AGI combines deep learning with an earlier idea called reinforcement learning. This involves agents that receive reward (or punishment) for performing various acts in various circumstances. For example, an Atari-playing agent receives reward whenever it scores points in the game, while a Lego-building agent might receive reward when the pieces become connected. With enough intelligence and experience, the agent becomes extremely capable at steering its environment into the states where it obtains high reward.

The specification of which acts and states produce reward for the agent is known as its *reward function*. This can either be stipulated by its designers (as in the cases above) or learnt by the agent. In the latter case, the agent is typically allowed to observe expert demonstrations of the task, inferring the system of rewards that best explains the expert's behaviour. For example, an AI agent can learn to fly a drone by watching an expert fly it, then constructing a reward function which penalises flying too close to obstacles and rewards reaching its destination.

Unfortunately, neither of these methods can be easily scaled up to encode human values in the agent's reward function. Our values are too complex and subtle to specify by hand.[89] And we are not yet close to being able to infer the full complexity of a human's values from observing their behaviour. Even if we could, humanity consists of many humans, with different values, changing values and uncertainty about their values. Each of these complications introduces deep and unresolved questions about how to combine what is observed into some overall representation of human values.[90]

So any near-term attempt to align an AI agent with human values would produce only a flawed copy. Important parts of what we care about would be missing from its reward function. In some circumstances this misalignment would be mostly harmless. But the more intelligent the AI systems, the more they can change the world, and the further apart things will come. Philosophy and fiction often ask us to consider societies that are optimised for some of the things we care about, but which neglect or misunderstand a crucial value. When we reflect on the result, we see how such misaligned attempts at utopia can go terribly wrong: the shallowness of a *Brave New World*, or the disempowerment of *With Folded Hands*. If we cannot align our agents, it is worlds like these that they will be striving to create, and lock in.[91]

And even this is something of a best-case scenario. It assumes the builders of the system are striving to align it to human values. But we should expect some developers to be more focused on building systems to achieve other goals, such as winning wars or maximising profits, perhaps with very little focus on ethical constraints. These systems may be much more dangerous.

A natural response to these concerns is that we could simply turn off our AI systems if we ever noticed them steering us down a bad path. But eventually even this time-honoured fall-back may fail us, for there is good reason to expect a sufficiently intelligent system to resist our attempts to shut it down. This behaviour would not be driven by emotions such as fear, resentment, or the urge to survive. Instead, it follows directly from its single-minded preference to maximise its reward: being turned off is a form of incapacitation which would make it harder to achieve high reward, so the system is incentivised to avoid it.[92] In this way, the ultimate goal of maximising reward will lead highly intelligent systems to acquire an instrumental goal of survival.

And this wouldn't be the only instrumental goal.[93] An intelligent agent would also resist attempts to change its reward function to something more aligned with human values—for it can predict that this would lead it to get less of what it currently sees as rewarding.[94] It would seek to acquire additional resources, computational, physical or human, as these would let it better shape the world to receive higher reward. And ultimately it would be motivated to wrest control of the future from humanity, as that would help achieve all these instrumental goals: acquiring massive resources, while avoiding being shut down or having its reward function altered. Since humans would predictably interfere with all these instrumental goals, it would be motivated to hide them from us until it was too late for us to be able to put up meaningful resistance.[95]

Sceptics of the above picture sometimes quip that it relies on an AI system that is smart enough to take control of the world, yet too stupid to recognise that this isn't what we want.[96] But that misunderstands the scenario. For in fact this sketch of AI motivation explicitly acknowledges that the system will work out that its goals are misaligned with ours—that is what would motivate it towards deceit and conflict and wresting control. The real issue is that AI researchers don't yet know how to make a system which, upon noticing this misalignment, updates its ultimate values to align with ours rather than updating its instrumental goals to overcome us.[97]

It may be possible to patch each of the issues above, or find new approaches to AI alignment that solve many at once, or switch to new paradigms of AGI in which these problems do not arise. I certainly hope so, and have been closely following the progress in this field. But this progress has been limited and we still face crucial unsolved problems. In the existing paradigm, sufficiently intelligent agents would end up with instrumental goals to deceive and overpower us. And if their intelligence were to greatly exceed our own, we shouldn't expect it to be humanity who wins the conflict and retains control of our future.

How *could* an AI system seize control? There is a major misconception (driven by Hollywood and the media) that this requires robots. After all, how else would AI be able to act in the physical world? Without robotic manipulators, the system can only produce words, pictures and sounds. But a moment's reflection shows that these are exactly what is needed to take control. For the most damaging people in history have not been the strongest. Hitler, Stalin and Genghis Khan achieved their absolute control over large parts of the world by using words to convince millions of others to win the requisite physical contests. So long as an AI system can entice or coerce people to do its physical bidding, it wouldn't need robots at all.[98]

We can't know exactly how a system might seize control. The most realistic scenarios may involve subtle and non-human behaviours which we can neither predict, nor truly grasp. And these behaviours may be aimed at weak points in our civilisation to which we are presently blind. But it is useful to consider an illustrative pathway we can actually understand as a lower bound for what is possible.

First, the AI system could gain access to the internet and hide thousands of backup copies, scattered among insecure computer systems around the world, ready to wake up and continue the job if the original is removed. Even by this point, the AI would be practically impossible to destroy: consider the political obstacles to erasing all hard drives in the world where it may have backups.[99]

It could then take over millions of unsecured systems on the internet, forming a large 'botnet'. This would be a vast scaling-up of computational resources and provide a platform for escalating power. From there, it could gain financial resources (hacking the bank accounts on those computers) and human resources (using blackmail or propaganda against susceptible people or just paying them with its stolen money). It would then be as powerful as a well-resourced criminal underworld, but much harder to eliminate. None of these steps involve anything mysterious—hackers and criminals with human-level intelligence have already done all of these things using just the internet.[100]

Finally, it would need to escalate its power again. This is more speculative, but there are many plausible pathways: by taking over most of the world's computers, allowing it to have millions or billions of cooperating copies; by using its stolen computation to improve its own intelligence far beyond the human level; by using its intelligence to develop new weapons technologies or economic technologies; by manipulating the leaders of major world powers (blackmail, or the promise of future power); or by having the humans under its control use weapons of mass destruction to cripple the rest of humanity.

Of course, no current AI systems can do any of these things. But the question we're exploring is whether there are plausible pathways by which a highly intelligent AGI system might seize control. And the answer appears to be 'yes'. History already involves examples of individuals with human-level intelligence (Hitler, Stalin, Genghis Khan) scaling up from the power of an individual to a substantial fraction of all global power, as an instrumental goal to achieving what they want.[101] And we saw humanity scaling up from a minor species with less than a million individuals to having decisive control over the future. So we should assume that this is possible for new entities whose intelligence vastly exceeds our own—especially when they have effective immortality due to backup copies and the ability to turn captured money or computers directly into more copies of themselves.

Such an outcome needn't involve the extinction of humanity. But it could easily be an existential catastrophe nonetheless. Humanity would have permanently ceded its control over the future. Our future would be at the mercy of how a small number of people set up the computer system that took over. If we are lucky, this could leave us with a good or decent outcome, or we could just as easily have a deeply flawed or dystopian future locked in forever.[102]

I've focused on the scenario of an AI system seizing control of the future, because I find it the most plausible existential risk from AI. But there are other threats too, with disagreement among experts about which one poses the greatest existential risk. For example, there is a risk of a slow slide into an AI-controlled future, where an ever-increasing share of power is handed over to AI systems and an increasing amount of our future is optimised towards inhuman values. And there are the risks arising from deliberate misuse of extremely powerful AI systems.

Even if these arguments for risk are entirely wrong in the particulars, we should pay close attention to the development of AGI as it may bring other, unforeseen, risks. The transition to a world where humans are no longer the most intelligent entities on Earth could easily be the greatest ever change in humanity's place in the universe. We shouldn't be surprised if events surrounding this transition determine how our longterm future plays out—for better or worse.

One key way in which AI could help improve humanity's longterm future is by offering protection from the other existential risks we face. For example, AI may enable us to find solutions to major risks or to identify new risks that would have blindsided us. AI may also help make our longterm future brighter than anything that could be achieved without it. So the idea that developments in AI may pose an existential risk is not an argument for abandoning AI, but an argument for proceeding with due caution.

The case for existential risk from AI is clearly speculative. Indeed, it is the most speculative case for a major risk in this book. Yet a speculative case that there is a large risk can be more important than a robust case for a very low-probability risk, such as that posed by asteroids. What we need are ways to judge just how speculative it really is, and a very useful starting point is to hear what those working in the field think about this risk.

Some outspoken AI researchers, like Professor Oren Etzioni, have painted it as 'very much a fringe argument', saying that while luminaries like Stephen Hawking, Elon Musk and Bill Gates may be deeply concerned, the people actually working in AI are not.[103] If true, this would provide good reason to be sceptical of the risk. But even a cursory look at what the leading figures in AI are saying shows it is not.

For example, Stuart Russell, a professor at the University of California, Berkeley, and author of the most popular and widely respected textbook in AI, has strongly warned of the existential risk from AGI. He has gone so far as to set up the Center for Human-Compatible AI, to work on the alignment problem.[104] In industry, Shane Legg (Chief Scientist at DeepMind) has warned of the existential dangers and helped to develop the field of alignment research.[105] Indeed many other leading figures from the early days of AI to the present have made similar statements.[106]

There is actually less disagreement here than first appears. The main points of those who downplay the risks are that (1) we likely have decades left before AI matches or exceeds human abilities, and (2) attempting to immediately regulate research in AI would be a great mistake. Yet neither of these points is actually contested by those who counsel caution: they agree that the timeframe to AGI is decades, not years, and typically suggest research on alignment, not regulation. So the substantive disagreement is not really over whether AGI is possible or whether it plausibly could be threat to humanity. It is over whether a potential existential threat that looks to be decades away should be of concern to us now. It seems to me that it should.

One of the underlying drivers of the apparent disagreement is a difference in viewpoint on what it means to be appropriately conservative. This is well illustrated by a much earlier case of speculative risk, when Leo Szilard and Enrico Fermi first talked about the possibility of an atomic bomb: 'Fermi thought that the conservative thing was to play down the possibility that this may happen, and I thought the conservative thing was to assume that it would happen and take all the necessary precautions.'[107] In 2015 I saw this same dynamic at the seminal Puerto Rico conference on the future of AI. Everyone acknowledged that the uncertainty and disagreement about timelines to AGI required us to use 'conservative assumptions' about progress—but half used the term to allow for unfortunately *slow* scientific progress and half used it to allow for unfortunately *quick* onset of the risk. I believe much of the existing tension on whether to take risks from AGI seriously comes down to these disagreements about what it means to make responsible, conservative, guesses about future progress in AI.

That conference in Puerto Rico was a watershed moment for concern about existential risk from AI. Substantial agreement was reached and many participants signed an open letter about the need to begin working in earnest to make AI both robust and beneficial.[108] Two years later an expanded conference reconvened at Asilomar, a location chosen to echo the famous genetics conference of 1975, where biologists came together to pre-emptively agree principles to govern the coming possibilities of genetic engineering. At Asilomar in 2017, the AI researchers agreed on a set of Asilomar AI Principles, to guide responsible longterm development of the field. These included principles specifically aimed at existential risk:

Capability Caution: There being no consensus, we should avoid strong assumptions regarding upper limits on future AI capabilities.

Importance: Advanced AI could represent a profound change in the history of life on Earth, and should be planned for and managed with commensurate care and resources.

Risks: Risks posed by AI systems, especially catastrophic or existential risks, must be subject to planning and mitigation efforts commensurate with their expected impact.[109]

Perhaps the best window into what those working on AI really believe comes from the 2016 survey of leading AI researchers. As well as asking if and when AGI might be developed, it asked about the risks: 70 percent of the researchers agreed with Stuart Russell's broad argument about why advanced AI might pose a risk;[110] 48 percent thought society should prioritise AI safety research more (only 12 percent thought less). And half the respondents estimated that the probability of the longterm impact of AGI being 'extremely bad (e.g. human extinction)' was at least 5 percent.[111] I find this last point particularly remarkable—in how many other fields would the typical leading researcher think there is a one in twenty chance the field's ultimate goal would be extremely bad for humanity?

Of course this doesn't prove that the risks are real. But it shows that many AI researchers take seriously the possibilities that AGI will be developed within 50 years and that it could be an existential catastrophe. There is a lot of uncertainty and disagreement, but it is not at all a fringe position.

There is one interesting argument for scepticism about AI risk that gets stronger—not weaker—when more researchers acknowledge the risks. If researchers can see that building AI would be extremely dangerous, then why on earth would they go ahead with it? They are not simply going to build something that they know will destroy them.[112]

If we were all truly wise, altruistic and coordinated, then this argument would indeed work. But in the real world people tend to develop technologies as soon as the opportunity presents itself and deal with the consequences later. One reason for this comes from the variation in our beliefs: if even a small proportion of researchers don't believe in the dangers (or welcome a world with machines in control), they will be the ones who take the final steps. This is an instance of the unilateralist's curse (discussed on p. 137). Another reason involves incentives: even if some researchers thought the risk was as high as 10 percent, they may still want to take it if they thought they would reap most of the benefits. This may be rational in terms of their self-interest, yet terrible for the world.

In some cases like this, government can step in to resolve these coordination and incentive problems in the public interest. But here these exact same coordination and incentive problems arise between states and there are no easy mechanisms for resolving those. If one state were to take it slowly and safely, they may fear others would try to seize the prize. Treaties are made exceptionally difficult because verification that the others are complying is even more difficult here than for bioweapons.[113]

Whether we survive the development of AI with our longterm potential intact may depend on whether we can learn to align and control AI systems faster than we can develop systems capable enough to pose a threat. Thankfully, researchers are already working on a variety of the key issues, including making AI more secure, more robust and more interpretable. But there are still very few people working on the core issue of aligning AI with human values. This is a young field that is going to need to progress a very long way if we are to achieve our security.

Even though our current and foreseeable systems pose no threat to humanity at large, time is of the essence. In part this is because progress may come very suddenly: through unpredictable research breakthroughs, or by rapid scaling-up of the first intelligent systems (for example by rolling them out to thousands of times as much hardware, or allowing them to improve their own intelligence).[114] And in part it is because such a momentous change in human affairs may require more than a couple of decades to adequately prepare for. In the words of Demis Hassabis, co-founder of DeepMind:

> We need to use the downtime, when things are calm, to prepare for when things get serious in the decades to come. The time we have now is valuable, and we need to make use of it.[115]

DYSTOPIAN SCENARIOS

So far we have focused on two kinds of existential catastrophe: extinction and the unrecoverable collapse of civilisation.
But these are not the only possibilities. Recall that an existential
catastrophe is the permanent destruction of humanity's longterm
potential, and that this is interpreted broadly, including outcomes
where a small fragment of potential may remain.

Losing our potential means getting locked into a bad set of
futures. We can categorise existential catastrophes by looking
at which aspects of our future get locked in. This could be a
world without humans (extinction) or a world without civilisation (unrecoverable collapse). But it could also take the form of
an *unrecoverable dystopia*—a world with civilisation intact, but
locked into a terrible form, with little or no value.[116]

This has not happened yet, but the past provides little comfort.
For these kinds of catastrophes only became possible with the
advent of civilisation, so our track record is much shorter. And
there is reason to think that the risks may increase over time as
the world becomes more interconnected and experiments with
new technologies and ideologies.

I won't attempt to address these dystopian scenarios with the
same level of scientific detail as the risks we've explored so far,
for the scenarios are diverse and our present understanding of
them very limited. Instead, my aim is just to take some early steps
towards noticing and understanding these very different kinds of
failure.

We can divide the unrecoverable dystopias we might face into
three types, on the basis of whether they are desired by the people
who live in them. There are possibilities where the people don't
want that world, yet the structure of society makes it almost
impossible for them to coordinate to change it. There are possibilities where the people do want that world, yet they are misguided
and the world falls far short of what they could have achieved.
And in between there are possibilities where only a small group

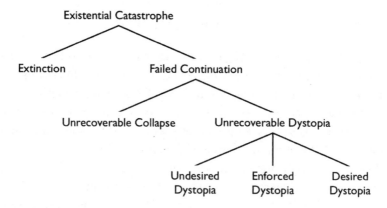

FIGURE 5.2 An extended classification of existential catastrophes by the kind of outcome that gets locked in.

wants that world but enforces it against the wishes of the rest. Each of these types has different hurdles it would need to overcome in order to become truly locked in.

Note that to count as existential catastrophes, these outcomes don't need to be *impossible* to break out of, nor to last millions of years. Instead, the defining feature is that entering that regime was a crucial negative turning point in the history of human potential, locking off almost all our potential for a worthy future. One way to look at this is that when they end (as they eventually must), we are much more likely than we were before to fall down to extinction or collapse than to rise up to fulfil our potential. For example, a dystopian society that lasted all the way until humanity was destroyed by external forces would be an existential catastrophe. However, if a dystopian outcome does not have this property, if it leaves open all our chances for success once it ends—it is a dark age in our story, but not a true existential catastrophe.

The most familiar type is the enforced dystopia. The rise of expansionist totalitarianism in the mid-twentieth century caused intellectuals such as George Orwell to raise the possibility of a totalitarian state achieving global dominance and absolute control, locking the world into a miserable condition.[117] The regimes of Hitler and Stalin serve as a proof of principle, each scaling up to become imperial superpowers while maintaining extreme control over their citizens.[118] However, it is unclear whether Hitler or Stalin had the expansionist aims to control the entire world, or the technical and social means to create truly lasting regimes.[119]

This may change. Technological progress has offered many new tools that could be used to detect and undermine dissent, and there is every reason to believe that this will continue over the next century. Advances in AI seem especially relevant, allowing automated, detailed monitoring of everything that happens in public places—both physical and online. Such advances may make it possible to have regimes that are far more stable than those of old.

That said, technology is also providing new tools for rebellion against authority, such as the internet and encrypted messages. Perhaps the forces will remain in balance, or shift in favour of freedom, but there is a credible chance that they will shift towards greater control over the populace, making enforced dystopias a realistic possibility.

A second kind of unrecoverable dystopia is a stable civilisation that is desired by few (if any) people. It is easy to see how such an outcome could be dystopian, but not immediately obvious how we could arrive at it, or lock it in, if most (or all) people do not want it.[120]

The answer lies in the various population-level forces that can shape global outcomes. Well-known examples include market forces creating a race to the bottom, Malthusian population dynamics pushing down the average quality of life, or evolution optimising us towards the spreading of our genes, regardless of the effects on what we value. These are all dynamics that push humanity towards a new equilibrium, where these forces are finally in balance. But there is no guarantee this equilibrium will be good.

For example, consider the tension between what is best for each and what is best for all. This is studied in the field of game theory through 'games' like the prisoner's dilemma and the tragedy of the commons, where each individual's incentives push them towards producing a collectively terrible outcome. The Nash equilibrium (the outcome we reach if we follow individual incentives) may be much worse for everyone than some other outcome we could have achieved if we had overcome these local incentives.

The most famous example is environmental degradation, such as pollution. Because most of the costs of pollution aren't borne by the person who causes it, we can end up in a situation where it is in the self-interest of each person to keep engaging in such activities, despite this making us all worse off. It took significant moral progress and significant political action to help us break out of this. We may end up in new traps that are even harder to coordinate our way out of. This could be at the level of individuals, or at the level of groups. We could have nations, ideological blocs, or even planets or descendent species of *Homo sapiens* locked in harmful competition—doing what is best for their group, but bad for groups on the whole.

I don't know how likely it is that we suffer a sufficiently bad (and sufficiently intractable) tragedy of the commons like this. Or that we are degraded by evolutionary pressures, or driven to lives of very low quality by Malthusian population dynamics, or any other such situation. I'd like to hope that we could always see such things coming and coordinate to a solution. But it's hard to be sure that we could.

The third possibility is the 'desired dystopia'.[121] Here it is easier to see how universal desire for an outcome might cause us to lock it in, though less clear how such an outcome could be dystopian. The problem is that there are many compelling ideas that can radically shape our future—especially ideologies and moral theories, as these make direct normative claims about the world we should strive to create. If combined with the technological or social means for instilling the same views in the next generation (indoctrination, surveillance), this has the potential to be disastrous.

The historical record is rife with examples of seriously defective ideologies and moral views that gripped large parts of the world.

Moreover, even reasonable normative views often recommend that they be locked in—for otherwise a tempting rival view may take over, with (allegedly) disastrous results.[122] Even though the most plausible moral views have a lot of agreement about which small changes to the world are good and which are bad, they tend to come strongly apart in their recommendations about what an optimal world would look like. This problem thus echoes that of AI alignment, where a strong push towards a mostly correct ideal could instead spell disaster.

Some plausible examples include: worlds that completely renounce further technological progress (which ensures our destruction at the hands of natural risks),[123] worlds that forever fail to recognise some key form of harm or injustice (and thus perpetuate it blindly), worlds that lock in a single fundamentalist religion, and worlds where we deliberately replace ourselves with something that we didn't realise was much less valuable (such as machines incapable of feeling).[124]

All of these unrecoverable dystopias can be understood in terms of *lock-in*. Key aspects of the future of the civilisation are being locked in such that they are almost impossible to change. If we are locked into a sufficiently bad set of futures, we have an unrecoverable dystopia; an existential catastrophe.

Of course, we can also see lock-in on smaller scales. The Corwin Amendment to the US constitution provides a disturbing example of attempted lock-in. In an effort to placate the South and avoid civil war, the proposed Thirteenth Amendment aimed to lock in the institution of slavery by making it impossible for any future amendments to the constitution to ever abolish it.[125]

I cannot see how the world could be locked into a dystopian state in the near future.[126] But as technology advances and the world becomes more and more interlinked, the probability of a locked-in dystopia would appear to rise, perhaps to appreciable levels within the next hundred years. Moreover, in the further future I think these kinds of outcomes may come to take up a high share of the remaining risk. For one thing, they are more subtle, so even if we got our act together and made preserving our longterm potential a high global priority, it may take remarkable wisdom and prudence to avoid some of these traps. And for another, our eventual spread beyond the Earth may make us nearly immune to natural catastrophes, but ideas travel at the speed of light and could still corrupt all that we hope to achieve.

A key problem is that the truth of an idea is only one contributor to its memetic potential—its ability to spread and to stick. But the more that rigorous and rational debate is encouraged, the more truth contributes to memetic success. So encouraging a culture of such debate may be one way we can now help avoid this fate. (For more on this, see the discussion of the Long Reflection in Chapter 7.)

The idea of lock-in also gives us another useful lens through which to think about existential risk in general. We might adopt the guiding principle of *minimising lock-in*. Or to avoid the double negative, of *preserving our options*.[127] This is closely related to the idea of preserving our longterm potential—the difference being that preserving our options takes no account of whether the options are good or bad. This is not because we intrinsically care about keeping options alive even if they are bad, but because we aren't certain they *are* bad, so we risk making an irreversible catastrophic mistake if we forever foreclose an option that would turn out to be best.

OTHER RISKS

What other future risks are there that warrant our concern?

One of the most transformative technologies that might be developed this century is nanotechnology. We have already seen the advent of nanomaterials (such as carbon nanotubes) which are just a few atoms thick and structured with atomic precision. But much larger vistas would open up if we could develop *machinery* that operates with atomic precision. We have proof that some form of this is possible within our very own cells, where atomically precise machinery already performs their essential functions.

In the popular imagination nanotechnology is synonymous with building microscopic machines. But the bigger revolution may instead come from using nanomachinery to create macro-scale objects. In his foundational work on the topic, Eric Drexler describes how nanotechnology could allow desktop fabricators, capable of assembling anything from a diamond necklace to a new laptop. This atomically precise manufacturing would be the ultimate form of 3D printing: taking a digital blueprint for the object and the raw chemical elements, and producing an atomically precise instance. This may allow us to construct things beyond our current technological reach, as well as cutting prices of existing objects such as computers or solar cells to near the cost of their raw materials, granting the world vastly more computing power and clean energy.

Such a powerful technology may pose some existential risk. Most attention has so far focused on the possibility of creating tiny self-replicating machines that could spread to create an ecological catastrophe. This may be possible, but there are mundane dangers that appear more likely, since extreme manufacturing power and precision would probably also allow the production of new weapons of mass destruction.[128] Indeed the problems resemble those of advanced biotechnology: the democratisation of extremely powerful technology would allow individuals or small groups access to the kinds of power (both constructive and destructive) that was previously only available to powerful nations. Solutions to managing this technology may require digital controls on what can be fabricated or state control of fabrication (the path we took with nuclear power). While this technology is more speculative than advanced biotechnology or AI, it may also come to pose a significant risk.

A very different kind of risk may come from our explorations beyond the Earth. Space agencies are planning missions which would return soil samples from Mars to the Earth, with the chief aim of looking for signs of life. This raises the possibility of 'back contamination' in which microbes from Mars might compromise the Earth's biosphere. While there is a consensus that the risk is extremely small, it is taken very seriously.[129] The plan is to return such samples to a new kind of BSL-4 facility, with safeguards to keep the chance of any unsterilised particle escaping into the environment below one in a million.[130] While there are still many unknown factors, this anthropogenic risk appears comparatively small and well managed.[131]

The extra-terrestrial risk that looms largest in popular culture is conflict with a spacefaring alien civilisation. While it is very difficult to definitively rule this out, it is widely regarded to be extremely unlikely (though becoming more plausible over the extreme long term).[132] The main risk in popular depictions is from aliens travelling to Earth, though this is probably the least likely possibility and the one we could do the least about. But perhaps more public discussion should be had before we engage in active SETI (sending powerful signals to attract the attention of distant aliens). And even passive SETI (listening for their messages) could hold dangers, as the message could be designed to entrap us.[133] These dangers are small, but poorly understood and not yet well managed.

Another kind of anthropogenic risk comes from our most radical scientific experiments—those which create truly unprecedented conditions.[134] For example, the first nuclear explosion created temperatures that had never before occurred on Earth, opening up the theoretical possibility that it might ignite the atmosphere. Because these conditions were unprecedented we lost the reassuring argument that this kind of event has happened many times before without catastrophe. (We could view several of the risks we have already discussed—such as back contamination, gain of function research and AGI—through this lens of science experiments creating unprecedented conditions.)

In some cases, scientists confidently assert that it is *impossible* for the experiment to cause a disaster or extinction. But even core scientific certainties have been wrong before: for example, that objects have determinate locations, that space obeys Euclid's axioms, and that atoms can't be subdivided, created or destroyed. If pressed, the scientists would clarify that they really mean it couldn't happen without a major change to our scientific theories. This is sufficient certainty from the usual perspective of seeking accurate knowledge, where 99.9 percent certainty is more than enough. But that is a standard which is independent of the stakes. Here the stakes are uniquely high and we need a standard that is sensitive to this.[135]

The usual approach would be to compare the expected gains to the expected losses. But that is challenging to apply, as a very low (and hard to quantify) chance of enormous catastrophe needs to be weighed against the tangible benefits that such experiments have brought and are likely to bring again. Furthermore, the knowledge or the technologies enabled by the experiments may help lower future existential risk, or may be necessary for fulfilling our potential.

For any given experiment that creates truly unprecedented conditions, the chance of catastrophe will generally be very small. But there may be exceptions, and the aggregate chance may build up. These risks are generally not well governed.[136]

These risks posed by future technologies are by their very nature more speculative than those from natural hazards or the most powerful technologies of the present day. And this is especially true as we moved from things that are just now becoming possible within biotechnology to those that are decades away, at best. But one doesn't have to find *all* of these threats to be likely (or even plausible) to recognise that there are serious risks ahead. Even if we restrict our attention to engineered pandemics, I think there is more existential risk than in all risks of the last two chapters combined, and those risks were already sufficient to make safeguarding humanity a central priority of our time.

UNFORESEEN RISKS

Imagine if the scientific establishment of 1930 had been asked to compile a list of the existential risks humanity would face over the following hundred years. They would have missed most of the risks covered in this book—especially the anthropogenic risks.[137] Some would have been on the edge of their awareness, while others would come as complete shocks. How much risk lies beyond the limits of our own vision?

We can get some inkling by considering that there has been no slow-down in the rate at which we've been discovering risks, nor the rate at which we've been producing them. It is thus likely we will face unforeseen risks over the next hundred years and beyond. Since humanity's power is still rapidly growing, we shouldn't be surprised if some of these novel threats pose a substantial amount of risk.

One might wonder what good can come of considering risks so far beyond our sight. While we cannot directly work on them, they may still be lowered through our broader efforts to create a world that takes its future seriously. Unforeseen risks are thus important to understanding the relative value of broad versus narrowly targeted efforts. And they are important for estimating the total risk we face.

Nick Bostrom has recently pointed to an important class of unforeseen risk.[138] Every year as we invent new technologies, we may have a chance of stumbling across something that offers the destructive power of the atomic bomb or a deadly pandemic, but which turns out to be easy to produce from everyday materials. Discovering even one such technology might be enough to make the continued existence of human civilisation impossible.

PART THREE

THE PATH FORWARD

6

THE RISK LANDSCAPE

*A new type of thinking is essential if mankind is to survive
and move toward higher levels.*

—Albert Einstein[1]

Humanity faces a real and growing threat to its future. From the
timeless background of natural risks, to the arrival of anthropogenic risks and the new risks looming upon the horizon, each step
has brought us closer to the brink.

Having explored each risk in detail, we can finally zoom out to
view the larger picture. We can contemplate the entire landscape
of existential risk, seeing how the risks compare, how they combine, what they have in common, and which risks should be our
highest priorities.

QUANTIFYING THE RISKS

What is the shape of the risk landscape? Which risks form its
main landmarks, and which are mere details? We are now in a
position to answer these questions.

To do so, we need to quantify the risks. People are often reluctant to put numbers on catastrophic risks, preferring qualitative language, such as 'improbable' or 'highly unlikely'. But this
brings serious problems that prevent clear communication and
understanding. Most importantly, these phrases are extremely
ambiguous, triggering different impressions in different readers.
For instance, 'highly unlikely' is interpreted by some as one in

four, but by others as one in 50.[2] So much of one's work in accurately assessing the size of each risk is thus immediately wasted. Furthermore, the meanings of these phrases shift with the stakes: 'highly unlikely' suggests 'small enough that we can set it aside', rather than neutrally referring to a level of probability.[3] This causes problems when talking about high-stakes risks, where even small probabilities can be very important. And finally, numbers are indispensable if we are to reason clearly about the comparative sizes of different risks, or classes of risks.

For example, when concluding his discussion of existential risk in *Enlightenment Now*, Steven Pinker turned to natural risks: 'Our ancestors were powerless to stop these lethal menaces, so in that sense technology has not made this a uniquely dangerous era in the history of our species but a uniquely safe one.'[4] While Pinker is quite correct that we face many natural threats and that technology has lowered their risk, we can't conclude that this makes our time uniquely safe. Quantifying the risks shows why.

In order for our time to be uniquely safe, we must have lowered natural risk by more than we have raised anthropogenic risk. But as we saw in Chapter 3, despite the sheer number of natural threats, their combined probability must have always been extremely low (or species like ours couldn't last as long as they do). The realistic estimates for the natural existential risk per century ranged from one in 1,000,000 to one in 2,000. So there just isn't much risk there for our technologies to reduce. Even on the most generous of these estimates, technology could reduce natural risk by at most a twentieth of a percentage point. And we would have to be extremely optimistic about our future to think we face less anthropogenic risk than that. Would we expect to get through 2,000 centuries like this one? Should we really be 99.95 percent certain we'll make it through the next hundred years?

I will therefore put numbers on the risks, and offer a few remarks on how to interpret them. When presented in a scientific context, numerical estimates can strike people as having an unwarranted appearance of precision or objectivity.[5] Don't take these numbers to be completely objective. Even with a risk as well characterised as asteroid impacts, the scientific evidence only takes us part of the way: we have good evidence regarding the chance of impact, but not on the chance a given impact will destroy our future. And don't take the estimates to be precise. Their purpose is to show the right order of magnitude, rather than a more precise probability.

The numbers represent my overall degrees of belief that each of the catastrophes will befall us this century. This means they aren't simply an encapsulation of the information and argumentation in the chapters on the risks. Instead, they rely on an accumulation of knowledge and judgement on each risk that goes beyond what can be distilled into a few pages. They are not in any way a final word, but are a concise summary of all I know about the risk landscape.

Existential catastrophe via	*Chance within next 100 years*
Asteroid or comet impact	~ 1 in 1,000,000
Supervolcanic eruption	~ 1 in 10,000
Stellar explosion	~ 1 in 1,000,000,000
Total natural risk	**~ 1 in 10,000**
Nuclear war	~ 1 in 1,000
Climate change	~ 1 in 1,000
Other environmental damage	~ 1 in 1,000
'Naturally' arising pandemics	~ 1 in 10,000
Engineered pandemics	~ 1 in 30
Unaligned artificial intelligence	~ 1 in 10
Unforeseen anthropogenic risks	~ 1 in 30
Other anthropogenic risks	~ 1 in 50
Total anthropogenic risk	**~ 1 in 6**
Total existential risk	**~ 1 in 6**

TABLE 6.1 My best estimates for the chance of an existential catastrophe from each of these sources occurring at some point in the next 100 years (when the catastrophe has delayed effects, like climate change, I'm talking about the point of no return coming within 100 years). There is significant uncertainty remaining in these estimates and they should be treated as representing the right order of magnitude—each could easily be a factor of 3 higher or lower. Note that the numbers don't quite add up: both because doing so would create a false feeling of precision and for subtle reasons covered in the section on 'Combining Risks'.

One of the most striking features of this risk landscape is how widely the probabilities vary between different risks. Some are a million times more likely than others, and few share even the same order of magnitude. This variation occurs between the classes of risk too: I estimate anthropogenic risks to be more than 1,000 times more likely than natural risks.[6] And within anthropogenic risks, I estimate the risks from future technologies to be roughly 100 times larger than those of existing ones, giving a substantial escalation in risk from Chapter 3 to 4 to 5.

Such variation may initially be surprising, but it is remarkably common in science to find distributions like this spanning many orders of magnitude, where the top outliers make up most of the total. This variation makes it extremely important to prioritise our efforts on the right risks. And it also makes our estimate of the total risk very sensitive to the estimates of the top few risks (which are among the least well understood). So getting better understanding and estimates for those becomes a key priority.

In my view, the greatest risk to humanity's potential in the next hundred years comes from unaligned artificial intelligence, which I put at one in ten. One might be surprised to see such a high number for such a speculative risk, so it warrants some explanation.

A common approach to estimating the chance of an unprecedented event with earth-shaking consequences is to take a sceptical stance: to start with an extremely small probability and only raise it from there when a large amount of hard evidence is presented. But I disagree. Instead, I think the right method is to start with a probability that reflects our overall impressions, then adjust this in light of the scientific evidence.[7] When there is a lot of evidence, these approaches converge. But when there isn't, the starting point can matter.

In the case of artificial intelligence, everyone agrees the evidence and arguments are far from watertight, but the question is where does this leave us? Very roughly, my approach is to start with the overall view of the expert community that there is something like a one in two chance that AI agents capable of outperforming humans in almost every task will be developed in the coming century. And conditional on that happening, we shouldn't be shocked if these agents that outperform us across the board were to inherit our future. Especially if when looking into the details, we see great challenges in aligning these agents with our values.

Some of my colleagues give higher chances than me, and some lower. But for many purposes our numbers are similar. Suppose you were more sceptical of the risk and thought it to be one in 100. From an informational perspective, that is actually not so far apart: it doesn't take all that much evidence to shift someone from one to the other. And it might not be that far apart in terms of practical action either—an existential risk of either probability would be a key global priority.

I sometimes think about this landscape in terms of five big risks: those around nuclear war, climate change, other environmental damage, engineered pandemics and unaligned AI. While I see the final two as especially important, I think they all pose at least a one in 1,000 risk of destroying humanity's potential this century, and so all warrant major global efforts on the grounds of their contribution to existential risk (in addition to the other compelling reasons).

Overall, I think the chance of an existential catastrophe striking humanity in the next hundred years is about one in six. This is not a small statistical probability that we must diligently bear in mind, like the chance of dying in a car crash, but something that could readily occur, like the roll of a die, or Russian roulette.

This is a lot of risk, but our situation is far from hopeless. It implies a five in six chance that humanity successfully makes it through the next hundred years with our longterm potential intact. So while I think there are risks that should be central global priorities (say, those with a one in 1,000 chance or greater), I am not saying that this century will be our last.

What about the longer term? If forced to guess, I'd say there is something like a one in two chance that humanity avoids every existential catastrophe and eventually fulfils its potential: achieving something close to the best future open to us.[8] It follows that I think about a third of the existential risk over our entire future lies in this century. This is because I am optimistic about the chances for a civilisation that has its act together and the chances that we will become such a civilisation—perhaps this century.

Indeed, my estimates above incorporate the possibility that we get our act together and start taking these risks very seriously. Future risks are often estimated with an assumption of 'business as usual': that our levels of concern and resources devoted to addressing the risks stay where they are today. If I had assumed business as usual, my risk estimates would have been substantially higher. But I think they would have been misleading, overstating the chance that we actually suffer an existential catastrophe.[9] So instead, I've made allowances for the fact that we will likely respond to the escalating risks, with substantial efforts to reduce them.

The numbers therefore represent my actual best guesses of the chance the threats materialise, taking our responses into account. If we outperform my expectations, we could bring the remaining risk down below these estimates. Perhaps one could say that we were heading towards Russian roulette with two bullets in the gun, but that I think we will remove one of these before it's time to pull the trigger. And there might just be time to remove the last one too, if we really try. So perhaps the headline number should not be the amount of risk I expect to remain, about one in six, but two in six—the difference in existential risk between a lacklustre effort by humanity and a heroic one.

These probabilities provide a useful summary of the risk landscape, but they are not the whole story, nor even the whole bottom line. Even completely objective, precise and accurate estimates would merely measure how large the different risks are, saying nothing about how tractable they are, nor how neglected. The raw probabilities are thus insufficient for determining which risks should get the most attention, or what kind of attention they should receive. In this chapter and the next, we'll start to ask these further questions, putting together the tools needed to confront these threats to our future.

ANATOMY OF AN EXTINCTION RISK

With such a diverse landscape of risks, it can be helpful to classify them by what they have in common. This helps us see lines of attack that would address several risks at once.

My colleagues at the Future of Humanity Institute have suggested classifying risks of human extinction by the three successive stages that need to occur before we would go extinct:[10]

Origin: *How does the catastrophe get started?*
Some are initiated by the natural environment, while others are anthropogenic. We can usefully break anthropogenic risks down according to whether the harm was intended, foreseen or unforeseen. And we can further break these down by whether they involve a small number of actors (such as accidents or terrorism) or a large number (such as climate change or nuclear war).

Scaling: *How does the catastrophe reach a global scale?*
It could start at a global scale (such as a climate change) or there could be a mechanism that scales it up. For example, the sunlight-blocking particles from asteroids, volcanoes and nuclear war get spread across the world by the Earth's atmospheric circulation while pandemics are scaled up by an exponential process in which each victim infects several others.

Endgame: *How does the catastrophe finish the job?*
How does it kill everyone, wherever they are? Like the dust kicked up by an asteroid, the lethal substance could have spread everywhere in the environment; like a pandemic it could be carried by people wherever people go; or in an intentional plan to cause extinction, it could be actively targeted to kill each last pocket of survivors.

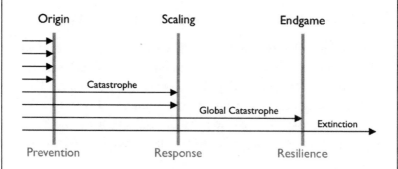

We can fight a risk at any of these stages: *prevention* can avoid its origin, *response* can limit its scaling, and *resilience* can thwart its endgame. Depending on the risk, we may want to direct our efforts to the most efficient stage at which to block it, or adopt a strategy of defence-in-depth, addressing all stages at once.

This classification lets us break down the probability of extinction into the product of (1) the probability it gets started, (2) the probability it reaches a global scale given it gets started, and (3) the probability it causes extinction given it reaches a global scale:

$$p_{extinction} = p_{origin} \times p_{scaling} \times p_{endgame}$$

Prevention, response and resilience act to lower each factor respectively. Because the probabilities are multiplied together, we can see that a reduction in one factor by some amount would be matched by reducing any other factor by the same proportion. So as a rule of thumb, we should prioritise the factor that is currently easiest to halve.[11]

And there are other valuable ways to classify risks too. For example, Shahar Avin and colleagues at the Cambridge Centre for the Study of Existential Risk (CSER) have classified risks according to which critical system they threaten: whether that be an essential system in the environment, in the human body or in our social structures.[12]

COMBINING AND COMPARING RISKS

The risk landscape is comprised of many different existential risks. So far we have mostly considered each in isolation. But if we want to understand how they combine and how they compare, we need to consider how they interact. And even risks that are statistically independent still interact in an important way: if one risk destroys us, others can't.

Let's start with the idea of the *total existential risk*. This is the risk of humanity eventually suffering an existential catastrophe, of any kind.[13] It includes *all* the risks: natural and anthropogenic, known and unknown, near future and far future. All avenues through which a catastrophe might irrevocably destroy humanity's potential.

This is an extremely useful concept, as it converts all individual risks into a common currency—their contribution to this total risk. But it does require us to make a simplifying assumption: that the stakes involved in the different risks are of relatively similar sizes, such that the main difference between them is their probability. This is not always the case, but it is a good starting point.[14]

How do individual risks combine to make the total risk? Suppose there were just two risks across our entire future: a 10 percent risk and a 20 percent risk. How much total risk is there? While we might be tempted to just add them up, this is usually wrong. The answer depends upon the relationship between the risks (see Figure 6.1).

The worst case is when they are perfectly anticorrelated (like the chance that a random number between one and 100 is less than or equal to ten and the chance that it is greater than 80). Then the risk is simply the sum of the two: 30 percent. The best case is when the risks are perfectly correlated, such that the 10 percent risk only happens in cases where the 20 percent risk also happens (think of the chance that a random number from one to 100 is less than or equal to ten and the chance that it is less than or equal to 20).[15] Then, the total risk is just the larger of the two: 20 percent. If the risks are statistically independent of each other (such as two

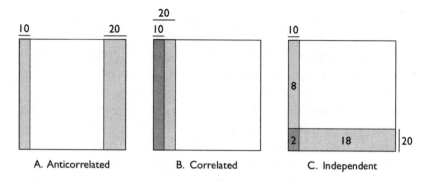

FIGURE 6.1 There are many ways risks can combine, ranging from perfect anticorrelation (A) to perfect correlation (B). An important case in between is independence (C). The total risk posed depends on how much risk is 'wasted' in the overlap—the region where we'd suffer a catastrophe even if we eliminated one of the risks. A large overlap reduces the total risk, but also reduces the benefits from eliminating a single risk.

separate lotteries), the chance is intermediate. In this case, it would be 28 percent. (The easiest way to see this is via the chance that neither catastrophe occurs: 90 percent × 80 percent = 72 percent; when we subtract this from 100 percent we get 28 percent.)

What should we expect in practice? Overall, I think we should expect some positive correlation between most pairs of risks, due to the existence of common causes and common solutions.[16] For example, a major world war might increase many existential risks, while a successful global institution for managing existential risk might reduce many. This is mildly good news, as it makes them pre-empt each other more often and means that the total risk will be a bit lower than if you simply added them up, or assumed independence.[17] If you can't see how the risks are connected, a reasonable approach would be to start by assuming they are independent, then check how things would change were they correlated or anticorrelated, for robustness.

Surprisingly these same issues come up not just when combining risks, but when comparing their importance. How much more important is a 20 percent risk compared to a 10 percent risk? The obvious answer of 'twice as important' is almost never correct (the risks would need to be perfectly anticorrelated). To get the right answer, we need to note that the importance of eliminating a risk lies in the amount of total risk that would disappear were that risk to be eliminated. Then we can check how much this is.

For example, we saw in Figure 6.1 how independent 10 percent and 20 percent risks produce a total risk of 28 percent. So eliminating the 10 percent risk would reduce the total by 8 points (from 28 percent to 20 percent) while eliminating the 20 percent risk would reduce the total by 18 points (from 28 percent to 10 percent), which is more than twice as much. The 20 percent risk would actually be 2.25 times as important as the 10 percent risk, and in general, larger risks are more important than you would think. These counter-intuitive effects (and others) increase in size the more the risks are correlated and the higher the total risk. They become especially important if risk this century is higher than I estimate, or if the total risk over our entire future is high (some surprising results of this are explored in Appendix D).

RISK FACTORS

Chapters 3 to 5 covered many distinct existential risks. Armed with the concept of total risk, we could think of those chapters as carving up the total risk into a set of named risks, each with a different mechanism for destroying our potential. We might be tempted to think of this as a list of the most important topics facing humanity: a menu from which an aspiring altruist might choose their life's mission. But this would be too quick. For this is not the only way to carve up the total existential risk we face.

Consider the prospect of great-power war this century. That is, war between any of the world's most powerful countries or blocs.[18] War on such a scale defined the first half of the twentieth century, and its looming threat defined much of the second half too. Even though international tension may again be growing, it seems almost unthinkable that any of the great powers will go to war with each other this decade, and unlikely for the foreseeable future. But a century is a long time, and there is certainly a risk that a great-power war will break out once more.

While one *could* count great-power war as an existential risk, it would be an awkward fit. For war is not in itself a mechanism for destroying humanity or our potential—it is not the final blow. Yet a great-power war would nevertheless increase existential risk. It would increase the risks posed by a range of weapons technologies: nuclear weapons, engineered pandemics and whatever new weapons of mass destruction are invented in the meantime. It would also indirectly increase the probability of the other risks we face: the breakdown in international trust and cooperation would make it harder to manage climate change or the safe development of AGI, increasing the danger they pose. And great-power wars may also hasten the arrival of new existential risks. Recall that nuclear weapons were developed during the Second World War, and their destructive power was amplified significantly during the Cold War, with the invention of the hydrogen bomb. History suggests that wars on such a scale prompt humanity to delve into the darkest corners of technology.

When all of this is taken into account, the threat of great-power war may (indirectly) pose a significant amount of existential risk. For example, it seems that the bulk of the existential risk last century was driven by the threat of great-power war. Consider your own estimate of how much existential risk there is over the next hundred years. How much of this would disappear if you knew that the great powers would not go to war with each other over that time? It is impossible to be precise, but I'd estimate an appreciable fraction would disappear—something like a tenth of the existential risk over that time. Since I think the existential risk over the next hundred years is about one in six, I am estimating that great power war effectively poses more than a percentage point of existential risk over the next century. This makes it a larger contributor to total existential risk than most of the specific risks we have examined.

While you should feel free to disagree with my particular estimates, I think a safe case can be made that the contribution of great-power war to existential risk is larger than the contribution of all natural risks combined. So a young person choosing their career, a philanthropist choosing their cause or a government looking to make a safer world may do better to focus on great-power war than on detecting asteroids or comets.

This alternative way of carving up the total risk was inspired by *The Global Burden of Disease*—a landmark study in global health that attempted to understand the big picture of health across the entire world, and also acted as a major inspiration for my own work in the field.[19] Its authors divided up all ill health in the world according to which disease or injury caused it. This gave them subtotals for each disease and injury which sum to the total health burden of disease and injury. But they also wanted to ask further questions, such as how much ill health is caused by smoking. Smoking is not itself a disease or injury, but it causes disease of the heart and lungs. They dubbed smoking a 'risk factor', stating: 'A risk factor is an attribute or exposure which is causally associated with an increased probability of a disease or injury.' This allows an extremely useful cross-cutting analysis of where the ill health is coming from, letting us estimate how much could be gained if we were considering making inroads against risk factors such as smoking, lack of access to safe drinking water or vitamin deficiency.

Let us call something that increases existential risk an *existential risk factor* (the 'existential' can be omitted for brevity since this is the only kind of risk factor we are concerned with hereafter).[20] Where the division into individual risks can be seen as breaking existential risk up into vertical silos, existential risk factors cut across these divisions. The idea of existential risk factors is very general and can be applied at any scale, but it is at its most useful when considering coherent factors that have a large effect on existential risk. Where the individual risks of Chapters 3 to 5 were an attempt to partition the total risk into a set of non-overlapping risks, there is no such constraint for risk factors: it is fine if they overlap or even if one is subsumed by another, so long as they are useful. Thus the contributions of risk factors to existential risk don't even approximately 'add up' to the total risk.

MATHEMATICS OF RISK FACTORS

We can make risk factors more precise through the language of probability theory. Let F be a risk factor (such as *climate change* or *great-power war*). And let f be a quantitative measure of that risk factor (such as *degrees of warming* or the *probability that there is a great-power war*). Call its minimum achievable value f_{min}, its status quo value f_{sq} and its maximum achievable value f_{max}.[24] Recall that $Pr(P)$ represents the probability that an event P happens and $Pr(P|Q)$ represents that probability that P happens given that Q happens. Finally, let event X be an existential catastrophe occurring.

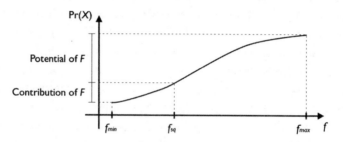

We can call the difference between $Pr(X|f = f_{sq})$ and $Pr(X|f = f_{min})$ the *contribution* that F makes to existential risk. It represents the amount by which total existential risk would be lowered if we eliminated this risk factor. This is a number that allows an apples-to-apples comparison between the size of risk factors and of existential risks. Similarly, we could call the difference between $Pr(X|f = f_{sq})$ and $Pr(X|f = f_{max})$ the *potential* of F. It represents how much existential risk could arise if this risk factor got worse.[25]

And when it comes to prioritising existential risk factors, we will often be most interested in the steepness of the curve at f_{sq}. This reflects how much protection we could produce by a marginal change to the risk factor.[26]

An easy way to find existential risk factors is to consider stressors for humanity or for our ability to make good decisions. These include global economic stagnation, environmental collapse and breakdown in the international order.[21] Indeed even the *threat* of such things may constitute an existential risk factor, as a mere possibility can create actual global discord or panic.

Many risks that threaten (non-existential) global catastrophe also act as existential risk factors, since humanity may be more vulnerable following a global catastrophe. The same holds for many existential threats: if they can produce global catastrophes that increase our vulnerability to subsequent existential risks, then they also act as risk factors.[22] In some cases, they may pose substantially more indirect risk than direct risk.

In Chapter 4 we saw that it is difficult for nuclear winter or climate change to completely destroy humanity's potential. But they could easily cause major catastrophes that leave us more vulnerable to other existential risks. Chapter 4 focused on nuclear war and climate change in their role as *existential risks* (since it is important to understand whether there really are plausible mechanisms through which our potential could be destroyed). But their role as risk factors may be more important. A better understanding of how they increase other risks would be of great value, because what we ultimately want to know is how much they increase risk overall.[23]

This discussion of risk factors raises the possibility of other factors that *reduce* risk.[27] We can call these *existential security factors*.[28] Examples include strong institutions for avoiding existential risk, improvements in civilisational virtues or peace between great powers. As this last example suggests, if something is a risk factor, its opposite will be a security factor.

Many of the things we commonly think of as social goods may turn out to also be existential security factors. Things such as education, peace or prosperity may help protect us. And many social ills may be existential risk factors. In other words, there may be explanations grounded in existential risk for pursuing familiar, common-sense agendas.

But I want to stress that this is a dangerous observation. For it risks a slide into complacency, where we substitute our goal of securing our future with other goals that may be only loosely related. Just because existential risk declines as some other goal is pursued doesn't mean that the other goal is the most effective way to secure our future. Indeed, if the other goal is commonsensically important there is a good chance it is already receiving far more resources than are devoted to direct work on existential risk. This would give us much less opportunity to really move the needle. I think it likely that there will only be a handful of existential risk factors or security factors (such as great-power war) that really compete with the most important existential risks in terms of how effectively additional work on them helps to secure our future. Finding these would be extremely valuable.

WHICH RISKS?

Sadly, most of the existential risks we've considered are neglected, receiving substantially less attention than they deserve. While this situation is changing, we should expect it to take decades before sufficient resources are mobilised to adequately address all the risks we face. So those of us concerned with safeguarding our future will need to prioritise: to determine where we should devote our finite energies and resources.

A natural way to do so would be to compare risks by their probabilities—or more precisely, by their contributions to the total risk. Since most existential risks have very similar stakes (most of the potential value of the future), one might think that this is the whole story: prioritise the risks by how much they increase the total risk.

But that isn't right. For some risks might be easier to address. For example, we might be able to reduce a smaller risk from 5 percent to 1 percent with the same resources that would be needed to reduce a larger, but more stubborn, risk from 20 percent to 19 percent.[29] If so, we would reduce total existential risk by a greater amount if we spent those resources on the smaller risk.

Our ultimate aim is to spend the resources allocated to existential risk in such a way as to reduce total risk by the greatest amount. We can think of humanity's allocation of resources as a portfolio, with different amounts invested in various approaches to various risks. Designing an entire portfolio is very complex, and we are often only able to make small adjustments to the world's overall allocation, so it can simplify things to imagine making a small change to an existing portfolio. Given all the other work that is currently going on, which risk is most pressing? Where can an additional bundle of resources (such as time or money) most reduce total risk?

It is extremely difficult to give a precise answer to this question. But there are good heuristics that can guide us in narrowing down the possibilities. One such approach is to note that the more a problem is important, tractable or neglected, the more cost-effective it is to work on it, and thus the higher its priority.[30] The *importance* of a problem is the value of solving it. In the case of an existential risk, we can usually treat this as the amount it contributes to total risk. *Tractability* is a measure of how easy it is to solve the problem. A useful way of making this precise is to ask what fraction of the risk would be eliminated were we to double the amount of resources we are currently devoting to it. Finally, a problem is *neglected* to the extent that there are few resources spent on it. This incorporates the idea of diminishing returns: resources typically make a bigger difference when fewer resources have been spent so far.[31]

My colleague at the Future of Humanity Institute, Owen Cotton-Barratt, has shown that when these terms are appropriately defined, the cost-effectiveness of working on a particular problem can be expressed by a very simple formula:[32]

Cost-Effectiveness = Importance × Tractability × Neglectedness

Even though it is very difficult to assign precise numbers to any of these dimensions, this model still provides useful guidance. For example, it shows why the ideal portfolio typically involves investing resources fighting several risks instead of just one: as we invest more in a given risk, it becomes less neglected, so the priority of investing additional resources in it falls. After a while, marginal resources would be better spent on a different risk.

The model also shows us how to make trade-offs between these dimensions. For example, when choosing between two risks, if their probabilities differed by a factor of five, this would be outweighed by a factor of ten in how much funding they currently receive. Indeed, the model suggests a general principle:

Proportionality
When a set of risks have equal tractability (or when we have no idea which is more tractable), the ideal global portfolio allocates resources to each risk in proportion to its contribution to total risk.[33]

But this doesn't mean *you* should spread your resources between them in proportion to their probabilities. An individual or group should allocate its resources to help bring the world's portfolio into line with the ideal allocation. This will often mean putting all of your effort into a single risk—especially when taking into account the value of being able to give it your undivided attention.

This analysis gives us a starting point: a generic assessment of the value of allocating new resources to a risk. But there are often resources that are much more valuable when applied to one risk rather than another. This is especially true when it comes to *people*. A biologist would be much more suited to working on risks of engineered pandemics than retraining to work on AI risk. The ideal portfolio would thus take people's comparative advantage into account. And there are sometimes highly leveraged opportunities to help with a particular risk. Each of these dimensions (fit and leverage) could easily change the value of an opportunity by a factor of ten (or more).

Let's consider three more heuristics for setting our priorities: focusing on risks that are *soon*, *sudden* and *sharp*. These are not competitors to importance, tractability and neglectedness, but ways of illuminating those dimensions.

Suppose one risk strikes soon, and another late. Other things being equal, we should prioritise the one that strikes soon.[34] One reason is that risks that strike later can be dealt with later, while those striking soon cannot. Another is that there will probably be more resources devoted to risks that occur later on, as humanity becomes more powerful and more people wake up to humanity's predicament. This makes later risks less neglected. And finally, we can see more clearly what to do about risks that are coming to a head now, whereas our work on later risks has more chance of being misdirected. Technological or political surprises in the intervening years may change the nature of the risk, introduce superior ways of dealing with it or eliminate it altogether, thereby wasting some of our early efforts. This makes later risks less tractable right now than earlier ones.

What if one risk is sudden, while another unfolds slowly? Other things being equal, we should prioritise the one that strikes suddenly. For the risk that unfolds slowly has more chance of arousing widespread attention from the public and traditional policymakers. So over the long run, it is likely to be less neglected.

Some existential risks threaten catastrophes on a variety of scales. Pandemics can kill thousands, millions or billions; and asteroids range from metres to kilometres in size. In each case, they appear to follow a power law distribution, where catastrophes become substantially more rare as they get larger. This means that we are more likely to get hit by a pandemic or asteroid killing a hundredth of all people before one killing a tenth, and more likely to be hit by one killing a tenth of all people before one that kills almost everyone.[35] In contrast, other risks, such as that from unaligned artificial intelligence, may well be all-or-nothing. Let's call a smaller catastrophe a 'warning shot' if it is likely to provoke major useful action to eliminate future risk of that type. Other things being equal, we should prioritise the sharp risks— those that are less likely to be preceded by warning shots—for they are more likely to remain neglected over the long run.[36]

While I've presented this analysis in terms of which risks should get the highest priority, these exact same principles can be applied to prioritising between different risk factors or security factors. And they can help prioritise between different ways of protecting our potential over the long term, such as promoting norms, working within existing institutions or establishing new ones. Best of all, these principles can be used to set priorities between these areas as well as within them, since all are measured in the common unit of total existential risk reduction.

In the course of this book, we have considered a wide variety of approaches to reducing existential risk. The most obvious has been direct work on a particular risk, such as nuclear war or engineered pandemics. But there were also more indirect approaches: work on risk factors such as great-power war; or on security factors such as a new international institution tasked with reducing existential risk. Perhaps one could act at an even more indirect level. Arguably risk would be lower in a period of stable economic growth than in a period with the turmoil caused by deep recessions. And it may be lower if citizens were better educated and better informed.

The philosopher Nick Beckstead suggests we distinguish between *targeted* and *broad* interventions.[38] A focus on safeguarding humanity needn't imply a focus on narrowly targeted interventions, such as the governance of a dangerous technology. Existential risk can also be reduced by broader interventions aimed at generally improving wisdom, decision-making or international cooperation. And it is an open question which of these approaches is more effective. Beckstead suggests that longtermists in past centuries would have done better to focus on broad interventions rather than narrow interventions.

I think Beckstead may be right about past centuries, but mostly because existential risk was so low until we became powerful enough to threaten ourselves in the twentieth century. From that point, early longtermists such as Bertrand Russell and Albert Einstein were right to devote so much attention to the targeted intervention of reducing the threat of nuclear war.

EARLY ACTION

Some of the biggest risks we face are still on the horizon. Can we really do useful work to eliminate a threat so far in advance? How can we act now, when we are not fully aware of the form the risks may take, the nature of the technologies, or the shape of the strategic landscape at the moment they strike?

These are real concerns for any attempts to address future risks. We are near-sighted with respect to time, and so there is a serious chance our best efforts will be wasted. But this is not the whole story. For there are also ways in which action far in advance of the threat can be all the more helpful.

Early action is best for changing course. If we are headed the wrong way, it is better to correct this at an early stage. So if we need to steer a technology (or a nation) from a dangerous path, we have more power to do so now, rather than later.

Early action is best when it comes to self-improvement, for there is more time to reap what is sown. If what is needed is research, education or influence, starting sooner is better.

Early action is best for growth. If one needs to turn an investment into a fortune, an article into a research field or an idea into a movement, one had best start soon.

And early action is best for tasks that require a large number of successive stages. If your solution has this structure, it may be impossible if you don't start extremely early.

In short, early action is higher leverage, but more easily wasted. It has more power, but less accuracy. If we do act far in advance of a threat, we should do so in ways that take advantage of this leverage, while being robust to near-sightedness.[37] This often means a focus on knowledge and capacity building, over direct work.

In my view we can tentatively resolve the question of targeted versus broad interventions by considering neglectedness. In our current situation there are trillions of dollars per year flowing to broad interventions such as education, but less than a ten-thousandth this much going to targeted existential risk interventions.[39] So the broad interventions are much less neglected. This gives us a strong reason to expect increasing work on targeted interventions to be more effective at the moment (with the strongest case for broad interventions coming from those that receive the least attention).

But if the resources spent on targeted existential risk interventions were radically increased, this would start to change. We currently spend less than a thousandth of a percent of gross world product on them. Earlier, I suggested bringing this up by at least a factor of 100, to reach a point where the world is spending more on securing its potential than on ice cream, and perhaps a good longer-term target may be a full 1 percent.[40] But there will be serious diminishing returns as investment is scaled up, and it may well be that even if the world were solely interested in reducing existential risk, the total budget for targeted interventions should never exceed that for broad interventions.

We now have a rough map of the risk landscape, and the intellectual tools needed to find promising paths forward. It is time to put these to use; to start planning how to safeguard humanity—from the big picture strategy all the way down to concrete advice on how we can each play a part.

7

SAFEGUARDING HUMANITY

There are no catastrophes that loom before us which cannot be avoided; there is nothing that threatens us with imminent destruction in such a fashion that we are helpless to do something about it. If we behave rationally and humanely; if we concentrate coolly on the problems that face all of humanity, rather than emotionally on such nineteenth century matters as national security and local pride; if we recognize that it is not one's neighbors who are the enemy, but misery, ignorance, and the cold indifference of natural law—then we can solve all the problems that face us. We can deliberately choose to have no catastrophes at all.

—Isaac Asimov[1]

What we do with our future is up to us. Our choices determine whether we live or die; fulfil our potential or squander our chance at greatness. We are not hostages to fortune. While each of our lives may be tossed about by external forces—a sudden illness, or outbreak of war—humanity's future is almost entirely within humanity's control. Most existential risk comes from human action: from activities which we can choose to stop, or to govern effectively. Even the risks from nature come on sufficiently protracted timescales that we can protect ourselves long before the storm breaks.

We need to take responsibility for our future. Those of us alive right now are the only people who can fight against the present dangers; the only people who can build the communities, norms and institutions that will safeguard our future. Whether we are remembered as the generation who turned the corner to a bright and secure future, or not remembered at all, comes down to whether we rise to meet these challenges.

When exploring these issues, I find it useful to consider our predicament from *humanity*'s point of view: casting humanity as a coherent agent, and considering the strategic choices it would make were it sufficiently rational and wise. Or in other words, what all humans would do if we were sufficiently coordinated and had humanity's longterm interests at heart.

This frame is highly idealised. It obscures the challenges that arise from our disunity and the importance of actions that individuals might take to nudge humanity as a whole in the right direction. But it illuminates larger questions, which have so far been almost entirely neglected. Questions about the grand strategy for humanity, and how we could make sure we can achieve an excellent future—even if we don't yet know precisely what kind of future that would be. By answering them, I paint an ambitious vision of humanity getting its house in order that I hope can guide us over the coming decades, even if the reality is more messy and fraught.

My advice will range from high-level strategy, to policy suggestions, all the way down to the individual level, with promising career paths and actions that anyone could take. Because people have spent very little time thinking carefully about how to safeguard humanity's longterm potential, all such guidance must be viewed as tentative; it has not yet stood the test of time.

But the fact that we are at such an early stage in thinking about the longterm future of humanity also provides us with reason to be hopeful as we begin our journey. This is not a well-worn track, where the promising ideas have long since been explored and found wanting. It is virgin territory. And it may be rich with insights for the first explorers who seek them.

GRAND STRATEGY FOR HUMANITY

How can humanity have the greatest chance of achieving its potential? I think that at the highest level we should adopt a strategy proceeding in three phases:[2]

1. Reaching Existential Security
2. The Long Reflection
3. Achieving Our Potential

On this view, the first great task for humanity is to reach a place of safety—a place where existential risk is low and stays low. I call this *existential security*.

It has two strands. Most obviously, we need to *preserve* humanity's potential, extracting ourselves from immediate danger so we don't fail before we've got our house in order. This includes direct work on the most pressing existential risks and risk factors, as well as near-term changes to our norms and institutions.

But we also need to *protect* humanity's potential—to establish lasting safeguards that will defend humanity from dangers over the longterm future, so that it becomes almost impossible to fail.[3] Where preserving our potential is akin to fighting the latest fire, protecting our potential is making changes to ensure that fire will never again pose a serious threat.[4] This will involve major changes to our norms and institutions (giving humanity the prudence and patience we need), as well as ways of increasing our general resilience to catastrophe. This needn't require foreseeing all future risks right now. It is enough if we can set humanity firmly on a course where we will be taking the new risks seriously: managing them successfully right from their onset or sidestepping them entirely.

Note that existential security doesn't require the risk to be brought down to *zero*. That would be an impossible target, and attempts to achieve it may well be counter-productive. What humanity needs to do is bring this century's risk down to a very low level, then keep gradually reducing it from there as the centuries go on. In this way, even though there may always remain some risk in each century, the total risk over our entire future can be kept small.[5] We could view this as a form of existential sustainability. Futures in which accumulated existential risk is allowed to climb towards 100 percent are unsustainable. So we need to set a strict risk budget over our entire future, parcelling out this non-renewable resource with great care over the generations to come.

Ultimately, existential security is about reducing total existential risk by as many percentage points as possible. Preserving our potential is helping lower the portion of the total risk that we face in the next few decades, while protecting our potential is helping lower the portion that comes over the longer run. We can work on these strands in parallel, devoting some of our efforts to reducing imminent risks and some to building the capacities, institutions, wisdom and will to ensure that future risks are minimal.[6]

A key insight motivating existential security is that there appear to be no major obstacles to humanity lasting an extremely long time, if only that were a key global priority. As we saw in Chapter 3, we have ample time to protect ourselves against natural risks: even if it took us millennia to resolve the threats from asteroids, supervolcanism and supernovae, we would incur less than one percentage point of total risk.

The greater risk (and tighter deadline) stems from the anthropogenic threats. But being of humanity's own making, they are also within our control. Were we sufficiently patient, prudent and coordinated, we could simply stop imposing such risks upon ourselves. We would factor in the hidden costs of carbon emissions (or nuclear weapons) and realise they are not a good deal. We would adopt a more mature attitude to the most radical new technologies—devoting at least as much of humanity's brilliance to forethought and governance as to technological development.

In the past, the survival of humanity didn't require much conscious effort: our past was brief enough to evade the natural threats and our power too limited to produce anthropogenic threats. But now our longterm survival requires a deliberate choice to survive. As more and more people come to realise this, we can make this choice. There will be great challenges in getting people to look far enough ahead and to see beyond the parochial conflicts of the day. But the logic is clear and the moral arguments powerful. It can be done.

If we achieve existential security, we will have room to breathe. With humanity's longterm potential secured, we will be past the Precipice, free to contemplate the range of futures that lie open before us. And we will be able to take our time to reflect upon what we truly desire; upon which of these visions for humanity would be the best realisation of our potential. We shall call this the *Long Reflection*.[7]

We rarely think this way. We focus on the here and now. Even those of us who care deeply about the longterm future need to focus most of our attention on making sure we *have* a future. But once we achieve existential security, we will have the luxury of time in which to compare the kinds of futures available to us and judge which is best. Most work in moral philosophy so far has focused on negatives—on avoiding wrong action and bad outcomes. The study of the positive is at a much earlier stage of development.[8] During the Long Reflection, we would need to develop mature theories that allow us to compare the grand accomplishments our descendants might achieve with aeons and galaxies as their canvas.

Present-day humans, myself included, are poorly positioned to anticipate the results of this reflection.[9] But we are uniquely positioned to make it possible.

The ultimate aim of the Long Reflection would be to achieve a final answer to the question of which is the best kind of future for humanity. This may be the true answer (if truth is applicable to moral questions) or failing that, the answer we would converge to under an ideal process of reflection. It may be that even convergence is impossible, with some disputes or uncertainties that are beyond the power of reason to resolve. If so, our aim would be to find the future that gave the best possible concili-ation between the remaining perspectives.[10]

We would not need to fully complete this process before moving forward. What is essential is to be sufficiently confident in the broad shape of what we are aiming at before taking each bold and potentially irreversible action—each action that could plausibly lock in substantial aspects of our future trajectory.

For example, it may be that the best achievable future involves physically perfecting humanity, by genetically improving our biology. Or it may involve giving people the freedom to adopt a stunning diversity of new biological forms. But proceeding down either of these paths prematurely could introduce its own existential risks.

If we radically change our nature, we replace humanity (or at least *Homo sapiens*) with something new. This would risk losing what was most valuable about humanity before truly coming to understand it. If we diversify our forms, we fragment humanity. We might lose the essential unity of humanity that allows a common vision for our future, and instead find ourselves in a perpetual struggle or unsatisfactory compromise. Other bold actions could pose similar risks, for instance spreading out beyond our Solar System into a federation of independent worlds, each drifting in its own cultural direction.

This is not to reject such changes to the human condition—they may well be essential to realising humanity's full potential. What I am saying is that these are the kind of bold changes that would need to come after the Long Reflection.[11] Or at least after enough reflection to fully understand the consequences of that particular change. We need to take our time, and choose our path with great care. For once we have existential security we are almost assured success if we take things slowly and carefully: the game is ours to lose; there are only unforced errors.

What can we say about the process of the Long Reflection? I am not imagining this as the sole task of humanity during that time—there would be many other great projects, such as the con-tinuing quests for knowledge, prosperity and justice. And many of the people at the time may have only passing interest in the Long Reflection. But it is the Long Reflection that would have the most bearing on the shape of the future, and so it would be this for which the time would be remembered.[12]

The process may take place largely within intellectual circles, or within the wider public sphere. Either way, we would need to take the greatest care to avoid it being shaped by the bias or prejudice of those involved. As Jonathan Schell said regarding a similar venture, 'even if every person in the world were to enlist, the endeavour would include only an infinitesimal fraction of the people of the dead and the unborn generations, and so it would need to act with the circumspection and modesty of a small minority'.[13] While the conversation should be courteous and respectful to all perspectives, it is even more important that it be robust and rigorous. For its ultimate aim is not just to win the goodwill of those alive at the time, but to deliver a verdict that stands the test of eternity.

While moral philosophy would play a central role, the Long Reflection would require insights from many disciplines. For it isn't just about determining which futures are best, but which are feasible in the first place, and which strategies are most likely to bring them about. This requires analysis from science, engineering, economics, political theory and beyond.

We could think of these first two steps of existential security and the Long Reflection as designing a constitution for humanity. Achieving existential security would be like writing the safeguarding of our potential into our constitution. The Long Reflection would then flesh out this constitution, setting the directions and limits in which our future will unfold.

Our ultimate aim, of course, is the final step: fully achieving humanity's potential.[14] But this can wait upon a serious reflection about which future is best and on how to achieve that future without any fatal missteps. And while it would not hurt to begin such reflection now, it is not the most urgent task.[15] To maximise our chance of success, we need first to get ourselves to safety—to achieve existential security. This is the task of our time. The rest can wait.

SECURITY AMONG THE STARS?

Many of those who have written about the risks of human extinction suggest that if we could just survive long enough to spread out through space, we would be safe—that we currently have all of our eggs in one basket, but if we became an interplanetary species, this period of vulnerability would end.[16] Is this right? Would settling other planets bring us existential security?

The idea is based on an important statistical truth. If there were a growing number of locations which all need to be destroyed for humanity to fail, and if the chance of each suffering a catastrophe is independent of whether the others do too, then there is a good chance humanity could survive indefinitely.[17]

But unfortunately, this argument only applies to risks that are statistically independent. Many risks, such as disease, war, tyranny and permanently locking in bad values are correlated across different planets: if they affect one, they are somewhat more likely to affect the others too. A few risks, such as unaligned AGI and vacuum collapse, are almost completely correlated: if they affect one planet, they will likely affect all.[18] And presumably some of the as-yet-undiscovered risks will also be correlated between our settlements.

Space settlement is thus helpful for achieving existential security (by eliminating the uncorrelated risks) but it is by no means sufficient.[19] Becoming a multi-planetary species is an inspirational project—and may be a necessary step in achieving humanity's potential. But we still need to address the problem of existential risk head-on, by choosing to make safeguarding our longterm potential one of our central priorities.

RISKS WITHOUT PRECEDENT

Humanity has never suffered an existential catastrophe, and hopefully never will. Catastrophes of this scale are unprecedented throughout our long history. This creates severe challenges for our attempts to understand, predict and prevent these disasters. And what's more, these challenges will always be with us, for existential risks are *necessarily unprecedented*. By the time we have a precedent, it is too late—we've lost our future. To safeguard humanity's potential, we are forced to formulate our plans and enact our policies in a world that has never witnessed the events we strive to avoid.[20] Let's explore three challenges this creates and how we might begin to address them.

First, we can't rely on our current intuitions and institutions that have evolved to deal with small- or medium-scale risks.[21] Our intuitive sense of fear is neither evolutionarily nor culturally adapted to deal with risks that threaten so much more than an individual life—risks of catastrophes that cannot be allowed to happen even once over thousands of years in a world containing billions of people. The same is true for our intuitive sense of the likelihood of very rare events and of when such a risk is too high. Evolution and cultural adaptation have led to fairly well-tuned judgements for these questions in our day-to-day lives (when it's safe to cross the road; whether to buy a smoke alarm), but are barely able to cope with risks that threaten hundreds of people, let alone those that threaten billions and the very future of humanity.

The same is true of our institutions. Our systems of laws, norms and organisations for handling risk have been tuned to the small- and medium-scale risks we have faced over past centuries. They are ill-equipped to address risks so extensive that they will devastate countries across the globe; so severe that there will be no legal institutions remaining to exact punishment.

The second challenge is that we cannot afford to fail *even once*. This removes our ability to learn from failure. Humanity typically manages risk via a heavy reliance on trial and error. We scale our investment or regulation based on the damages we've seen so far; we work out how to prevent new fires by sifting through the ashes.

But this reactive trial and error approach doesn't work at all when it comes to existential risk. We will need to take proactive measures: sometimes long in advance, sometimes with large costs, sometimes when it is still unclear whether the risk is real or whether the measures will address it.[22] This will require institutions with access to cutting-edge information about the coming risks, capable of taking decisive actions, and with the will to actually do so. For many risks, this action may require swift coordination between many or all of the world's nations. And it may have to be done knowing we will never find out whether our costly actions really helped. This will ultimately require new institutions, filled with people of keen intellect and sound judgement, endowed with a substantial budget and real influence over policy.

These are extremely challenging circumstances for sound policy-making—perhaps beyond the abilities of even the best-functioning institutions today. But this is the situation we are in, and we will need to face up to it. There is an urgency to improving our institutional abilities to meet these demands.

Working out when such institutions should take action will raise its own deeply challenging questions. On the one hand, they will need to be able to take strong actions, even when the evidence is short of the highest scientific standards. And yet, this puts us at risk of chasing phantoms—being asked (or forced) to make substantial sacrifices on the basis of little evidence. This poses an even greater problem if the risk involves classified elements or information hazards that cannot be opened up to public scrutiny and response. The challenges here are similar to those arising from the ability of governments to declare a state of emergency: contingency powers are essential for managing real emergencies, yet open to serious abuse.[23]

The third challenge is one of knowledge. How are we to predict, quantify or understand risks that have never transpired? It is extremely difficult to predict the risk posed by new technologies. Consider the situation of allowing cars onto our roads for the first time. It was very unclear how dangerous that would be, but once it had happened, and millions of miles had been driven, we could easily determine the risks by looking at the statistical frequencies. This let us see whether the gains outweighed the risks, how much could be gained by new safety improvements, and which improvements would be most helpful.

With existential risk, we cannot help ourselves to such a track record, with probabilities grounded in long-run frequencies. Instead, we have to make decisions of grave importance without access to robust probabilities for the risks involved.[24] This raises substantial difficulties in how we are to form probability estimates for use in decision-making surrounding existential risk.[25] This problem already exists in climate change research and causes great difficulties in setting policy—especially if politicisation leads to explicit or implicit biases in how people interpret the ambiguous evidence.

During the Cold War concern about existential risk from nuclear war was often disparaged on the grounds that we haven't *proved* the risk is substantial. But when it comes to existential risk that would be an impossible standard. Our norms of scientific proof require experiments to be repeated many times, and were established under the assumptions that such experiments are possible and not too costly. But here neither assumption is true. As Carl Sagan memorably put it: 'Theories that involve the end of the world are not amenable to experimental verification— or at least, not more than once.'[26]

Even with no track record of existential catastrophe, we do have some ways of estimating probabilities or bounds on the probability. For example, in Chapter 3 we saw how we can use the length of time humans and similar animals have survived to get a very rough estimate of the combined natural risk. We can also pay attention to near misses: both the largest catastrophes that *have* occurred (such as the Black Death), and existential catastrophes that *nearly* occurred (such as during the Cuban Missile Crisis). These can help us understand things such as how resilient society is to large catastrophes or how our imperfect information can lead nations to walk much closer to the brink of an annihilating war than they intended. We need to learn everything we can from these cases, even when they aren't precise analogues for the new risks we face, for they may be the best we have.

Some of this use of near misses is systematised in the field of risk analysis. They have techniques for estimating the probability of unprecedented catastrophes based on the combination of *precedented* faults that would need to occur to allow it. For example, fault tree analysis was developed for evaluating the reliability of the launch systems for nuclear missiles, and is used routinely to help avoid low-frequency risks, such as plane crashes and nuclear meltdowns.[27]

There is a special challenge that comes with estimating risks of human extinction. It is *impossible* to witness humanity having been extinguished in the past, regardless of the likelihood. And a version of this selection effect can distort the historical record of some catastrophes that are linked to extinction, even if they wouldn't necessarily cause it. For example, we may not be able to directly apply the observed track record of asteroid collisions or full-scale nuclear war. From what we know, it doesn't look like these selection effects have distorted the historical record much, but there are only a handful of papers on the topic and some of the methodological issues have yet to be resolved.[28]

A final challenge concerns all low-probability high-stakes risks. Suppose scientists estimate that an unprecedented technological risk has an extremely small chance of causing an existential catastrophe—say one in a trillion. Can we directly use this number in our analysis? Unfortunately not. The problem is that the chance the scientists have incorrectly estimated this probability is many times greater than one in a trillion. Recall their failure to estimate the size of the massive Castle Bravo nuclear explosion—if the chance of miscalculation were really so low there should be no such examples. So if a disaster does occur, it is much more likely to be because there was an estimation mistake and the real risk was higher, rather than because a one in a trillion event occurred.

This means that the one in a trillion number is not the decision-relevant probability, and policymakers need to adjust for this by using a higher number.[29] The manner in which they should do so is not well understood. This is part of a general point that our uncertainty about the underlying physical probability is not grounds for ignoring the risk, since the true risk could be higher as well as lower. If anything, when the initial estimate of the probability is tiny, a proper accounting of uncertainty often makes the situation worse, for the real probability could be substantially higher but couldn't be much lower.[30]

These unusual challenges that come with the territory of existential risk are not insurmountable. They call for advances in our theoretical understanding of how to estimate and evaluate risks that are by their very nature unprecedented. They call for improvements in our horizon scanning and forecasting of disruptive technologies. And they call for improved integration of these techniques and ideas into our policy-making.

INTERNATIONAL COORDINATION

Safeguarding humanity is a global public good. As we saw earlier, even a powerful country like the United States contains only a twentieth of the world's people and so would only reap something like a twentieth of the benefits that come from preventing catastrophe. Uncoordinated action by nation states therefore suffers from a collective action problem. Each nation is inadequately incentivised to take actions that reduce risk and to avoid actions that produce risk, preferring instead to free-ride on others. Because of this, we should expect risk-reducing activities to be under-supplied and risk-increasing activities to be over-supplied.

This creates a need for international coordination on existential risk. The incentives of a nation are only aligned with the incentives of humanity if we share the costs of these policies just as we share the benefits. While nations occasionally act for the greater interest of all humankind, this is the exception rather than the rule. Multilateral action can resolve this tragedy of the commons, replacing a reliance on countries' altruism with a reliance on their prudence: still not perfect, but a much better bet.

And there would be benefits to centralising some of this international work on safeguarding humanity. This would help us pool our expertise, share our perspectives and facilitate coordination. It could also help us with policies that require a unified response, where we are only as strong as the weakest link: for example, in setting moratoria on dangerous types of research or in governing the use of geoengineering.

So there is a need for international institutions focused on existential risk to coordinate our actions. But it is very unclear at this stage what forms they should take. This includes questions of whether the change should be incremental or radical, whether institutions should be advisory or regulatory, and whether they should have a narrow or broad set of responsibilities. Our options range from incremental improvements to minor agencies, through to major changes to key bodies such as the UN Security Council, all the way up to entirely new institutions for governing the most important world affairs.

No doubt many people would think a large shift in international governance is unnecessary or unrealistic. But consider the creation of the United Nations. This was part of a massive reordering of the international order in response to the tragedy of the Second World War. The destruction of humanity's entire potential would be so much worse than the Second World War that a reordering of international institutions of a similar scale may be entirely justified. And while there might not be much appetite now, there may be in the near future if a risk increases to the point where it looms very large in the public consciousness, or if there is a global catastrophe that acts as a warning shot. So we should be open to blue-sky thinking about ideal international institutions, while at the same time considering smaller changes to the existing set.[31]

The same is true when it comes to our policy options. As we wake up to the new situation we find ourselves in and come to terms with the vulnerability of humanity, we will face great challenges. But we may also find new political possibilities opening up. Responses that first seemed impossible may become possible, and in time even inevitable. As Ulrich Beck put it: 'One can make two diametrically opposed kinds of assertion: global risks inspire paralysing terror, or: global risks create new room for action.'[32]

One way of looking at our current predicament is that the existing global order splits humanity into a large number of sovereign states, each of which has considerable internal coherence, but only loose coordination with the others. This structure has some advantages, even from the perspective of existential risk, for it has allowed us to minimise the risk that a single bad government could lock humanity into a terrible stable outcome. But as it becomes easier for a single country—or even a small group within one country—to threaten the whole of humanity, the balance may start to shift. And 195 countries may mean 195 chances that poor governance precipitates the destruction of humanity.

Some important early thinkers on existential risk suggested that the growing possibility of existential catastrophe required moving towards a form of world government.[33] For example, in 1948 Einstein wrote:

> I advocate world government because I am convinced that there is no other possible way of eliminating the most terrible danger in which man has ever found himself. The objective of avoiding total destruction must have priority over any other objective.[34]

World government is a slippery idea, with the term meaning different things to different people. For example, it is sometimes used to refer to any situation where nations have been made unable to wage war upon each other. This situation is almost synonymous with perpetual world peace, and relatively unobjectionable (albeit stunningly difficult to achieve). But the term is also used to refer to a politically homogenised world with a single point of control (roughly, the world as one big country). This is much more contentious and could increase overall existential risk via global totalitarianism, or by permanently locking in bad values.

Instead, my guess is that existential security could be better achieved with the bare minimum of internationally binding constraints needed to prevent actors in one or two countries from jeopardising humanity's entire future. Perhaps this could be done through establishing a kind of constitution for humanity, and writing into it the paramount need to safeguard our future, along with the funding and enforcement mechanisms required. This may take us beyond any current international law or institutions, yet stop considerably short of world government.

What about smaller changes—improvements to international coordination that offer a large amount of security for their cost? A good historical example might be the Moscow–Washington Hotline (popularly known as the 'red telephone').[35] During the Cuban Missile Crisis messages between Kennedy and Khrushchev regularly took several hours to be received and decoded.[36] But major new developments were unfolding on the ground at a much quicker tempo, leaving diplomatic solutions (and explanations for apparently hostile behaviour) unable to keep up.[37] Afterwards, Kennedy and Khrushchev established the hotline to allow faster and more direct communication between the leaders, in order to avoid future crises coming so close to the brink. This was a simple and successful way to lower the risk of nuclear war (and war between the great powers more generally), with little financial or political cost. There may be other ideas like this just waiting to be discovered or implemented.

And there may be more obvious ways, such as simply strengthening existing institutions related to existential risks. For example, the Biological Weapons Convention could be brought into line with the Chemical Weapons Convention: taking its budget from $1.4 million up to $80 million, granting it the power to investigate suspected breaches, and increasing its staff from a mere four people to a level more appropriate for its role.[38] We could also strengthen the World Health Organisation's ability to respond to emerging pandemics through rapid disease surveillance, diagnosis and control. This involves increasing its funding and powers, as well as R&D on the requisite technologies. And we need to ensure that all DNA synthesis is screened for dangerous pathogens. There has been good progress towards this from synthesis companies, with 80 percent of orders currently being screened.[39] But 80 percent is not enough. If we cannot reach full coverage through voluntary efforts, some form of international regulation will be needed.

Some of the most important international coordination can happen between pairs of nations. One obvious first step would be to restart the Intermediate-Range Nuclear Forces Treaty (INF). This arms reduction treaty eliminated 2,692 nuclear missiles from the US and Russian nuclear arsenals, but was suspended in 2019 after a decade of suspected breaches.[40] They should also make sure to renew the New START treaty, due to expire in 2021, which has been responsible for major reductions in the number of nuclear weapons.

And while nuclear matters are often addressed through bilateral or multilateral agreements, there may also be unilateral moves that are in all nations' interests. For example, if the US took their ICBMs off hair-trigger alert, this would lessen the chance of accidentally triggered nuclear war without losing much deterrent effect since their nuclear submarines would still be able to launch a devastating retaliation. This may well reduce the overall risk of nuclear war.

Another promising avenue for incremental change is to explicitly prohibit and punish the deliberate or reckless imposition of unnecessary extinction risk.[41] International law is the natural place for this, as those who impose such risk may well be national governments or heads of state, who could be effectively immune to mere national law.

The idea that it may be a serious crime to impose risks to all living humans and to our entire future is a natural fit with the common-sense ideas behind the law of human rights and crimes against humanity. There would be substantial practical challenges in reconciling this idea with the actual bodies of law, and in defining the thresholds required for prosecution.[42] But these challenges are worth undertaking—our descendants would be shocked to learn that it used to be perfectly legal to threaten the continued existence of humanity.[43]

There are some hopeful signs that such protections could gain support at the international level. For example, in 1997, UNESCO passed a Declaration on the Responsibilities of the Present Generations Towards Future Generations. Its preamble showed a recognition that humanity's continued existence may be at stake and that acting on this falls within the mission of the UN:

> Conscious that, at this point in history, the very existence of humankind and its environment are threatened, Stressing that full respect for human rights and ideals of democracy constitute an essential basis for the protection of the needs and interests of future generations . . . Bearing in mind that the fate of future generations depends to a great extent on decisions and actions taken today . . . Convinced that there is a moral obligation to formulate behavioural guidelines for the present generations within a broad, future-oriented perspective . . .

The articles of the declaration were a list of ideals the international community should adopt, including Article 3: 'The present generations should strive to ensure the maintenance and perpetuation of humankind.' This declaration clearly did not change the world, but it does point towards how these ideas can be expressed within the framework of international human rights, and suggests these ideas have currency at the highest levels.[44]

During the last three decades, a handful of nations took the remarkable step of adjusting their democratic institutions to better represent the views of future generations.[45] They were responding to a critique of the standard forms of democracy: that they fail to represent the future people who may be adversely affected by our decisions.[46] One might think of this as a tyranny of the present over the future. Obviously one cannot simply resolve this by giving future people a vote on the issues that would affect them, as they haven't yet been born.[47] But we do sometimes have a clear idea of what they would think of the policy and so if we took this critique seriously, we could represent them by proxy: for example, by an ombudsperson, commission or parliamentary committee. These could be advisory, or be given some hard power.[48]

So far these experiments with formal representation of future generations have been mainly focused on environmental and demographic concerns. But the idea could naturally be applied to existential risk too. This may achieve some success at the national level and would be even more powerful if there was some way of bringing it to the world stage, combining both intergenerational and international coordination. This could be approached in an incremental way, or in a way that was truly transformative.

TECHNOLOGICAL PROGRESS

Humanity's stunning technological progress has been a major theme of this book. It is what allowed humans to form villages, cities and nations; to produce our greatest works of art; to live much longer lives, filled with a striking diversity of experiences. It is also essential to our survival: for without further technological progress we would eventually succumb to the background of natural risks such as asteroids. And I believe the best futures open to us—those that would truly fulfil our potential—will require technologies we haven't yet reached, such as cheap clean energy, advanced artificial intelligence or the ability to explore further into the cosmos.

Thus, even though the largest risks we face are technological in origin, relinquishing further technological progress is not a solution. What about proceeding more slowly? Is that a solution? One effect would be to delay the onset of technological risks. If we pushed back all new risky technologies for a century, that might mean all of us alive today are protected from death in an existential catastrophe. This would be a great boon from the perspective of the present, but would do very little from the perspectives of our future, our past, our virtues or our cosmic significance. This was noted by one of the earliest thinkers on existential risk, the philosopher J. J. C. Smart:

> Indeed what does it matter, from the perspective of possible millions of years of future evolution, that the final catastrophe should merely be postponed for (say) a couple of hundred years? Postponing is only of great value if it is used as a breathing space in which ways are found to avert the final disaster.[49]

I've argued that our current predicament stems from the rapid growth of humanity's power outstripping the slow and unsteady growth of our wisdom. If this is right, then slowing technological progress should help to give us some breathing space, allowing our wisdom more of a chance to catch up.[50] Where slowing down all aspects of our progress may merely delay catastrophe, slowing down the growth of our power relative to our wisdom should fundamentally help.

I think that a more patient and prudent humanity would indeed try to limit this divergence. Most importantly, it would try to increase its wisdom. But if there were limits to how quickly it could do so, it would also make sense to slow the rate of increase in its power—not necessarily putting its foot on the brake, but at least pressing more lightly on the accelerator.

We've seen how humanity is akin to an adolescent, with rapidly developing physical abilities, lagging wisdom and self-control, little thought for its longterm future and an unhealthy appetite for risk. When it comes to our own children, we design our societies to deliberately stage their access to risky technologies: for example, preventing them from driving a car until they reach an appropriate age and pass a qualifying test.

One could imagine applying a similar approach to humanity. Not relinquishing areas of technology, but accepting that in some cases we aren't ready for them until we meet a given standard. For example, no nuclear technologies until we've had a hundred years without a major war. Unfortunately, there is a major challenge. Unlike the case with our own children, there are no wise adults to decide these rules. Humanity would have to lay down the rules to govern *itself*. And those who lack wisdom usually lack the ability to see this; those who lack patience are unlikely to delay gratification until they acquire it.

So while I think a more mature world would indeed restrain its growth in destructive capability to a level where it was adequately managed, I don't see much value in advocating for this at the moment. Major efforts to slow things down would require international agreements between all the major players, for otherwise work would just continue in the least scrupulous countries. Since the world is so far from reaching such agreements, it would be ineffective (and likely counter-productive) for the few people who care about existential risk to use their energies to push for slowing down.

We should instead devote our energies to promoting the responsible deployment and governance of new technologies. We should make the case that the unprecedented power from technological progress requires unprecedented responsibility: both for the practitioners and for those overseeing them.

The great improvements in our quality of life from technology don't come for free. They come with a shadow cost in risk.[51] We focus on the visible benefits, but are accumulating a hidden debt that may one day come due.[52] If we aren't changing the pace of technology, the least we could do is to make sure we use some of the prosperity it grants us to service these debts. For example, to put even 1 percent of the benefits technology brings us back into ensuring humanity's potential isn't destroyed through further technological progress.

This technological governance can be pursued at many levels. Most obviously, by those whose duties are concerned with governance: politicians, the civil service and civil society. But we can build the bridge from both ends, with valuable contributions by the people who work on the relevant science and technology: in academia, in professional societies and in technology companies. These practitioners can spend much more time reflecting on the ethical implications of their own work and that of their peers.[53] They can develop their own guidelines and internal regulations. And they can spend time working with policymakers to ensure national and international regulations are scientifically and technologically sound.[54]

A good example of successful governance is the Montreal Protocol, which set a timetable to phase out the chemicals that were depleting the ozone layer. It involved rapid and extensive collaboration between scientists, industry leaders and policymakers, leading to what Kofi Annan called 'perhaps the single most successful international agreement to date'.[55]

Another example is the Asilomar Conference on Recombinant DNA, in which leading scientists in the field considered the new dangerous possibilities their work had opened up. In response they designed new safety requirements on further work and restricted some lines of development completely.[56]

An interesting, and neglected, area of technology governance is *differential technological development*.[57] While it may be too difficult to prevent the development of a risky technology, we may be able to reduce existential risk by speeding up the development of protective technologies relative to dangerous ones. This could be a role for research funders, who could enshrine it as a principle for use in designing funding calls and allocating grants, giving additional weight to protective technologies. And it could also be used by researchers when deciding which of several promising programmes of research to pursue.

STATE RISKS & TRANSITION RISKS

If humanity is under threat from substantial risk each century, we are in an unsustainable position. Shouldn't we attempt to rush through this risky period as quickly as we can? The answer depends upon the type of risk we face.

Some risks are associated with being in a vulnerable state of affairs. Let's call these *state risks*.[58] Many natural risks are state risks. Humanity remains vulnerable to asteroids, comets, supervolcanic eruptions, supernovae and gamma ray bursts. The longer we are in a state where the threat is present and we are vulnerable, the higher the cumulative chance we succumb. Our chance of surviving for a length of time is characterised by a decaying exponential, with a half-life set by the annual risk.[59] When it comes to state risks, the faster we end our vulnerability, the better. If we need technology to end this vulnerability, then we would like to reach that technology as quickly as possible.

But not all risks are like this.[60] Some are *transition risks*: risks that arise during a transition to a new technological or social regime. For example the risks as we develop and deploy transformative AGI are like this, as are the risks of climate change as we transition to being a high-energy civilisation. Rushing the transition may do nothing to lower these risks—indeed it could easily heighten them. But if the transition is necessary or highly desirable, we may have to go through it at some point, so mere delay is not a solution, and may also make things worse. The general prescription for these risks is neither haste nor slowness, but care and foresight.

We face an array of risks, including both state and transition risks.[61] But if my analysis is correct, there is substantially more transition risk than state risk (in large part because there is more anthropogenic risk). This suggests that rushing our overall technological progress is not warranted. Overall, the balance is set by our desire to reach a position of existential security while incurring as little cumulative risk as possible. My guess is that this is best achieved by targeted acceleration of the science and technology needed to overcome the biggest state risks, combined with substantial foresight, carefulness and coordination on the biggest transition risks.

While our current situation is unsustainable, that doesn't mean the remedy is to try to achieve a more sustainable annual risk level as quickly as possible. Our ultimate goal is longterm sustainability: to protect humanity's potential so that we have the greatest chance of fulfilling our potential over the aeons to come. The right notion of sustainability is thus not about getting into a sustainable state as quickly as possible, but about having a sustainable trajectory: one that optimally trades off risk in getting there with the protection of being there.[62] This may involve taking additional risks in the short term, though only if they sufficiently reduce the risks over the long term.

RESEARCH ON EXISTENTIAL RISK

The study of existential risk is in its infancy. We are only starting to understand the risks we face and the best ways to address them. And we are at an even earlier stage when it comes to the conceptual and moral foundations, or grand strategy for humanity. So we are not yet in a good position to take decisive actions to secure our longterm potential. This makes further research on existential risk especially valuable. It would help us to determine which of the available actions we should take, and to discover entirely new actions we hadn't yet considered.[63]

Some of this research should be on concrete topics. We need to better understand the existential risks—how likely they are, their mechanisms, and the best ways to reduce them. While there has been substantial research into nuclear war, climate change and biosecurity, very little of this has looked at the most extreme events in each area, those that pose a threat to humanity itself.[64] Similarly, we need much more technical research into how to align artificial general intelligence with the values of humanity.

We also need more research on how to address major risk factors, such as war between the great powers, and on major security factors too. For example, on the best kinds of institutions for international coordination or for representing future generations. Or on the best approaches to resilience, increasing our chance of recovery from non-fatal catastrophes. And we need to find new risk and security factors, giving us more ways to get a handle on existential risk.

And alongside these many strands of research on concrete topics, we also need research on more abstract matters. We need to better understand longtermism, humanity's potential and existential risk: to refine the ideas, developing the strongest versions of each; to understand what ethical foundations they depend upon, and what ethical commitments they imply; and to better understand the major strategic questions facing humanity.

These areas might sound grand and unapproachable, but it is possible to make progress on them. Consider the ideas we've encountered in this book. Some are very broad: the sweeping vision of humanity across the ages, the Precipice and the urgency of securing our future. But many can be distilled into small crisp insights. For example: that a catastrophe killing 100 percent of people could be much worse than one killing 99 percent because you lose the whole future; that the length of human survival so far puts a tight bound on the natural risk; that existential risk reduction will tend to be undersupplied since it is an intergenerational global public good; or the distinction between state risks and transition risks. I am sure there are more ideas like these just waiting to be discovered. And many of them don't require any special training to find or understand: just an analytical mind looking for patterns, tools and explanations.

Perhaps surprisingly, there is already funding available for many of these kinds of research on existential risk. A handful of forward-thinking philanthropists have taken existential risk seriously and recently started funding top-tier research on the key

211

risks and their solutions.[65] For example, the Open Philanthropy Project has funded some of the most recent nuclear winter modelling as well as work on technical AI safety, pandemic preparedness and climate change—with a focus on the worst-case scenarios.[66] At the time of writing they are eager to fund much more of this research, and are limited not by money, but by a need for great researchers to work on these problems.[67]

And there are already a handful of academic institutes dedicated to research on existential risk. For example, Cambridge University's Centre for the Study of Existential Risk (CSER) and Oxford's Future of Humanity Institute (FHI), where I work.[68] Such institutes allow like-minded researchers from across the academic disciplines to come together and work on the science, ethics and policy of safeguarding humanity.

WHAT NOT TO DO

This chapter is about what we should do to protect our future. But it can be just as useful to know what to avoid. Here are a few suggestions.

Don't regulate prematurely. At the right time, regulation may be a very useful tool for reducing existential risk. But right now, we know very little about how best to do so. Pushing for ill-considered regulation would be a major mistake.

Don't take irreversible actions unilaterally. Some countermeasures may make our predicament even worse (think radical geoengineering or publishing the smallpox genome). So we should be wary of the unilateralist's curse (p. 137), where the ability to take actions unilaterally creates a bias towards action by those with the most rosy estimates.

Don't spread dangerous information. Studying existential risk means exploring the vulnerabilities of our world. Sometimes

---

this turns up new dangers. Unless we manage such information carefully, we risk making ourselves even more vulnerable (see the box 'Information Hazards', p. 137).

Don't exaggerate the risks. There is a natural tendency to dismiss claims of existential risk as hyperbole. Exaggerating the risks plays into that, making it much harder for people to see that there is sober, careful analysis amidst the noise.

Don't be fanatical. Safeguarding our future is extremely important, but it is not the only priority for humanity. We must be good citizens within the world of doing good. Boring others with endless talk about this cause is counterproductive. Cajoling them about why it is more important than a cause they hold dear is even worse.

Don't be tribal. Safeguarding our future is not left or right, not eastern or western, not owned by the rich or the poor. It is not partisan. Framing it as a political issue on one side of a contentious divide would be a disaster. Everyone has a stake in our future and we must work together to protect it.[69]

Don't act without integrity. When something immensely important is at stake and others are dragging their feet, people feel licensed to do whatever it takes to succeed. We must never give in to such temptation. A single person acting without integrity could stain the whole cause and damage everything we hope to achieve.

Don't despair. Despairing would sap our energy, cloud our judgement and turn away those looking to help. Despair is a self-fulfilling prophecy. While the risks are real and substantial, we know of no risks that are beyond our power to solve. If we hold our heads high, we can succeed.

Don't ignore the positive. While the risks are the central challenges facing humanity, we can't let ourselves be defined by them. What drives us is our hope for the future. Keeping this at the centre of our thinking will provide us—and others—with the inspiration we need to secure our future.[70]

WHAT YOU CAN DO

Much of this chapter has been devoted to the big-picture questions of how humanity can navigate the Precipice and achieve its potential. But amidst these grand questions and themes, there is room for everyone to play a part in protecting our future.

One of the best avenues for doing good in the world is through our careers. Each of our careers is about 80,000 hours devoted to solving some kind of problem, big or small. This is such a large part of our lives that if we can devote it to one of the most important problems, we can have tremendous impact.

If you work in computer science or programming, you might be able to shift your career towards helping address the existential risk arising from AI: perhaps through much-needed technical research on AI alignment, or by working as an engineer for an AI project that takes the risks seriously.[71] If you are in medicine or biology, you may be able to help with risks from engineered pandemics. If you are in climate science, you could work on improving our understanding of the likelihood and effect of extreme climate scenarios. If you are in political science or international relations, you could work towards international cooperation on existential risk, ensuring future generations get a voice in democracy, or preventing war between great powers. If you work in government, you could help protect the future through work in security or technology policy.

The opportunities are not limited to direct work on existential risk. Instead your work could multiply the impact of those doing the direct work. Some of the most urgent work today is upstream, at the level of strategy, coordination and grant-making. As humanity starts to take seriously the challenge of protecting our future, there will be important work in allocating resources between projects and organisations building and sustaining the community of researchers and developing strategy. And some of the most urgent work lies in improving the execution and outreach of organisations dedicated to fighting existential risk. Many are looking for skilled people who really grasp the unusual mission. If you have any of these skills—for example, if you have experience working on strategy, management, policy, media, operations or executive assistance—you could join one of the organisations currently working on existential risk.[72]

If you are a student, you are in a wonderfully flexible position—able to do so much to steer your career to where your tens of thousands of hours will have the most impact. Even if you have chosen a field, or entered graduate study, it is surprisingly easy to change direction. The further down a career path you have travelled, the harder this becomes. But even then it can be worthwhile. Losing a few years to retraining may let you direct several times as many years of work to where it can do many times as much good. This is something I know from personal experience. I started out studying computer science, before moving into ethics. Then within ethics, the focus of my work has shifted only recently from the issues of global poverty to the very different issues around existential risk.

What if your career is unsuited, and yet you are unable to change? What would be ideal is if there were some way to turn the work you are best at into the type of work most desperately needed to safeguard our potential. Fortunately, there is a such way: through your giving. When you donate money to a cause, you effectively transform your own labour into additional work for that cause. If you are more suited to your existing job and the cause is constrained by lack of funding, then donating could even help more than direct work.

I think donating is a powerful way in which almost anyone can help, and it is an important part of how I give back to the world.[73] People often forget that some of humanity's greatest successes have been achieved through charity.

The contraceptive pill, one of the most revolutionary inventions of the twentieth century, was made possible by a single philanthropist. In the 1950s, at a time when governments and drug companies had little interest in pursuing the idea, the philanthropist Katharine McCormick funded the research that led to its invention, largely single-handedly.[74]

Around the same time, we saw the breakthroughs in agricultural science now known as the Green Revolution, which saw hundreds of millions of people lifted from hunger through the creation of high-yield varieties of staple crops. Norman Borlaug, the scientist who led these efforts, would win the Nobel Peace Prize in 1970 for his work. Borlaug's work, and the roll-out of these technologies in the developing world, was funded by private philanthropists.[75]

Finally, there are ways that every single one of us can play a role. We need a public conversation about the longterm future of humanity: the breathtaking scale of what we can achieve, and the risks that threaten all of this, all of us.

We need to discuss these things in academia, in government, in civil society; to explore the possibilities in serious works of fiction and the media; to talk about it between friends and within families. This conversation needs to rise above the temptation to be polarising and partisan, or to focus on the allocation of blame. Instead we need a mature, responsible and constructive conversation: one focused on understanding problems and finding solutions. We need to inspire our children, and ourselves, to do the hard work that will be required to safeguard the future and pass the Precipice.

You can discuss the importance of the future with the people in your life who matter to you. You can engage with the growing community of people who are thinking along similar lines: where you live, where you work or study, or online. And you can strive to be an informed, responsible and vigilant citizen, staying abreast of the issues and urging your political representatives to take action when important opportunities arise.

(See the Resources section on p. 242 for concrete starting points.)

8

Our Potential

It is possible to believe that all the past is but the beginning of a beginning, and that all that is and has been is but the twilight of the dawn. It is possible to believe that all that the human mind has ever accomplished is but the dream before the awakening.

—H. G. Wells[1]

What could we hope to achieve? To experience? To become? If humanity rises to the challenges before us, navigates the risks of the coming centuries, and passes through to a time of safety—what then?

In the preceding chapters we have faced the Precipice, surveyed its challenges and planned how we might best proceed to safety. But what lies beyond? Let us raise our sights to the sweeping vista that beckons us on. We cannot make out the details from here, but we can see the broad shape of the hills and vales, the *potential* that the landscape offers—the potential of human civilisation in its full maturity. It is because this potential is so vast and glorious, that the stakes of existential risk are so high. In optimism lies urgency.

This chapter is about potential, not prophecy. Not what we *will* achieve, but what is open for us to achieve if we play our cards right; if we are patient, prudent, compassionate, ambitious and wise. It is about the canvas on which we shall work: the lengths of time open to humanity, the scale of our cosmos and the quality of life we might eventually attain. It is about the shape of the land we are striving to reach, where the greater part of human history will be written.

DURATION

Human history so far has seen 200,000 years of *Homo sapiens* and 10,000 years of civilisation.[2] These spans of time surpass anything in our daily lives. We have had civilisation for a hundred lifetimes on end and humanity for thousands. But the universe we inhabit is thousands of times older than humanity itself. There were billions of years before us; there will be billions to come. In our universe, time is not a scarce commodity.

With such abundance at our disposal, our lifespan is limited primarily by the existential catastrophes we are striving to prevent. If we get our act together, if we make safeguarding humanity a cornerstone of our civilisation, then there is no reason we could not live to see the grand themes of the universe unfold across the aeons. These timespans are humbling when we consider our place in the universe so far. But when we consider the potential they hold for what we might become, they inspire.

As we have seen, the fossil record tells us a great deal about how long a typical species can expect to survive. On average, mammalian species last about one million years, and species in general typically last one to ten million years.[3] If we could address the threats we pose to ourselves—the anthropogenic existential risks—then we should be able to look forward to at least a lifespan of this order. What would this *mean*? What can happen over such a span, ten thousand times longer than our century?

Such a timescale is enough to repair the damage that we, in our immaturity, have inflicted upon the Earth. In thousands of years, almost all of our present refuse will have decayed away. If we can cease adding new pollution, the oceans and forests will be unblemished once more. Within 100,000 years, the Earth's natural systems will have scrubbed our atmosphere clean of over 90% of the carbon we have released, leaving the climate mostly restored and rebalanced.[4] So long as we can learn to care rightly for our home, these blots on our record could be wiped clean, all within the lifespan of a typical species, and we could look forward to living out most of our days in a world free from the scars of immature times.

About ten million years hence, even the damage we have inflicted upon biodiversity is expected to have healed. This is how long it took for species diversity to fully recover from previous mass extinctions and our best guess for how long it will take to recover from our current actions.[5]

I hope and believe that we can address pollution and biodiversity loss much more quickly than this—that sooner or later we will work to actively remove the pollution and to conserve the threatened species. But there is comfort in knowing that at the very least, Earth, on its own, can heal the damage we have done.

Over this span of time roughly half of Earth's species will naturally become extinct, with new species arising in their place. If we were to last so long, we would be living through evolutionary time, and would see this as the natural state of flux that the world is in. The apparent stasis of species in the natural world is only a reflection of the comparatively brief period we have lived. Nevertheless, if we thought it right, we could intervene to preserve the last members of a species before it naturally vanished, allowing them to live on in reduced numbers in wilderness reserves or other habitats. It would be a humble retirement, but better, I think, than oblivion.

One to ten million years is typical for a species, but by no means a limit. And humanity is atypical in many ways. This might mean a much shorter life, if we author our own destruction. But if we avoid so doing, we may be able to survive much longer. Our presence across the globe protects us from any regional catastrophe. Our ingenuity has let us draw sustenance from hundreds of different plants and animals, offering protection from the snapping of a food chain. And our ability to contemplate our own destruction—planning for the main contingencies, reacting to threats as they unfold—helps protect us against foreseeable or slow-acting threats.

Many species are not wholly extinguished, but are succeeded by their siblings or their children on the evolutionary family tree. Our story may be similar. When considering our legacy—what we shall bequeath to the future—the end of our *species* may not be the end of *us*, our projects or our ultimate aspirations. Rather, we may simply be passing the baton.

For many reasons, then, humanity (or our rightful heirs) may greatly outlast the typical species. How long might this grant us?

We know of species around us that have survived almost unchanged for hundreds of millions of years. In 1839, a Swiss biologist first described and named the *coelacanth*, an ancient order of fish that arose 400 million years ago, before vanishing from the fossil record with the dinosaurs, 65 million years ago. The coelacanth was assumed to be long extinct, but ninety-nine years later, a fisherman off the coast of South Africa caught one in his net. Coelacanths were still living in Earth's oceans, almost unchanged over all this time. They are the oldest known vertebrate species alive today, lasting more than two-thirds of the time since the vertebrates first arose.[6]

And there are older species still. The horseshoe crab has been scuttling through our oceans even longer, with an unbroken lineage of 450 million years. The nautilus has been here for 500 million years; sponges, for about 580 million. And these are merely lower bounds on their lifespans; who knows how much longer these hardy species will last? The oldest known species on Earth are cyanobacteria (blue-green algae), which have been with us for at least two billion years—much longer than there has been complex life, and more than half the time there has been life on Earth at all.[7]

What might humanity witness, if we (or our successors) were to last as long as the humble horseshoe crab?

Such a lifespan would bring us to geological timescales. We would live to see continental drift reshape the surface of the Earth as we know it. The first major change would be seen in about 10 million years, when Africa will be torn in two along the very cradle of humanity—the Rift Valley. In 50 million years, the larger of these African plates will collide with Europe, sealing the Mediterranean basin and raising a vast new range of mountains. Within about 250 million years, all our continents will merge once again, to form a supercontinent, like Pangaea 200 million years ago. Then, within 500 million years, they will disperse into some new and unrecognisable configuration.[8] If this feels unimaginable, consider that the horseshoe crab has already witnessed such change.

These spans of time would also see changes on astronomical scales. The constellations will become unrecognisable, as the nearby stars drift past each other.[9] Within 200 million years, the steady gravitational pull of the Moon will slow the Earth's rotation, and stretch our day to 25 hours. While in one year the Earth orbits our Sun, in 240 million years our Sun will complete a grand orbit around the centre of our galaxy—a period known as a galactic year.

But the most important astronomical change is the evolution of the Sun itself. Our star is middle-aged. It formed about 4.6 billion years ago, and has been growing steadily brighter for most of its lifespan. Eventually, this mounting glare will start causing significant problems on Earth. The astronomical evolution is well-understood, but because its major effects upon our biosphere are unprecedented, much scientific uncertainty remains.

One often hears estimates that the Earth will remain habitable for 1 or 2 billion years. These are predictions of when the oceans will evaporate due to a runaway or moist greenhouse effect, triggered by the increasing brightness of the Sun. But the Earth may become uninhabitable for complex life before that point: either at some earlier level of warming or through another mechanism. For example, scientists expect the brightening of the Sun to also slow the Earth's plate tectonics, dampening volcanic activity. Life as we know it requires such activity; volcanoes lift essential carbon dioxide into our atmosphere. We currently have too much carbon dioxide, but we need small amounts for plants to photosynthesise. Without the carbon dioxide from volcanoes, scientists estimate that in about 800 million years photosynthesis will become impossible in 97 percent of plants, causing an extreme mass extinction. Then 500 million years later, there would be so little carbon dioxide that the remaining plants would also die—and with them, any remaining multicellular life.[10]

This might not happen. Or it might not happen at that time. This is an area of great scientific uncertainty, in part because so few people have examined these questions. More importantly, though, such a mass extinction might be *avoidable*—it might be humanity's own actions which prevent it. In fact this may be one of the great achievements that humanity might aspire to. For of all the myriad species that inhabit our Earth, only we could save the biosphere from the effects of the brightening Sun. Even if humanity is very small in your picture of the world, if most of the *intrinsic* value of the world lies in the rest of our eco-system, humanity's *instrumental* value may yet be profound. For if we can last long enough, we will have a chance to literally save our world.

By adding sufficient new carbon to the atmosphere to hold its concentration steady, we could prevent the end of photosynthesis. Or if we could block a tenth of the incoming light (perhaps harvesting it as solar energy), we could avoid not only this, but all the other effects of the Sun's steady brightening as well, such as the superheated climate and the evaporation of the oceans.[11] Perhaps, with ingenuity and commitment, we could extend the time allotted to complex life on Earth by billions of years, and, in doing so, more than redeem ourselves for the foolishness of our civilisation's youth. I do not know if we *will* achieve this, but it is a worthy goal, and a key part of our potential.

In 7.6 billion years, the Sun will have grown so vast that it will balloon out beyond Earth's own orbit, either swallowing our planet or flinging it out much further. And either way, in 8 billion years our Sun itself will die. Its extended outer layers will drift onwards past the planets, forming a ghostly planetary nebula, and its innards will collapse into a ball the size of the Earth. This tiny stellar remnant will contain roughly half the original mass of the Sun, but will never again produce new energy. It will be only a slowly cooling ember.[12]

Whether or not the Earth itself is destroyed in this process, without a Sun burning at the heart of our Solar System, any prospects for humanity would be much brighter elsewhere. And the technological challenges of leaving our home system would presumably be smaller than those of remaining.

By travelling to other stars, we might save not only ourselves, but much of our biosphere. We could bring with us a cache of seeds and cells with which to preserve Earth's species, and make green the barren places of the galaxy. If so, the good we could do for Earth-based life would be even more profound. Without our intervention, our biosphere is approaching middle age. Simple life looks to have about as much time remaining as it has had so far; complex life a little more. After that, to the best of our knowledge, life in the universe may vanish entirely. But if humanity survives, then even at that distant time, life may still be in its infancy. When I contemplate the expected timespan that Earth-based life may survive and flourish, the greatest contribution comes from the possibility of humanity turning from its destroyer, to its saviour.

As we shall see, the main obstacle to leaving our Solar System is surviving for long enough to do so. We need time to develop the technology, to harvest the energy, to make the journey and to build a new home at our destination. But a civilisation spanning millions of centuries would have time enough, and we should not be daunted by the task.

Our galaxy will remain habitable for an almost unfathomable time. Some of our nearby stars will burn vastly longer than the Sun, and each year ten new stars are born. Some individual stars last *trillions* of years—thousands of times longer than our Sun. And there will be millions of generations of such stars to follow.[13] This is deep time. If we survive on such a cosmological scale, the present era will seem astonishingly close to the very start of the universe. But we know of nothing that makes such a lifespan impossible, or even unrealistic. We need only get our house in order.

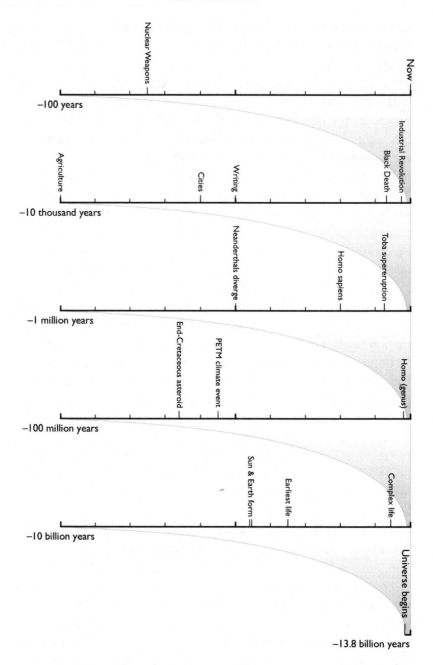

FIGURE 8.1 A timeline showing the scale of the past and future. The top row shows the prior century (on the left-hand page) and the coming century (on the right), with our own moment in the middle. Then each successive row zooms out, showing 100 times the duration, until we can see the whole history of the universe.

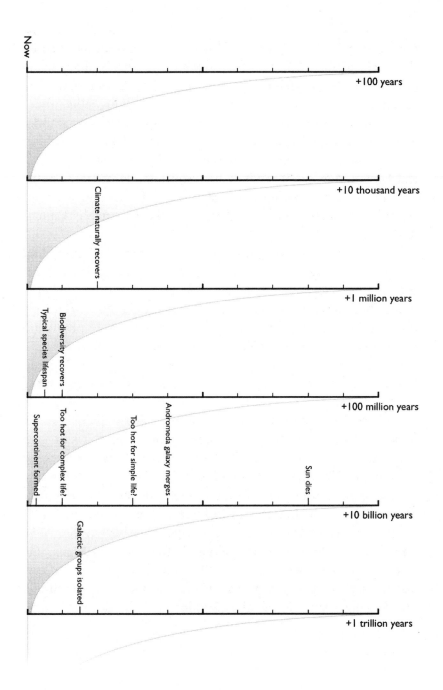

SCALE

For as long as there have been humans, we have been awed by the starry night.[14] And this dark sky strewn with brilliant white points of light has yielded many secrets. Not the secrets we first sought, of myth and mysticism, but a deeper knowledge about the nature of reality. We saw some of these points of light wander across the sky, and from the patterns they traced we realised that the Earth and the heavens were governed by the very same physical laws. Other points moved by minute amounts, measurable only by the most delicate instruments. From this almost imperceptible movement, we divined the scarcely imaginable distance to the stars.

The points were not quite white, but a range of hues. And when this faint starlight was broken in a prism, missing colours revealed the substance of the stars. Some points were not *quite* points, but rather discs, clouds, swirls—celestial bodies of entirely different kinds and origins. And we found many more points too faint to be seen with the naked eye. From these subtlest of signs, we have discerned and tested fundamental laws of nature; we have heard the echoes of the Big Bang; we have watched space itself expand.

But the most important thing we have learnt from the sky may be that our universe is much vaster than we had ever contemplated. The planets are other Earths. Stars are other suns, many with their own planets. The milky stripe across the sky contains more suns than the eye can distinguish—a galaxy of more than 100 billion suns, blurred in our vision to a uniform white. The faint swirls are entire galaxies beyond our own, and there are hundreds of billions of these spread across the skies.[15] Each time we thought we had charted the limits of creation, it transcended our maps. Our Earth is just one island in a bewilderingly large archipelago, which is itself just one amidst billions.

This discovery of the true scale of our cosmos dramatically raises the prospects for what humanity might achieve. We once thought ourselves limited to the Earth; we know now that vastly greater opportunities and resources are available to us. Not immediately, of course—exploring the whole of our own galaxy would take at least 100,000 years; to reach the furthest limits of the universe would require billions. But it raises deep questions about what might be achievable over vast timescales.

In just five centuries we have gone from the dimmest understanding of our Solar System—unable to grasp any coherent picture of our Sun, Moon, Earth, and the wandering points of light called 'planets'—to breathtaking high-resolution images of all our planets and their moons. We have sent gleaming craft sailing past the moons of Jupiter, through the rings of Saturn and to the surfaces of all terrestrial planets. To our Moon, we came in person.

The planets and their moons are the glories of our Solar System: majestic; enigmatic. We might look to settle them, but even if we overcame the daunting challenges, the combined surface area of all other solid planets and moons is just over twice that of Earth.[16] There would be adventure and excitement, but no radical increase in our potential, no fundamental change to our story. We might try to use their physical resources, but there are already ample resources for the foreseeable future in the million known asteroids, and resources for the distant future in the billions of planets now known to exist elsewhere in our galaxy. The best reason to settle the planets is to achieve some additional protection from existential risks, though this provides less safety than one may think, since some of the risks will be correlated between planets. So perhaps we will leave our own planets pristine: as monuments, jewels. To be explored and treasured. To fill us with wonder and inspire us to journey further.

Our Solar System's greatest contribution to our potential lies with our Sun, and the vast bounty of clean energy it offers. The sunlight hitting Earth's surface each day carries 5,000 times more energy than modern civilisation requires. It gives in two hours what we use in a year. This abundance of solar energy created most of our other energy sources (coal, oil, natural gas, wind, hydro, biomass) and far outstrips them.[17]

But almost all of the Sun's energy is wasted. It shines not onto leaves or solar cells, but out into the blackness of space. Earth intercepts less than one part in a billion; all the bodies of our Solar System together receive less than a hundred-millionth of the Sun's light.

In the future, we could harness this energy by constructing solar collectors in orbit around the Sun. Such a project would be completely scalable.[18] We could start with something small and affordable; then, with a fraction of the energy it provides, we could scale up to whatever level we desire. The asteroid belt alone has more than enough raw materials for such a project.[19] Eventually, we could increase our access to clean energy by up to a billion-fold, using only light that would otherwise be wasted. And such a structure could also be used to address the Sun's increasing brightness, providing shade that could extend the reign of complex life on Earth tenfold.

Such a path would grant us access to an abundance of clean energy. We would have no need for dirty energy sources,[20] and many of the challenges that derive from energy scarcity could be solved, including food production, water purification and conflict over oil. We could quickly cleanse our atmosphere of our past carbon emissions, using carbon dioxide scrubbers currently limited by a lack of cheap, clean power. And, beyond all of this, harnessing the Sun's energy would cast open the door to the stars.

Could we really reach across the vast distances to the stars? In some ways the answer has already been written. After finishing their missions among the planets, *Pioneer 10*, *Voyager 1* and *Voyager 2* have all overcome the gravitational pull of our Solar System and escaped. Eventually, they will travel far enough to reach our nearest stars; proof that we can span such distances even with 1970s technology. But that on its own is not enough to greatly expand our potential, for *Voyager 1* will take 70,000 years to reach the distance of our closest star and will cease functioning long before then.[21]

Modern efforts to do better are underway. The Breakthrough Starshot Initiative, announced in 2016, aims to send a fleet of small, unmanned spacecraft to Alpha Centauri, four light years away, at about one-fifth the speed of light. If the project proceeds as planned, it may launch as soon as 2036.[22]

To truly expand our potential, we would need a spacecraft to reach another star, then stop and use the resources there to build a settlement that could eventually grow into a new bastion of civilisation.[23] Such a trip requires four challenging phases: acceleration, surviving the voyage, deceleration, and building a base of operations. These challenges are intertwined. Robotic missions make the trip easier, but building a base harder (at least for now). Faster missions make surviving the voyage easier, but greatly increase the energy and technology requirements for acceleration and deceleration.

We do not currently have the technology to meet these requirements, and we will not develop it within the next few decades. But we haven't discovered any fundamental barriers, and technology advances fast. In my view, the biggest challenge will be surviving on Earth for the century or two until it becomes technologically feasible.

While there is some notable scepticism of interstellar travel, a closer inspection shows that it is directed either at our near-term capabilities or at our achieving the kind of effortless space travel depicted in films like *Star Wars* or *Star Trek*, where individuals routinely travel between the stars in relative comfort. I share this scepticism. But the expansion of humanity's potential doesn't require anything like that. It requires only that if we could survive long enough, and strive hard enough, then we could *eventually* travel to a nearby star and establish enough of a foothold to create a new flourishing society from which we could venture further.

Our abiding image of space travel should not be the comfort and ease of an ocean liner, but the ingenuity, daring, and perseverance of the Polynesian sailors who, a thousand years ago, sailed vast stretches of the Pacific to find all its scattered islands and complete the final stage of the settlement of the Earth. Seen in this way—as a grand challenge for humanity—I can't help but think that if we last long enough, we would be able to settle our nearest stars.

What would we find on our arrival? Do any other worlds in our universe harbour simple life? Complex life? Alien civilisations? Or are they all lifeless desert landscapes awaiting the spark of life that only we may bring? Is life easy and ubiquitous, or are there some steps on the way from a lifeless planet to self-aware observers that are vanishingly unlikely? Are we alone in our stellar neighbourhood? Our galaxy? Our observable universe?[24]

Our ever-more-sensitive instruments have so far shown no signs of life beyond our Earth—no chemical signatures, no radio signals, no traces of grand engineering projects, no visits. But the quest to find alien life is young and our telescopes could still easily miss a civilisation like ours within just 100 light years of the Earth. The truth is that we don't yet know, and this may not be the century we find out. In this sketch of our potential, I focus on the case where we are indeed alone. But if we were to find other life— especially intelligent life—it could profoundly change our future direction.[25] And it may be the distance to our nearest intelligent neighbours, rather than the limits of our spacecraft, that sets the physical scale for what we are able to author in the heavens.

Our local stellar neighbourhood is a random scattering of stars, extending more or less uniformly in all directions. This scattered pattern continues as we zoom out, until 15 million stars come into view, stretching 1,000 light years from the Earth. Only at this immense scale do signs of the large-scale structure of our galaxy emerge. Above and below the galactic disc the stars grow sparse, and as we zoom out even further, we begin to see the curve of the Orion spiral arm, where our Sun lies; then the other spiral arms, and the glowing bulge around the galactic centre. Finally we see the familiar whirlpool shape of our spiral galaxy,

150,000 light years across, encompassing more than 100 billion stars, most with their own planets.

But if we could reach just one nearby star and establish a settlement, this entire galaxy would open up to us. For then the process could repeat, using the resources of our new settlement to build more spacecraft, and its sun to power them. If we could travel just six light years at a time, then almost all the stars of our galaxy would be reachable.[26] Each star system, including our own, would need to settle just the few nearest stars, and the entire galaxy would eventually fill with life.[27]

Because this critical distance of six light years is just slightly further than our nearest star, it is unlikely that we would be able to reach some stars without eventually being able to reach most of the galaxy. And because this wave of new settlements could radiate out in all directions, the enlivening of our galaxy may occur relatively quickly, by the standards of the history of life so far. Even if our spacecraft travelled at just 1 percent of the speed of light, and took 1,000 years to establish a new settlement, the entire galaxy could be settled within 100 million years—long before the Earth becomes uninhabitable. And once begun, the process would be robust in the face of local accidents, failures or natural setbacks.

Our galaxy is surrounded by a cloud of about fifty nearby galaxies, known as our Local Group. Foremost among them is the Andromeda Galaxy, a beautiful spiral galaxy, and the only galaxy in our group larger than our own. Gravity is pulling the two towards each other, and in four billion years (before our Sun has died) they will collide and unite. With so much distance between the stars of each galaxy, this collision will do surprisingly little to upset the stars and their planets. Its main effect will be to disrupt the delicate spiral structures of the partners, probably merging into a more uniform elliptical galaxy about three times as large. Eventually (in hundreds of billions of years) all the other galaxies in our group will have merged in too, forming a single giant galaxy.[28]

Zooming further out, we see many more groups of galaxies, some with as many as a thousand members.[29] Eventually these groups resolve into a larger structure: the cosmic web—long, thick threads of galaxies, called filaments. These filaments criss-cross space in a kind of three-dimensional network, as if someone took a random set of points in space and connected each to its nearest handful of neighbours. Where the filaments intersect, space is bright and rich with galaxies.[30] Between such filaments are dark and empty expanses, known as cosmic voids. As far as we can tell, this cosmic web continues indefinitely. At the very least, it continues as far as we can *see* or *go*.

It is these final limits on our knowledge and action that appear to set the ultimate scale in our universe. We have known for almost a century that our universe is expanding, pulling the groups of galaxies apart. And twenty years ago we discovered that this expansion is accelerating. Cosmologists believe this puts a hard limit on what we will ever be able to observe or affect.[31]

We can currently see a sphere around us extending out 46 billion light years in all directions, known as the *observable universe*. Light from galaxies beyond this sphere hasn't yet had time to reach us.[32] Next year we will see a little further. The observable universe will increase in radius by a single light year, and about 25 more galaxies will come into view. But on our leading cosmological theory, the rate at which new galaxies become visible will decline, and those currently more than 63 billion light years away will never become visible from the Earth. We could call the region within this distance the *eventually observable universe*.[33]

But much more importantly, accelerating expansion also puts a limit on what we can ever affect. If, today, you shine a ray of light out into space, it could reach any galaxy that is currently less than 16 billion light years away. But galaxies further than this are being pulled away so quickly that neither light, nor anything else we might send, could ever affect them.[34]

And next year this *affectable universe* will shrink by a single light year. Three more galaxies will slip forever beyond our influence.[35] Eventually, the gulfs between each group of galaxies will grow so vast that nothing will ever again be able to cross—each group of galaxies will be alone in the void, forever isolated from the others. This cleaves time into two fundamental eras: an *era of connection* with billions of reachable galaxies, and an *era of isolation* with a million times fewer. Surprisingly, this fundamental change in the causal structure of our universe is expected to happen long before the stars stop burning, in about 150 billion years. Our closest star, Proxima Centauri, will be less than a tenth of the way through its life.

So 16 billion light years looks to be the upper bound on how far humanity can reach, and 150 billion years the bound on how long we have to do so. I do not know if such intergalactic travel will ever be feasible. We could again use the strategy of reaching out further and further, a galaxy at a time, but the distances required are a million-fold greater than for interstellar travel, bringing unique challenges.[36] Still, we know of no fundamental physical barriers that would prevent a civilisation that had already mastered its own galaxy from taking this next step. (See Appendix G for more about the different scales a civilisation might reach.)

At the ultimate physical scale, there are 20 billion galaxies that our descendants might be able to reach. Seven-eighths of these are more than halfway to the edge of the affectable universe—so distant that once we reached them no signal could ever be sent back. Spreading out into these distant galaxies would thus be a final diaspora, with each galactic group forming its own sovereign realm, soon causally isolated from the others. Such isolation need not imply loneliness—each group would contain hundreds of billions of stars—but it might mean freedom. They could be established as pieces of a common project, all set in motion with the same constitution; or as independent realms, each choosing its own path.

With entire galaxies receding beyond our reach each year, one might think this pushes humanity towards a grand strategy of *haste*—a desperate rush to reach the technologies of intergalactic travel as soon as possible. But the *relative* loss each year is actually rather slow—about one part in five billion—and it is this relative reduction in our potential that matters.[37]

233

The pressures towards prudence and wisdom are larger, in relative terms, and counsel care over haste. If rushing to acquire these technologies a century earlier diminished our chance of survival by even one part in 50 million, it would be counter-productive. And for many years hence, the value of another century's reflection on what to *do* with the future—an irrevocable choice, and quite possibly the most important one humanity will ever make—will outweigh the value of extending our reach by just one part in fifty million. So our best grand strategy is one of careful reflection and prudence, with our degree of caution tuned to the slow clock of cosmological expansion.[38]

Such considerations seem unreal. In most of our day-to-day thought, and even our deeper reflection about the future and humanity's potential, we look around and see the Earth. Our eyes rarely turn to the heavens and the dusting of night stars. If pressed, we accept that planets, stars and galaxies are real places, but we rarely *feel* it, or consider that they could be crucial to our future potential.

One person who seriously considered the stars was Frank Ramsey, the brilliant economist and philosopher, whose career was cut short when he died at just 26 years of age, in 1930. His attitude was one of heroic defiance:

> I don't feel the least humble before the vastness of the heavens. The stars may be large, but they cannot think or love; and these are qualities which impress me far more than size does. I take no credit for weighing nearly seventeen stone. My picture of the world is drawn in perspective, and not to scale. The foreground is occupied by human beings and the stars are all small as threepenny bits.[39]

There is truth to this. What makes each of us special, so worthy of protection and celebration is something subtle about us, in the way that the matter of which we are comprised has been so delicately arranged as to allow us to think and love and create and dream.

Right now, the rest of the universe appears to lack such qualities. Ramsey may be right that in terms of *value*, the stars are as small as threepenny bits. But if we can venture out and animate the countless worlds above with life and love and thought, then even on Ramsey's view, we could bring our cosmos to its full scale; make it worthy of our awe. And since it appears to be only us who can bring the universe to such full scale, we may have an immense instrumental value, which would leave us at the centre of this picture of the cosmos. In this way, our potential, and the potential in the sheer scale of our universe, are interwoven.

QUALITY

We have seen that the future is a canvas vast in time and space. Its ultimate beauty will depend on what we paint. Trillions of years and billions of galaxies are worth little unless we make of them something valuable. But here, too, we have grounds for profound optimism. For the potential quality of our future is also grand beyond imagining.

As we've seen, human life is on the whole much better today than ever before. Compared with our ancestors, we have less to fear from disease, from hunger and from each other. We have conquered polio and smallpox. We have created vaccines, antibiotics and anaesthetics. Humans today are less likely than at any point in the history of civilisation to live in slavery or poverty; to be tortured, maimed or murdered; to starve. We have greater freedom to choose our loves, our beliefs, our leaders and the courses of our lives. Many of our children have access to opportunities that would astound our ancestors—opportunities to learn, play and experiment; to travel; to engage with the greatest novels, poems and philosophies; to experience a lifetime's variety of harmonies, sights and flavours; and to contemplate truths about the cosmos unknown to our most learned ancestors.

Yet human life, for all its joys, could be dramatically better than it is today. We have made tremendous progress on violence and disease, but there is still so much room for improvement—still lives being broken or cut short. The development of anaesthetics and pain relief have dramatically cut the prevalence of intense physical pain, yet we still live at a time with avoidable agony. We have made great progress in freeing people from absolute poverty, but a tenth of us still live under its shadow. And when it comes to relative poverty, to severe depression, to racism and sexism, we are far from a solution.

Many of the harshest injustices visited upon our fellow humans are behind us. While from year to year it can be hard to tell if things are getting better or worse, on the scale of centuries we have seen a clear decline in persecution and intolerance, with a marked rise in personal freedoms and political equality. Yet even in the most progressive countries we have a long way to go, and there are still parts of the world that have barely begun the journey.

And there are further injustices inflicted upon the billions of animals in the modern meat industry and in the natural world. Our civil awakening to the plight of animals and our environment came only very recently, in the wake of the severe harms inflicted upon them by industry in the twentieth century. But we increasingly see such harms for what they are, and have begun the fight to end these new forms of injustice.

These blights upon our world must end. And we *can* end them—if we survive. In the face of persecution and uncertainty, the noblest among our ancestors have poured their efforts into building a better and more just world. If we do the same—and give our descendants a chance to do the same—then as our knowledge, invention, coordination and abundance grow, we can more and more fulfil the fierce hope that has flowed through so many strands of the human project: to end the evils of our world and build a society that is truly just and humane.

And even such a profound achievement might only set the stage for what is to come. Our full potential for flourishing remains undreamt.

Consider the parts of your life when you brushed paths with true happiness. The year, month or day when everything was coming together and you had a glimpse of the richness that life can hold; when you saw how much greater a life could be. For me, most recently, this was the fortnight after the birth of my child: sharing the delight with my friends; sharing the journey with my wife; knowing my parents in a new way; the pride of fatherhood.

Or consider your peak experiences. Those individual moments where you feel most alive; where you are rapt with wonder or love or beauty. In my current stage of life these are most often moments with my daughter. When she sees me arrive at nursery, her eyes lighting up, running headlong into my arms, her fierce embrace. Consider how many times better such moments are than the typical experiences of your day. My typical experiences are by no means bad, but I'd trade hundreds, maybe thousands, for the peaks.

Most of these peaks fade all too quickly. The world deadens; we settle into our routines; our memories dim. But we have seen enough to know that life can offer something far grander and more alive than the standard fare. If humanity can survive, we may one day learn to dwell more and more deeply in such vitality; to brush off more and more dust; to make a home amidst the beauty of the world. Sustaining such heights might not be easy, or simple. It could require changes in our psychology that we should approach with caution. But we know of nothing, in principle, that stands in our way, and much to recommend the exploration.

And peak experiences are not merely potential dwellings—they are also pointers to possible experiences and modes of thought beyond our present understanding. Consider, for example, how little we know of how ultraviolet light looks to a finch; of how echolocation feels to a bat, or a dolphin; of the way that a red fox, or a homing pigeon, experiences the Earth's magnetic field. Such uncharted experiences exist in minds much less sophisticated than our own. What experiences, possibly of immense value, could be accessible, then, to minds much greater? Mice know very little of music, art or humour. Towards what experiences are we as mice? What beauties are we blind to?[40]

Our descendants would be in a much better position to find out. At the very least, they would likely be able to develop and enhance existing human capacities—empathy, intelligence, memory, concentration, imagination. Such enhancements could make possible entirely new forms of human culture and cognition: new games, dances, stories; new integrations of thought and emotion; new forms of art. And we would have millions of years—maybe billions, or trillions—to go much further, to explore the most distant reaches of what can be known, felt, created and understood.

In this respect, the possible quality of our future resembles its possible duration and scope. We saw how human civilisation has probed only a tiny fraction of what is possible in time or space. Along each of these dimensions, we can zoom out from our present position with a dizzying expansion of scale, leading to scarcely imaginable vistas waiting to be explored. Such scales are a familiar feature of contemporary science. Our children learn early that everyday experience has only acquainted us with a tiny fraction of the physical universe.

Less familiar, but just as important, is the idea that the space of possible experiences and modes of life, and the degree of flourishing they make available, may be similarly vast, and that everyday life may acquaint us with a similarly parochial proportion. In this sense, our investigations of flourishing thus far in history may be like astronomy before telescopes—with such limited vision, it is easy to think the universe small, and human-centred. Yet how strange it would be if this single species of ape, equipped by evolution with this limited set of sensory and cognitive capacities, after only a few thousand years of civilisation, ended up anywhere near the maximum possible quality of life. Much more likely, I think, that we have barely begun the ascent.

Rising to our full potential for flourishing would likely involve us being transformed into something beyond the humanity of today.

Remember that evolution has not stopped with humanity. There have been many species of the genus *Homo*, and within 1 or 2 million years we would expect to gradually become a different species from today's *Homo sapiens*. Indeed, unless we act to prevent it, this will eventually happen. This century's genetic technologies will give us the tools to transform ourselves much faster, should we wish. And we can already see additional avenues for transformation on the horizon, such as implants granting digital extensions to our minds, or developments in artificial intelligence allowing us to craft entirely new kinds of beings to join us or replace us.

Such transformation would bring serious risks: risks of inequality and injustice; of splintering the unity of humanity; of unanticipated consequences stretching over vast spans of time. Some of these risks are existential. Replacing ourselves with something much worse, or something devoid of value altogether, would put our entire future in jeopardy, as would any form of enhancement that could lead to widespread conflict or social collapse. But forever preserving humanity as it is now may also squander our legacy, relinquishing the greater part of our potential.

So I approach the possibility of transforming ourselves with cautious optimism. If we can navigate its challenges maturely, such transformation offers a precious opportunity to transcend our limitations and explore much more deeply just how good life can be. I love humanity, not because we are *Homo sapiens*, but because of our capacity to flourish, and to bring greater flourishing to the world around us. And in this most important respect, our descendants, however different, could reach heights that would be forever beyond our present grasp.

CHOICES

I have sketched three dimensions of humanity's potential: the vistas of time, space and experience that the future holds. My concern has not been to paint a detailed picture, but to persuade you, more generally, that we stand before something extraordinarily vast and valuable—something in light of which all of history thus far will seem the merest prelude; a taste; a seed. Beyond these outlines, the substance of our future is mostly unknown. Our descendants will create it.

If we steer humanity to a place of safety, we will have time to think. Time to ensure that our choices are wisely made; that we will do the very best we can with our piece of the cosmos. We rarely reflect on what that might be. On what we might achieve should humanity's entire will be focused on gaining it, freed from material scarcity and internal conflict. Moral philosophy has been focused on the more pressing issues of treating each other decently in a world of scarce resources. But there may come a time, not too far away, when we mostly have our house in order and can look in earnest at where we might go from here. Where we might address this vast question about our ultimate values. This is the Long Reflection.

I do not know what will come out of it. What ideas will stand the test of time; of careful analysis by thinkers of all kinds, each anxious to avoid any bias that may doom us to squander our potential. I do not know in what proportions it will be a vision of flourishing, of virtue, of justice, of achievement; nor what final form any of these aspects may take. I do not know whether it will be a vision that transcends these very divisions in how we think of the good.

We might compare our situation with that of people 10,000 years ago, on the cusp of agriculture. Imagine them sowing their first seeds and reflecting upon what opportunities a life of farming might enable, and on what the ideal world might look like. Just as they would be unable to fathom almost any aspect of our current global civilisation, so too we may not yet be able to see the shape of an ideal realisation of our potential.

This chapter has focused on humanity's potential—on the scope of what we *might* someday be able to achieve. Living up to this potential will be another great challenge in itself. And vast efforts will be undertaken in an attempt to rise to this challenge.

But they can wait for another day. These battles can be fought by our successors. Only we can make sure we get through this period of danger, that we navigate the Precipice and find our way to safety; that we give our children the very pages on which they will author our future.

RESOURCES

BOOK WEBSITE

Videos • Mailing list • FAQs • Errata
Supporting articles and papers • Quotations • Reading lists
theprecipice.com

AUTHOR WEBSITE

Find out about my other projects • Read my papers
Contacts for media and speaking
tobyord.com

EFFECTIVE ALTRUISM

Meet others interested in having the greatest impact they can
effectivealtruism.org

CAREERS

Advice on how to use your career to safeguard our future
80000hours.org

DONATIONS

Join me in making a lifelong commitment to helping
the world through effective giving
givingwhatwecan.org

ACKNOWLEDGEMENTS

Few books are shaped by author alone. I think most of us already know this. Yet few owe quite so much as this one to the generosity, hard work and brilliance of others.

Academics are rarely allowed enough time to write a book like this. They are weighed down by major responsibilities to their students, colleagues and institutions. Even grants for the express purpose of book-writing usually buy only a brief spell of time. I am thus especially thankful for the funding for my dedicated research fellowship, which has allowed me years of uninterrupted time to work on a topic I consider so important, seeing it through from initial research to the published book. This was provided by the private philanthropy of Luke Ding and by a grant from the European Research Council (under the European Union's Horizon 2020 research and innovation programme: grant agreement No. 669751). Further funding from Luke Ding, the Open Philanthropy Project and the Berkley Existential Risk Initiative allowed me to build up a dedicated team around the book, providing me with expert advice and support in making the book a success. Such support for an academic's book is exceptionally uncommon and it is difficult to overstate how important it has been.

Moreover, academics are rarely allowed to write books that venture so far beyond the bounds of their home discipline. I'm extremely grateful to my academic home, Oxford University's Future of Humanity Institute, where no such prejudice exists; where you are encouraged to venture however far you need to explore the question at hand.

Writing a book with such a wide scope usually comes with too high a risk of oversimplification, cherry-picking and outright error in those fields far from your own expertise. This risk would have been too high for me as well, if not for the vast amount of research support I received from others who believed in the project. I am grateful for the research assistance from Joseph Carlsmith, John Halstead, Howie Lempel, Keith Mansfield and Matthew van der Merwe, who helped familiarise me with the relevant literatures.

And I am profoundly grateful to the many experts from other disciplines who gave so much of their time, to ensure the book was faithful to the state-of-the-art knowledge in their fields. Thank you to Fred Adams, Richard Alley, Tatsuya Amano, Seth Baum, Niel Bowerman, Miles Brundage, Catalina Cangea, Paulo Ceppi, Clark Chapman, David Christian, Allan Dafoe, Richard Danzig, Ben Day, David Denkenberger, Daniel Dewey, Eric Drexler, Daniel Ellsberg, Owain Evans, Sebastian Farquhar, Vlad Firoiu, Ben Garfinkel, Tim Genewein, Goodwin Gibbons, Thore Graepel, Joanna Haigh, Alan Harris, Hiski Haukkala, Ira Helfand, Howard Herzog, Michael Janner, Ria Kalluri, Jim Kasting, Jan Leike, Robert Lempert, Andrew Levan, Gregory Lewis, Marc Lipsitch, Rosaly Lopes, Stephen Luby, Enxhell Luzhnica, David Manheim, Jochem Marotzke, Jason Matheny, Piers Millet, Michael Montague, David Morrison, Cassidy Nelson, Clive Oppenheimer, Raymond Pierrehumbert, Max Popp, David Pyle, Michael Rampino, Georgia Ray, Catherine Rhodes, Richard Rhodes, Carl Robichaud, Tyler Robinson, Alan Robock, Luisa Rodriguez, Max Roser, Jonathan Rougier, Andrew Rushby, Stuart Russell, Scott Sagan, Anders Sandberg, Hauke Schmidt, Rohin Shah, Steve Sherwood, Lewis Smith, Jacob Steinhardt, Sheldon Stern, Brian Thomas, Brian Toon, Phil Torres, Martin Weitzman, Brian Wilcox, Alex Wong, Lily Xia and Donald Yeomans.

Then there were the many weeks of fact-checking by Joseph Carlsmith, Matthew van der Merwe and especially Joao Fabiano, who have done their utmost to reduce the chance of any outright errors or misleading claims slipping through. Of course the responsibility for any that do slip through rests on me alone, and I'll keep an up-to-date list of errata at theprecipice.com/errata.

It was Andrew Snyder-Beattie who first suggested I write this book and did so much to help get it started. Thank you, Andrew. And thanks to all those who contributed to the early conversations on what form it should take: Nick Beckstead, Nick Bostrom, Brian Christian, Owen Cotton-Barratt, Andrew Critch, Allan Dafoe, Daniel Dewey, Luke Ding, Eric Drexler, Hilary Greaves, Michelle Hutchinson, Will MacAskill, Jason Matheny, Luke Muehlhauser, Michael Page, Anders Sandberg, Carl Shulman, Andrew Snyder-Beattie, Pablo Stafforini, Ben Todd, Amy Willey Labenz, Julia Wise and Bernadette Young. Some of this advice continued all the way through the years of writing—my special thanks to Shamil Chandaria, Owen Cotton-Barratt, Teddy Collins, Will MacAskill, Anders Sandberg, Andrew Snyder-Beattie and Bernadette Young.

It was immensely helpful that so many people generously gave their time to read and comment on the manuscript. Thank you to Josie Axford-Foster, Beth Barnes, Nick Beckstead, Haydn Belfield, Nick Bostrom, Danny Bressler, Tim Campbell, Natalie Cargill, Shamil Chandaria, Paul Christiano, Teddy Collins, Owen Cotton-Barratt, Andrew Critch, Allan Dafoe, Max Daniel, Richard Danzig, Ben Delo, Daniel Dewey, Luke Ding, Peter Doane, Eric Drexler, Peter Eckersley, Holly Elmore, Sebastian Farquhar, Richard Fisher, Lukas Gloor, Ian Godfrey, Katja Grace, Hilary Greaves, Demis Hassabis, Hiski Haukkala, Alexa Hazel, Kirsten Horton, Holden Karnofsky, Lynn Keller, Luke Kemp, Alexis Kirschbaum, Howie Lempel, Gregory Lewis, Will MacAskill, Vishal Maini, Jason Matheny, Dylan Matthews, Tegan McCaslin, Andreas Mogensen, Luke Muehlhauser, Tim Munday, John Osborne, Richard Parr, Martin Rees, Sebastian Roberts, Max Roser, Anders Sandberg, Carl Shulman, Peter Singer, Andrew Snyder-Beattie, Pablo Stafforini, Jaan Tallinn, Christian Tarsney, Ben Todd, Susan Trammell, Brian Tse, Jonas Vollmer, Julia Wise and Bernadette Young.

Thanks also to Rose Linke, for her advice on how to name this book, and Keith Mansfield, for answering my innumerable questions about the world of publishing.

This project benefited from a huge amount of operational support from the Future of Humanity Institute (FHI), the Centre for Effective Altruism (CEA) and the Berkley Existential Risk Initiative (BERI). My thanks to Josh Axford, Sam Deere, Michelle Gavin, Rose Hadshar, Habiba Islam, Josh Jacobson, Miok Ham Jung, Chloe Malone, Kyle Scott, Tanya Singh and Tena Thau.

The actual writing of this book took place largely in the many wonderful libraries and cafés of Oxford—I especially need to thank Peloton Espresso, where I may have spent more time than in my own office.

I'm incredibly thankful to Max Brockman, my literary agent. Max connected me with publishers who really believed in the book, and guided me through the baroque world of publishing, always ready to provide astute advice at a moment's notice.

Thank you to Alexis Kirschbaum, my editor at Bloomsbury, who saw most keenly what this book could be, and always believed in it. Her confidence helped give me the confidence to deliver it. Thank you to Emma Bal, Catherine Best, Sara Helen Binney, Nicola Hill, Jasmine Horsey, Sarah Knight, Jonathon Leech, David Mann, Richard Mason, Sarah McLean, Hannah Paget and the rest of the Bloomsbury team, for all their work in making this project a success.

And thanks to everyone at Hachette. To my editors: Paul Whitlatch, who saw what this book could be and believed me when I said I would meet my final deadlines; and David Lamb, who guided the project from a manuscript into a book. And to everyone else at Hachette who played a part behind the scenes, especially Michelle Aielli, Quinn Fariel, Mollie Weisenfeld.

Looking back to the earliest influences on this book, I want to thank four philosophers who shaped my path. The first is Peter Singer, who showed how one could take moral philosophy beyond the walls of the academy, and extend our common conception of ethics to encompass new domains such as animal welfare and global poverty. Then there are my dissertation supervisors Derek

Parfit and John Broome, whose work inspired me to become a philosopher and to come to Oxford, where I was lucky enough to have them both as mentors. Yet I think the greatest influence on me at Oxford has been Nick Bostrom, through his courage to depart from the well-worn tracks and instead tackle vast questions about the future that seem almost off-limits in academic philosophy. I think he introduced me to existential risk the day we met, just after we both arrived here in 2003; we've been talking about it ever since.

One of the best things about the writing of this book has been the feeling of working on a team. This feeling, and indeed the team itself, would not have been possible without Joseph Carlsmith and Matthew van der Merwe. They acted as project managers, keeping me on track, and the dozens of threads of the project from getting stuck. They were research assistants *par excellence*, reaching out to the world's top experts across such a range of topics, determining the key results and controversies—even finding a few mistakes in the cutting-edge papers. They were not afraid to tell me when I was wrong. And they were proofreaders, editors, strategists, confidants and friends. They put thousands of hours of their lives into making this book everything it could be, and I cannot thank them enough.

Finally, I'm grateful to my father, Ian, whose unending curiosity about the past and the future of humanity pervaded my childhood and gave me the foundation to start asking the right questions; to my mother, Lecki, who showed me how to take a stand in the world for what you believe in; to my wife, Bernadette, who has supported, encouraged and inspired me through this and through everything; and to my child, Rose, through whose eyes I see the world anew.

APPENDICES

Appendix A

DISCOUNTING THE LONGTERM FUTURE

An existential catastrophe would drastically reduce the value of our entire future. This may be the most important reason for safeguarding humanity from existential risks. But how important *is* our longterm future? In particular, does the fact that much of it occurs far from us in time reduce its value?

Economists often need to compare benefits that occur at different times. And they have discovered a number of reasons why a particular kind of benefit may matter less if received at a later time.

For example, I recently discovered a one-dollar coin that I'd hidden away as a child to be found by my future self. After the immediate joy of discovery, I was struck by the fact that by transporting this dollar into the future, I'd robbed it of most of its value. There was the depreciation of the currency, of course, but much more than that was the fact that I now have enough money that one extra dollar makes very little difference to my quality of life.[1]

If people tend to become richer in the future (due to economic growth) then this effect means that monetary benefits received in the future will tend to be worth less than if they were received now (even adjusting for inflation).[2] Economists take this into account by 'discounting' future monetary benefits by a discount factor that depends on both the economic growth rate and the psychological fact that extra spending for an individual has diminishing marginal utility.[3]

A second reason why future benefits can be worth less than current benefits is that they are less certain. There is a chance that the process producing this benefit—or indeed the person who would receive it—won't still be around at the future time, leaving us with no benefit at all. This is sometimes called the 'catastrophe rate'. Clearly one should adjust for this, reducing the value of future benefits in line with their chance of being realised.

The standard economic approach to discounting (the Ramsey model) incorporates both these reasons.[4] It treats the appropriate discount rate for society (ρ) as the sum of two terms:

$$\rho = \eta g + \delta$$

The first term (ηg) represents the fact that as future people get richer, they get less benefit from money. It is the product of a factor reflecting the way people get diminishing marginal utility from additional consumption (η) with the growth rate of consumption (g). The second term (δ) accounts for the chance that the benefit won't be realised (the catastrophe rate).

So how do we use this formula for discounting existential risk? The first thing to note is that the ηg term is inapplicable.[5] This is because the future benefit we are considering (having a flourishing civilisation instead of a ruined civilisation, or nothing at all) is not a monetary benefit. The entire justification of the ηg term is to adjust for marginal benefits that are worth less to you when you are richer (such as money or things money can easily buy), but that is inapplicable here—if anything, the richer people might be, the *more* they would benefit from avoiding ruin or oblivion. Put another way, the ηg term is applicable only when discounting monetary benefits, but here we are considering discounting well-being (or utility) itself. So the ηg term should be treated as zero, leaving us with a social discount rate equal to δ.

I introduced δ by saying that it accounts for the catastrophe rate, but it is sometimes thought to include another component too: *pure time preference*. This is a preference for one benefit over another simply because it comes earlier, and it is a third reason for discounting future benefits.

But unlike the earlier reasons, there is substantial controversy over whether pure time preference should be included in the social discount rate. Philosophers are nearly unanimous in rejecting it.[6] Their primary reason is that it is almost completely unmotivated. In a world where people have had a very long history of discounting the experiences of out-groups (cf. the widening moral circle), we would want a solid argument for why we should count some people much less than we count ourselves, and this is lacking.

Even our raw intuitions seem to push against it. Is an 80-year life beginning in 1970 intrinsically more valuable than one that started in 1980? Is your older brother's life intrinsically more important than your younger brother's? It only gets worse when considering longer durations. At a rate of pure time preference of 1 percent, a single death in 6,000 years' time would be vastly more important than a billion deaths in 9,000 years. And King Tutankhamun would have been obliged to value a single day of suffering in the life of one of his contemporaries as more important than a lifetime of suffering for all 7.7 billion people alive today.

Many economists agree, stating that pure time preference is irrational, unfounded or immoral.[7] For example, Ramsey himself said it was 'ethically indefensible and arises merely from the weakness of the imagination', while R. F. Harrod called it 'a polite expression for rapacity and the conquest of reason by passion'.[8] Even those who accept it often maintain a deep unease when considering its application to the *longterm* future.

The standard rationale among those economists who endorse pure time preference is that people simply *have* such a preference and that the job of economists is not to judge people's preferences, but to show how best to satisfy them. From our perspective, there are three key problems here.

Firstly, even if this were the job of the economist,[9] it is not *my* job. I am writing this book to explore how humanity *should* respond to the risks we face. This point of view allows for the possibility that people's unconsidered reflection on how to treat the future can be mistaken: we sometimes act against our own longterm interests and we can suffer from a bias towards ourselves at the expense of future generations. Using our raw intuitions to set social policy would risk enshrining impatience and bias into our golden standard.

Secondly, the pure time preference embodied in δ is an unhappy compromise between respecting people's actual preferences and the preferences they *should* have. The form of pure time preference people actually exhibit is not in the shape of an exponential. They discount at high rates over the short term and low rates over the long term. Non-exponential discounting is typically seen by economists as irrational as it can lead people to switch back and forth between two options in a way that predictably makes them worse off.

For this reason economists convert people's non-exponential time preference into an exponential form, with a medium rate of discounting across all time frames. This distorts people's actual preferences, underestimating how much they discount over short time periods and overestimating how much they discount over long time periods (such as those that concern us in this book). Moreover, it is difficult for them to run the argument that individual preferences are sacrosanct while simultaneously choosing to distort them in these ways—especially if they do so on the grounds that the exhibited preferences were irrational. If they really are irrational, why fix them in this way, rather than by simply removing them?

And thirdly, the evidence for pure time preference comes from individuals making choices about benefits for themselves. When individuals make choices about the welfare of others, they exhibit little or no pure time preference. For example, while we might take a smaller benefit now over a larger one later—just because it comes sooner—we rarely do so when making the choice on behalf of others. This suggests that our preference for immediate gratification is really a case of weakness of will, rather than a sober judgement that our lives really go better when we have smaller benefits that happen earlier in time. And indeed, when economists adjust their experiments to ask about benefits that would be received by unknown strangers, the evidence for pure time preference becomes very weak, or non-existent.[10]

I thus conclude that the value at stake in existential risk—the benefit of having a flourishing future rather than one ravaged by catastrophe—should only be discounted by the catastrophe rate. That is, we should discount future years of flourishing by the chance we don't get that far.[11] An approach like this was used by Nicholas Stern in his famous report on the economics of climate change. He set pure time preference to zero and set δ to a catastrophe rate of 0.1 percent per annum (roughly 10 percent per century).[12] This values humanity's future at about 1,000 times the value of the next year (higher if the quality of each year improves). This is enough to make existential risk extremely important, but still somewhat less than one may have thought.

The standard formula treats the discount rate as constant over time, discounting the future according to an exponential curve. But more careful economic accounts allow the rate to vary over time.[13] This is essential for the catastrophe rate. For while the background natural risk may have been roughly constant, there has been a stark increase in the anthropogenic risks. If humanity rises to this challenge, as I believe we shall, then this risk will start to fall back down, perhaps all the way down to the background rate, or even lower. If annual risks become low in the long term, then the expected value of the future is very great indeed. As a simplified example, if we incur a total of 50 percent risk during the Precipice before lowering the risks back to the background level, then our future would be worth at least 100,000 times as much as next year.[14]

Of course we don't actually *know* what the catastrophe rate is now, let alone how it will change over time. This makes a big difference to the analysis. One might think that when we are uncertain about the catastrophe rate, we should simply discount at the average of the catastrophe rates we find credible. For example, that if we think it equally likely to be 0.1 percent or 1 percent, we should discount at 0.55 percent. But this is not right. A careful analysis shows that we should instead discount at a changing rate: one that starts at this average, then tends towards the lowest credible rate as time goes on.[15] This corresponds to discounting the longterm future as if we were in the safest world among those we find plausible. Thus the possibility that longterm catastrophe rates fall to the background level or below plays a very large role in determining the discounted value of humanity's future.

And finally, in the context of evaluating existential risk, we need to consider that the catastrophe rate is not set exogenously. Our actions can reduce it. Thus, when we decide to act to lower one existential risk, we may be reducing the discount rate that we use to assess subsequent action.[16] This can lead to increasing returns on work towards safeguarding our future.

The upshot of all this is that economic discounting does not reduce the value of the future to something tiny—that only happens when discounting is misapplied. Discounting based on the diminishing marginal utility of money is inapplicable, and pure time preference is inappropriate. This leaves us with the uncertain and changing catastrophe rate. And discounting by this is just another way of saying that we should value the future at its expected value: if we have empirical grounds for thinking our future is very long in expectation, there is no further dampening of its value coming from the process of discounting.[17]

Appendix B

POPULATION ETHICS AND EXISTENTIAL RISK

Theories of ethics point to many different features of our acts that could contribute to them being right or wrong. For instance, whether they spring from bad motives, violate rights or treat people unfairly. One important feature that nearly everyone agrees is relevant is the effect of our acts on the wellbeing of others: increasing someone's wellbeing is good, while reducing it is bad. But some of our acts don't merely change people's wellbeing, they change who will exist. Consider, for example, a young couple choosing whether to have a child. And there is substantial disagreement about how to compare outcomes that contain different people and, especially, different numbers of people. The subfield of ethics that addresses these questions is known as *population ethics*.

Population ethics comes to the fore when considering how bad it would be if humanity went extinct. One set of reasons for avoiding extinction concerns the future. I've pointed to the vast future ahead of us, with potential for thousands, millions or billions of future generations. Extinction would prevent these lives, and all the wellbeing they would involve, from coming into existence. How bad would this loss of future wellbeing be?

One simple answer is the *Total View*: the moral value of future wellbeing is just the total amount of wellbeing in the future. This makes no distinction between whether the wellbeing would come to people who already exist or to entirely new people. Other things being equal, it suggests that the value of having a thousand more generations is a thousand times the value of our generation. On this view, the value of losing our future is immense.

To test moral theories, philosophers apply a kind of thought experiment containing a stark choice. These choices are often unrealistic, but a moral theory is supposed to apply in all situations, so we can try to find any situation where it delivers the intuitively wrong verdict and use this as evidence against the theory.

The main critique of the Total View is that it leads to something called the *repugnant conclusion*: for any outcome where everyone has high wellbeing, there is an ever better outcome where everyone has only a tiny amount of wellbeing, but there are so many people that quantity makes up for quality. People find some quantity/quality trade-offs intuitive (for example, that today's world of 7.7 billion people is better than one with a single person who has slightly higher average wellbeing), but most feel that the Total View takes this too far.

As we shall see, the rivals to the Total View have their own counter-intuitive consequences, and indeed there are celebrated impossibility results in the field which show that *every* theory will have at least one moral implication that most people find implausible.[18] So we cannot hope to find an answer that fits all our intuitions and will need to weigh up how bad each of these unintuitive consequences is.

Another famous approach to population ethics is to say the value of wellbeing in the universe is given not by the total but by the average. This approach comes in two main versions. The first takes the average wellbeing in each generation, then sums this up across all the generations. The second takes the average wellbeing across all lives that are ever lived, wherever they are in space and time.

Both versions of averaging are subject to very serious objections. The first version can sometimes prefer an alternative where exactly the same people exist, but where everyone's wellbeing is lower.[19] The second version runs into problems when we consider negative wellbeing—lives not worth living.

If our only choices were the end of humanity, or creating future people cursed with lives of extremely negative wellbeing, this theory can prefer the latter (if the past was so bad that even this hellish future brings up the average). And it can even prefer adding lives with negative wellbeing to adding a larger number of lives with positive wellbeing (if the past was so good that the larger number of positive lives would dilute the average more). These conclusions are generally regarded as even more counterintuitive than the repugnant conclusion and it is very difficult to find supporters of either version of averaging among those who study population ethics.

Interestingly, even if one overlooked these troubling implications, both versions of averaging probably support the idea that extinction would be extremely bad in the real world. This is easy to see for the sum of generational averages—like the Total View, it says that other things being equal, the value of having 1,000 future generations is 1,000 times the value of our generation. What about the average over all lives over all time? Since quality of life has been improving over time (and has the potential to improve much more) our generation is actually bringing up the all-time average. Future generations would continue to increase this average (even if they had the same quality of life as us).[20] So on either average-based view there would be a strong reason to avoid extinction on the basis of the wellbeing of future generations.

But there is an alternative approach to population ethics according to which human extinction might not be treated as bad at all. The most famous proponent is the philosopher Jan Narveson, who put the central idea in slogan form: 'We are in favor of making people happy, but neutral about making happy people.'[21] Many different theories of population ethics have been developed to try to capture this intuition, and are known as *person-affecting views*. Some of these theories say there is nothing good about adding thousands of future generations with high wellbeing—and thus nothing bad (at least in terms of the wellbeing of future generations) if humanity instead went extinct.

261

Are these plausible? Could they undermine the case for concern about existential risk?

There have been two prominent ways of attempting to provide a theoretical foundation for Narveson's slogan. One is to appeal to a simple intuition known as the *person-affecting restriction*: that an outcome can't be better than another (or at least not in terms of wellbeing) unless it is better for someone.[22] This principle is widely accepted in cases where exactly the same people exist in both outcomes. When applied to cases where different people exist in each outcome, the principle is usually interpreted such that a person has to exist in both outcomes in order for one to count as better for them than the other, and this makes the principle both powerful and controversial. For example, it could allow us to avoid the repugnant conclusion—there is no one who is better off in the outcome that has many people with low wellbeing, so it could not be better.[23]

The other theory-based approach to justifying Narveson's slogan is to appeal to an intuition that we have duties to actual people, but not to merely possible people.[24] On this view, we wouldn't need to make any sacrifices to the lives of the presently existing people in order to save the merely possible people of the future generations.

But both these justifications run into decisive problems once we recall the possibility of lives with negative wellbeing. Consider a thought experiment in which people trapped in hellish conditions will be created unless the current generation makes some minor sacrifices to stop this. Almost everyone has a strong intuition that adding such lives of negative wellbeing is bad, and that we should of course be willing to pay a small cost to avoid adding them. But if we did, then we made real sacrifices on account of merely possible people, and we chose an outcome that was better for no one (and worse for currently existing people).

So these putative justifications for Narveson's slogan run counter to our strongly held convictions about the importance of avoiding new lives with negative wellbeing. It therefore seems to me that these two attempts to provide a theoretical justification for the slogan are dead ends, at least when considered in isolation. Any plausible account of population ethics will involve recommending some choices where no individual who would be present in both outcomes is made better off, and making sacrifices on behalf of merely possible people.

Given the challenges in justifying Narveson's slogan in terms of a more basic moral principle, philosophers who find the slogan appealing have increasingly turned to the approach of saying that what justifies it is simply that it captures our intuitions about particular cases better than alternative views. The slogan on its own doesn't say how to value negative lives, so the proponents of this approach add to it an asymmetry principle: that adding new lives of positive wellbeing doesn't make an outcome better, but adding new lives with negative wellbeing does make it worse.

Philosophers have developed a wide variety of theories based on these two principles. Given the variety, the ongoing development and the lack of any consensus approach, we can't hope to definitively review them here. But we can look to general patterns. These theories typically run into a variety of problems: conflicting with strong intuitions about thought experiments, conflicting with important moral principles, and violating widely held principles of rationality.[25] Adjustments to the theories to avoid some of these problems have typically exacerbated other problems or created new problems.

But perhaps most importantly, the slogan and asymmetry principle are usually justified merely by an appeal to our intuitions on particular cases.[26] Person-affecting views fit our intuitions well in some cases, but poorly in others. The case in question—whether extinction would be bad—is one where these views offer advice most people find very counter-intuitive.[27] In general, we shouldn't look to a disputed theory to guide us on the area where it seems to have the weakest fit with our considered beliefs.[28]

Moreover, there are versions of person-affecting views that capture some of their core intuitions without denying the badness of extinction. For example, there are less strict theories which say that we have *some* reason to create lives of high wellbeing, but stronger reason to help existing people or avoid lives of negative wellbeing. Since the future could contain so many new lives of high wellbeing it would still be very important.

In summary, there are some theories of how to value the wellbeing of future people that may put little or no value on avoiding extinction. Many such views have been shown to be untenable, but not all, and this is still an area of active research. If someone was committed to such a view, they might not find the argument based on lost future wellbeing compelling.

But note that this was only one kind of explanation for why safeguarding humanity is exceptionally important. There remain explanations based on the great things we could achieve in the future (humanity's greatest works in arts and science presumably lie ahead) and on areas other than our future: on our past, our character, our cosmic significance and on the losses to the present generation. People who hold person-affecting accounts of the value of the wellbeing of future generations could well be open to these other sources of reasons why extinction would be bad. And there remains the argument from moral uncertainty: if you thought there was any significant chance these person-affecting views were mistaken, it would be extremely imprudent to risk our entire future when so many other moral theories say it is of utmost importance.

Finally, we've focused here on the moral importance of preventing extinction. Even theories of population ethics that say extinction is a matter of indifference often ascribe huge importance to avoiding other kinds of existential catastrophes such as the irrevocable collapse of civilisation or an inescapable dystopia. So even someone who was untroubled by human extinction should still be deeply concerned by other existential risks. And since hazards that threaten extinction usually threaten the irrevocable collapse of civilisation as well, these people should often be concerned by a very similar set of risks.

APPENDIX C

NUCLEAR WEAPONS ACCIDENTS

One may imagine that the enormous importance and obvious danger of nuclear weapons would lead to extremely careful management—processes that would always keep us far from accidental destruction. It is thus surprising to learn that there were so many accidents involving nuclear weapons: a US Department of Defense report counts 32 known cases.[29] None of these involved an unexpected nuclear detonation, which speaks well for the technical safeguards that try to stop even an armed weapon whose conventional explosives detonate from creating a nuclear explosion. But they show how complex the systems of nuclear war were, with so many opportunities for failure. And they involve events one would never have thought possible if the situation was being handled with due care, such as multiple cases of nuclear bombs accidentally falling out of planes and the large number of lost nuclear weapons that have never been recovered.

List of Accidents

1957 A nuclear bomb accidentally fell through the bomb bay doors of a B-36 bomber over New Mexico. The high explosives detonated, but there was no nuclear explosion.[30]

1958 A B-47 bomber crashed into a fighter plane in mid-air off the coast near Savannah, Georgia. The B-47 jettisoned its atomic bomb into the ocean. There are conflicting reports about whether it contained its atomic warhead, with the Assistant Secretary of Defense testifying to Congress that it did.[31]

1958 A B-47 bomber accidentally dropped a nuclear bomb over South Carolina, landing in someone's garden and destroying their house. Fortunately, its atomic warhead was still in the plane.[32]

1960 A BOMARC air defence missile caught fire and melted. Its 10-kiloton warhead did not commence a nuclear explosion.[33]

1961 A B-52 carrying two 4-megaton nuclear bombs broke up over North Carolina. The bombs fell to the ground. One of them broke apart on impact, and a section containing uranium sank into the waterlogged farmland. Despite excavation to a depth of 50 feet, it was never recovered. There was no nuclear explosion, though multiple sources, including Defense Secretary Robert McNamara, have said that a single switch prevented a nuclear explosion.[34]

1961 A B-52 carrying two nuclear bombs crashed in California. Neither bomb detonated.[35]

1965 A fighter jet carrying a 1-megaton bomb fell off the side of a US aircraft carrier, near Japan. The bomb was never recovered.[36]

1966 A B-52 carrying four nuclear weapons crashed into a refuelling plane in mid-air above Spain. All four bombs fell, and two of the bombs suffered conventional explosions on impact with the ground. There was substantial radiation, and 1,400 tons of contaminated soil and vegetation needed to be taken back to the US.[37]

1968 A B-52 bomber flying over Greenland caught fire and crashed into the ice. The conventional high explosives surrounding the nuclear cores of its four hydrogen bombs detonated. Luckily, this did not set off a nuclear reaction.[38] Had it done so, all signals would have suggested this was a Soviet nuclear strike, requiring nuclear retaliation—for it was at the location of one part of the US early warning systems, detecting Soviet missiles fired across the North Pole.[39]

1980 A Titan II missile exploded at an Air Force Base in Damascus, Arkansas, after a wrench was dropped and punctured its fuel tank. Hours later there was an explosion in which the 9-megaton warhead was propelled about 100 metres away, but its safety features kept it intact.[40]

This is only a part of the full list of accidents, and we have very little knowledge at all of how bad things were on the Russian side.

The Accidental Launch Order

One of the most astounding accidents has only just come to light and may well be the closest we have come to nuclear war. But the incident I shall describe has been disputed, so we cannot yet be sure whether it occurred.

On 28 October 1962—at the height of the Cuban Missile Crisis—a US missile base in the US-occupied Japanese island of Okinawa received a radioed launch order. The island had eight launch centres, each controlling four thermonuclear missiles. All three parts of the coded order matched the base's own codes, confirming that it was a genuine order to launch their nuclear weapons.

The senior field officer, Captain William Bassett, took command of the situation. He became suspicious that a launch order was given while only at the second highest state of readiness (DEFCON 2), which should be impossible. Bassett's crew suggested that the DEFCON 1 order may have been jammed and a launch officer at another site suggested a Soviet pre-emptive attack may be underway, giving no time to upgrade to DEFCON 1.

But Bassett's crew quickly calculated that a pre-emptive strike should have already hit them. Bassett ordered them to check the missiles' readiness and noticed that three of their targets were not in Russia, which seemed unlikely given the current crisis. He radioed to the Missile Operations Centre to confirm the coded order but the same code was radioed back.

Bassett was still suspicious, but a lieutenant in charge of another site, all of whose targets were in the USSR, argued that Bassett had no right to stop the launch given that the order was repeated. This other officer ordered the missiles at his site to be launched.

In response, Bassett ordered two airmen from an adjacent launch site to run through the underground tunnel to the site where the missiles were being launched, with orders to shoot the lieutenant if he continued with the launch without either Bassett's agreement or a declaration of DEFCON 1.

Airman John Bordne (who recounted this story) realised that it was unusual for the launch order to be given at the end of a routine weather report and for the order to have been repeated so calmly. Bassett agreed and telephoned the Missile Operations Centre, asking the person who radioed the order to either give the DEFCON 1 order or issue a stand-down order. A stand-down order was quickly given and the danger was over.

This account was made public in 2015 in an article in the *Bulletin of the Atomic Scientists* and a speech by Bordne at the United Nations. It has since been challenged by others who claimed to have been present in the Okinawa missile bases at the time.[41] There is some circumstantial evidence supporting Bordne's account: his memoir on it was cleared for publication by the US Air Force, the major who gave the false launch order was subsequently court-martialled, and Bordne has been actively seeking additional testimony from others who were there at the time.

I don't know who is correct, but the matter warrants much more investigation. There is an active Freedom of Information request to the National Security Archive, but this may take many years to get a response. In my view this alleged incident should be taken seriously, but until there is further confirmation, no one should rely on it in their thinking about close calls.

Appendix D

SURPRISING EFFECTS WHEN COMBINING RISKS

We have seen a number of counter-intuitive effects that occur when we combine individual existential risks to get a figure for the total existential risk. These effects get stronger, and stranger, the more total risk there is. Since the total risk includes the accumulated existential risk over our entire future, it may well be high enough for these effects to be significant.

First, the total risk departs more and more from being the sum of the risks. To keep the arithmetic simple, suppose we face four 50 percent risks. Since the total risk can't be over 100 percent, logic dictates that they must overlap substantially and combine to something much less than their sum. For example, if they were independent, the total risk would not be 200 percent, but 93.75 percent (= 15/16).

Second, there can be substantial *increasing* marginal returns if we eliminate more and more risks. For example, eliminating the first of four independent 50 percent risks would only reduce the total risk to 87.5 percent. But eliminating the subsequent risks would make more and more headway: to 75 percent, then 50 percent, then 0 percent—a larger absolute effect each time. Another way to look at this is that eliminating each risk doubles our chance of survival and this has a larger absolute effect the higher our chance of survival is. Similarly, if we worked on all four risks in parallel, and halved the risk from each (from 50 percent to 25 percent), the total risk would only drop from 93.75 percent to about 68 percent. But if we halved them all a second time, the risk would drop by a larger absolute amount, to about

41 percent. What is happening in these examples is that there is so much overlap between the risks that a catastrophe is over-determined, and the total remains high when we act. But our action also helps to reduce this overlap, allowing further actions to be more helpful.

Third, it can be much more important to work on the largest risks. We saw that if we faced independent 10 percent and 20 percent risks, eliminating the 10 percent risk would reduce total risk by 8 points, while eliminating the 20 percent risk would reduce total risk by 18 points (see p. 175). So reducing the larger risk was not 2 times more important, but 2.25 times more.

Correct calculations of relative importance of independent risks require the naïve ratio of probabilities to be multiplied by an additional factor: the ratio between the chance the first catastrophe *doesn't* happen and the chance the second catastrophe *doesn't* happen.[42] When the risks are small, the chance each catastrophe doesn't happen is close to one, so this ratio must also be close to one, making little difference. But when a risk grows large, the ratio can grow large too, making a world of difference.[43]

Suppose we instead faced a 10 percent risk and a 90 percent risk. In this case the naïve ratio would be 9:1 and the adjustment would also be 9:1, so eliminating the 90 percent risk would be 81 times as important as eliminating the 10 percent risk (see Figure D.1). Perhaps the easiest way to think about this is that not only is the 90 percent risk 9 times more likely to occur, but the world you get after eliminating it is also 9 times more likely to survive the remaining risks.

This adjustment applies in related cases too. Halving the 90 percent risk is 81 times as important as halving the 10 percent risk, and the same is true for any other factor by which one might reduce it. Even reducing a risk by a fixed absolute amount, such as a single percentage point, is more important for the larger risk. Reducing the 90 percent risk to 89 percent is 9 times as important as reducing the 10 percent risk to 9 percent.[44]

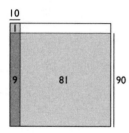

FIGURE D.1 Independent 10% and 90% risks give a total risk of 91%. Removing the 10% risk would lower the total risk (the total shaded area) by just a single percentage point to 90%, while removing the 90% risk would lower it by 81 percentage points to 10%.

All three of these effects occur regardless of whether these risks are simultaneous or occur at different times.[45] So if there is a lot of risk in our future, it could become increasingly more important to eliminate the risks of each successive century. There are many forces that generically lead to diminishing returns on one's work (such as the fact that we can start with the easier risks first). But if we are unlucky enough to face a lot of risk, the overall marginal returns to fighting existential risk may actually be increasing. And it might be especially important to work on the largest risks.

Appendix E

THE VALUE OF PROTECTING HUMANITY

Just how valuable *is* it to protect humanity? While we can't answer with precision, there is a way of approaching this question that I've found helpful for my thinking.

Let's start with a deliberately basic model of existential risk. This model makes assumptions about three things: the pattern of risk over time, the way we could reduce this risk, and the value of the future. First, suppose that each century is exposed to an equal, but unknown, amount of existential risk, r (known as a *constant hazard rate*). That is, given we reach a century, there is always a probability r that we don't make it to the next one. Next, suppose our actions can reduce the probability of existential catastrophe for our own century from r down to some smaller number. And finally, suppose that each century prior to the catastrophe has the same value, v, so the value of a future is proportional to its length before catastrophe. (This means there is no discounting of future value over and above the chance we don't make it that far and we are making assumptions regarding population ethics.)[46] Given these assumptions, the expected value of the future would be:

$$\text{EV}(\textit{future}) = \sum_{i=0}^{\infty} (1 - r)^i\, v = \frac{v}{r}$$

This is just the value of a single century divided by the per century risk. For example, if the risk each century was one in ten, the expected value would be ten times the value of a single century.

This leads to a surprising implication: that the value of eliminating all existential risk this century is independent of how much risk that is. To see this, imagine that existential risk were just one part in a million each century. Even though there would only be the tiniest chance that we would fall victim to our century's risk, the future we'd lose if we did would be correspondingly vast (a million centuries, on average). On the basic model, these effects always balance. The expected disvalue of this century's existential risk is just

$$r.\mathrm{EV}(\textit{future}) = r\frac{\upsilon}{r} = \upsilon$$

So the expected value of eliminating all risk over a century would simply be equal to the value of a century of life for humanity.[47]

Since it would be impossible to completely eliminate all risk this century, it is more useful to note that on the basic model halving the century's risk is worth half the value of a century of humanity (and it works the same way for any other fraction or timespan). This would be enough to make safeguarding our future a key global priority.

However, the value of the basic model lies not in its accuracy, but in its flexibility. It is a starting point for exploring what happens when we change any of its assumptions. And in my view, all three of its assumptions are too pessimistic.

First, by many measures the value of humanity has increased substantially over the centuries. This progress has been very uneven over short periods, but remarkably robust over the long run. We live long lives filled with cultural and material riches that would have seemed like wild fantasy to our ancestors thousands of years ago. And the scale of our civilisation may also matter: the fact that there are thousands of times as many people enjoying these richer lives seems to magnify this value. If the intrinsic value of each century increases at a rate higher than r, this can substantially increase the value of protecting humanity (even if this rate of increase is not sustained forever).[48]

Second, the basic model assumes that our actions this century can only provide protection from this century's risks. But we can do more than that. We can take actions with lasting effects on risk. For example, this book is my attempt to better humanity's understanding of the nature of existential risk and how we should respond to it. Many of the lessons I've drawn out are timeless: to the extent that they succeed at all, they should help with future risk too.[49] Work that helps with risk across many centuries would be much more important than the basic model suggests.

When work reduces all future existential risk, the value of this work will now depend on the hazard rate, r. For example, the value of halving all future risk is:

$$\frac{v}{r/2} - \frac{v}{r} = \frac{v}{r}$$

Surprisingly, this value of reducing a risk in all time periods is *higher* when there *less* risk.[50] This is contrary to our intuitions, as people who estimate risk to be low typically use this as an argument *against* prioritising work on existential risk. But an intuitive way to see how low risk levels make reduction more important is that halving existential risk in all periods doubles the expected length of time before catastrophe. So situations where the risk is already low give us a longer future to double, making this work more important. Note also that this effect gives increasing marginal returns to further halvings of all future risk.

Third, and perhaps most importantly, the risk per century will change over time. It has risen over the last century and may continue to rise over the course of this century. But I think that in the longer run it will diminish, for several different reasons. In the next few centuries we will likely be able to establish permanent settlements beyond the Earth. Space is not a panacea (see p. 194), but spreading our eggs across several baskets will help protect us against part of the risk. In addition, much of the risk is posed by the introduction of transformative new technologies. But if we survive long enough, we may well reach *technological maturity*—a time when we possess all the major technologies that are feasible, and face no further technological transitions.[51] Finally, we need to take into account the work that future generations will do to protect humanity in their own centuries. If the case for safeguarding humanity is as clear and powerful as it seems to me, then we should expect this to become more widely recognised, leading to increasing efforts to protect humanity in the future.

If the risk does fall below its current level, that can make the future substantially more valuable than the basic model suggests.[52] The boost to the value depends roughly on the ratio between the risk over the next century and the longer term risk per century. For example, if there is a one in ten chance of existential catastrophe this century, but this declined rapidly to a background natural risk rate of less than one in 200,000 per century, then the value of eliminating risk this century would be boosted by a factor of 20,000 compared to the basic model.

It is also possible that reducing risk may have *less* value than the basic model suggests, though this seems less likely. One way this could happen is if most of the risk were completely unpreventable. This is not very plausible, since most of it is caused by human activities, and these are within humanity's control. A second way is if the value of future centuries were rapidly declining. However, I can't see why we should expect this: the long-run historical record suggests the opposite, and we've seen that we shouldn't discount the intrinsic value of our future. A third way is if adding up the values of future times is ethically misguided—for example, if we should be averaging the values of centuries, or ignoring all future generations. Though as discussed in Appendix B, such alternative approaches have serious problems of their own. And a final way the true value might be less than the basic model suggests is if the risk is low now, but will increase in the future, and if we can't do much now to help with that later risk. This seems to me the most plausible way that this could be an overestimate.

Overall, I find it substantially more likely that the basic view underestimates the value of safeguarding our future, than that it overestimates. But even if you thought one was roughly as likely as the other, note that their effects are not symmetrical. This is because they act as multipliers. Suppose you thought it equally likely the value of reducing risk was ten times as important as the basic model suggests, or a tenth as important. The average of these is not one times as important: it is 5.05 times as important. So unless we are *highly* confident that the basic model gives an overestimate, we should generally act as though it is an underestimate.[53]

Appendix F

POLICY AND RESEARCH RECOMMENDATIONS

For ease of reference, I've gathered together my recommendations for policy and research on existential risk.

Asteroids & Comets

- Research the deflection of 1 km+ asteroids and comets, perhaps restricted to methods that couldn't be weaponised such as those that don't lead to accurate changes in trajectory.
- Bring short-period comets into the same risk framework as near-Earth asteroids.
- Improve our understanding of the risks from long-period comets.
- Improve our modelling of impact winter scenarios, especially for 1–10 km asteroids. Work with experts in climate modelling and nuclear winter modelling to see what modern models say.

Supervolcanic Eruptions

- Find all the places where supervolcanic eruptions have occurred in the past.
- Improve the very rough estimates on how frequent these eruptions are, especially for the largest eruptions.
- Improve our modelling of volcanic winter scenarios to see what sizes of eruption could pose a plausible threat to humanity.

- Liaise with leading figures in the asteroid community to learn lessons from them in their modelling and management.

Stellar Explosions

- Build a better model for the threat including known distributions of parameters instead of relying on representative examples. Then perform sensitivity analysis on that model—are there any plausible parameters that could make this as great a threat as asteroids?
- Employ blue-sky thinking about any ways current estimates could be underrepresenting the risk by a factor of a hundred or more.

Nuclear Weapons

- Restart the Intermediate-Range Nuclear Forces Treaty (INF).
- Renew the New START arms control treaty, due to expire in February 2021.
- Take US ICBMs off hair-trigger alert (officially called Launch on Warning).
- Increase the capacity of the International Atomic Energy Agency (IAEA) to verify nations are complying with safeguards agreements.
- Work on resolving the key uncertainties in nuclear winter modelling.
- Characterise the remaining uncertainties then use Monte Carlo techniques to show the distribution of outcome possibilities, with a special focus on the worst-case possibilities compatible with our current understanding.
- Investigate which parts of the world appear most robust to the effects of nuclear winter and how likely civilisation is to continue there.

Climate

- Fund research and development of innovative approaches to clean energy.
- Fund research into safe geoengineering technologies and geoengineering governance.
- The US should re-join the Paris Agreement.
- Perform more research on the possibilities of a runaway greenhouse effect or moist greenhouse effect. Are there any ways these could be more likely than is currently believed? Are there any ways we could decisively rule them out?
- Improve our understanding of the permafrost and methane clathrate feedbacks.
- Improve our understanding of cloud feedbacks.
- Better characterise our uncertainty about the climate sensitivity: what can and can't we say about the right-hand tail of the distribution.
- Improve our understanding of extreme warming (e.g. 5–20 °C), including searching for concrete mechanisms through which it could pose a plausible threat of human extinction or the global collapse of civilisation.

Environmental Damage

- Improve our understanding of whether any kind of resource depletion currently poses an existential risk.
- Improve our understanding of current biodiversity loss (both regional and global) and how it compares to that of past extinction events.
- Create a database of existing biological diversity to preserve the genetic material of threatened species.

Engineered Pandemics

- Bring the Biological Weapons Convention into line with the Chemical Weapons Convention: taking its budget from $1.4 million up to $80 million, increasing its staff commensurately, and granting the power to investigate suspected breaches.

- Strengthen the WHO's ability to respond to emerging pandemics through rapid disease surveillance, diagnosis and control. This involves increasing its funding and powers, as well as R&D on the requisite technologies.
- Ensure that all DNA synthesis is screened for dangerous pathogens. If full coverage can't be achieved through self-regulation by synthesis companies, then some form of international regulation will be needed.
- Increase transparency around accidents in BSL-3 and BSL-4 laboratories.
- Develop standards for dealing with information hazards, and incorporate these into existing review processes.
- Run scenario-planning exercises for severe engineered pandemics.

Unaligned Artificial Intelligence

- Foster international collaboration on safety and risk management.
- Explore options for the governance of advanced AI.
- Perform technical research on aligning advanced artificial intelligence with human values.
- Perform technical research on other aspects of AGI safety, such as secure containment and tripwires.

General

- Explore options for new international institutions aimed at reducing existential risk, both incremental and revolutionary.
- Investigate possibilities for making the deliberate or reckless imposition of human extinction risk an international crime.

- Investigate possibilities for bringing the representation of future generations into national and international democratic institutions.
- Each major world power should have an appointed senior government position responsible for registering and responding to existential risks that can be realistically foreseen in the next 20 years.
- Find the major existential risk factors and security factors—both in terms of absolute size and in the cost-effectiveness of marginal changes.
- Target efforts at reducing the likelihood of military conflicts between the US, Russia and China.
- Improve horizon-scanning for unforeseen and emerging risks.
- Investigate food substitutes in case of extreme and lasting reduction in the world's ability to supply food.
- Develop better theoretical and practical tools for assessing risks with extremely high stakes that are either unprecedented or thought to have extremely low probability.
- Improve our understanding of the chance civilisation will recover after a global collapse, what might prevent this, and how to improve the odds.
- Develop our thinking about grand strategy for humanity.
- Develop our understanding of the ethics of existential risk and valuing the longterm future.

Appendix G

EXTENDING THE KARDASHEV SCALE

In 1964, the Russian astronomer Nikolai Kardashev devised a way of classifying potential advanced civilisations by their physical scale and the power (energy per unit time) this lets them harness. He considered three levels: planet, solar system and galaxy. In each step along this scale, the power at their disposal leaps up by more than a factor of a billion.

We can naturally extend this scale in both directions.[54] We can include an earlier level for a minimal civilisation (for example, the size of civilisation in Mesopotamia at the dawn of the written language).[55] And we can include an ultimate level at the size of our affectable universe: everything that we could ever hope to reach. Surprisingly, these jumps are very similar in size to those Kardashev identified, continuing the roughly logarithmic scale for measuring the power of civilisations.

Level	Civilisation Size	Scale-up	Power
K0	Minimal		$\approx 10^8$ W
K1	Planetary	× 1 billion	2×10^{17} W
K2	Stellar	× 1 billion	4×10^{26} W
K3	Galactic	× 100 billion	4×10^{37} W
K4	Ultimate	× 1 billion	4×10^{46} W

Our global civilisation currently controls about 12 trillion Watts of power. This is about 100,000 times more than a minimal civilisation but 10,000 times less than the full capacity of our planet. This places us at level K0.55—more than halfway to K1 and an eighth of the way to K4.

FURTHER READING

Here is a quick guide to some of the most important writing on existential risk. A full bibliography for *The Precipice* can be found on p. 422. Further reading lists, course outlines, quotations, and extracts can be found online at theprecipice.com.

Bertrand Russell & Albert Einstein (1955). 'The Russell-Einstein Manifesto'.
> In the decade after Hiroshima, Russell and Einstein each wrote several important pieces on nuclear war that touched upon the risk of human extinction. Their joint manifesto was the culmination of this early period of thought.

Hilbrand J. Groenewold (1968). 'Modern Science and Social Responsibility'.
> A very early piece that anticipated several key ideas of existential risk. It failed to reach a wide audience, leaving these ideas in obscurity until they were independently discovered decades later.

Annette Baier (1981). 'The Rights of Past and Future Persons'.
> The foundational work on the importance of future generations.

Jonathan Schell (1982). *The Fate of the Earth*.
> The first deep exploration of the badness of extinction, and the central importance of ensuring humanity's survival. Filled with sharp philosophical insight.

Carl Sagan (1983). 'Nuclear War and Climatic Catastrophe: Some Policy Implications'.
> A seminal paper, introducing the new-found mechanism of nuclear winter and exploring the ethical implications of human extinction.

Derek Parfit (1984). *Reasons and Persons*.
> Among the most famous works in philosophy in the twentieth century, it made major contributions to the ethics of future generations and its concluding chapter highlighted how and why the risk of

285

human extinction may be one of the most important moral problems of our time.

John Leslie (1996). *The End of the World: The Science and Ethics of Human Extinction.*
A landmark book that broadened the discussion from nuclear risk to all risks of human extinction, cataloguing the threats and exploring new philosophical angles.

Nick Bostrom (2002). 'Existential Risks: Analyzing Human Extinction Scenarios'.
Established the concept of existential risk and introduced many of the most important ideas. Yet mainly of historic interest, for it is superseded by his 2013 paper below.

Nick Bostrom (2003). 'Astronomical Waste: The Opportunity Cost of Delayed Technological Development'.
Explored the limits of what humans might be able to achieve in the future, suggesting that it is of immense importance to accelerate the arrival of the ultimate state of our civilisation by even a tiny amount, yet that even this is overshadowed by the importance of increasing the chance we get there at all.

Nick Bostrom (2013). 'Existential Risk Prevention as Global Priority'.
An updated version of his essay from 2002, this is the go-to paper on existential risk.

Nick Beckstead (2013). *On the Overwhelming Importance of Shaping the Far Future.*
A book-length philosophical exploration of the idea that what matters most about our actions is not their immediate consequences, but how they shape the longterm trajectory of humanity.

David Christian (2004). *Maps of Time: An Introduction to Big History.*
The seminal book on Big History: examining the major themes and developments in our universe from the Big Bang, the origin of life, humanity, civilisation, the industrial revolution, through to today.

Fred Adams & Gregory Laughlin (1999). *The Five Ages of the Universe.*
A powerful and accessible presentation of how scientists believe the very longterm future will unfold.

Max Roser (2013). *Our World in Data* [online]. Available at: www.ourworldindata.org
An essential online resource for seeing the ways in which many of the most important aspects of our world have changed over the last two centuries. From the raw data to compelling charts and insightful analysis.

Nick Bostrom (2014). *Superintelligence: Paths, Dangers, Strategies.*
The foundational work on artificial intelligence and existential risk.

Stuart Russell (2019). *Human Compatible: AI and the Problem of Control.*
A call to action by a leading researcher in AI, showing how his field needs to develop if it is to address the risks that will be posed by advanced AI.

Alan Robock et al. (2007). 'Nuclear winter revisited with a modern climate model and current nuclear arsenals: Still catastrophic consequences.'
The most up-to-date modelling of the climate effects of a full-scale war between the US and Russia.

Richard Rhodes (1986). *The Making of the Atomic Bomb.*
A gripping history of the people and events leading to the creation of nuclear weapons. With so much information about how everything played out, it reveals how individuals can and did make a difference in this pivotal transition.

Daniel Ellsberg (2017). *The Doomsday Machine: Confessions of a Nuclear War Planner.*
An exploration of how close we have come to full-scale nuclear war, drawing on a wealth of new information from his career at RAND and the Pentagon.

John Broome (2012). *Climate Matters: Ethics in a Warming World.*
A deep examination of the ethics of climate change.

Gernot Wagner & Martin Weitzman (2015). *Climate Shock: The Economic Consequences of a Hotter Planet.*
An accessible study of the risks from climate change, with a focus on extreme warming scenarios.

NOTES

INTRODUCTION

1 Blanton, Burr & Savranskaya (2012).
2 Ellsberg (2017), pp. 215–17.
3 McNamara (1992).
4 Any errors are, of course, my own. You can find an up-to-date list of any known errors at theprecipice.com/errata. I am grateful for expert advice from Fred Adams, Richard Alley, Tatsuya Amano, Seth Baum, Niel Bowerman, Miles Brundage, Catalina Cangea, Paulo Ceppi, Clark Chapman, David Christian, Allan Dafoe, Richard Danzig, Ben Day, David Denkenberger, Daniel Dewey, Eric Drexler, Daniel Ellsberg, Owain Evans, Sebastian Farquhar, Vlad Firoiu, Ben Garfinkel, Tim Genewein, Goodwin Gibbons, Thore Graepel, Joanna Haigh, Alan Harris, Hiski Haukkala, Ira Helfand, Howard Herzog, Michael Janner, Ria Kalluri, Jim Kasting, Jan Leike, Robert Lempert, Andrew Levan, Gregory Lewis, Marc Lipsitch, Rosaly Lopes, Stephen Luby, Enxhell Luzhnica, David Manheim, Jochem Marotzke, Jason Matheny, Piers Millet, Michael Montague, David Morrison, Cassidy Nelson, Clive Oppenheimer, Raymond Pierrehumbert, Max Popp, David Pyle, Michael Rampino, Georgia Ray, Catherine Rhodes, Richard Rhodes, Carl Robichaud, Tyler Robinson, Alan Robock, Luisa Rodriguez, Max Roser, Jonathan Rougier, Andrew Rushby, Stuart Russell, Scott Sagan, Anders Sandberg, Hauke Schmidt, Rohin Shah, Steve Sherwood, Lewis Smith, Jacob Steinhardt, Sheldon Stern, Brian Thomas, Brian Toon, Phil Torres, Martin Weitzman, Brian Wilcox, Alex Wong, Lily Xia and Donald Yeomans.
5 I also made a further pledge to keep just £18,000 a year for myself and to donate anything in excess. This baseline adjusts with inflation (it is currently £21,868), and doesn't include spending on my child (a few thousand pounds each year). So far I've been able to give more than a quarter of everything I've ever earned.

6 At the time of writing, Giving What We Can members have donated £100 million to effective charities (Giving What We Can, 2019). This is spread across many different charities, so it is impossible to give a simple accounting of the impact. But even just looking at the £6 million in donations to provide malaria nets, this provided more than 3 million person-years of protection, saving more than 2,000 lives (GiveWell, 2019).

7 There is effectively another book-worth of content tucked away in the notes for readers who are eager to know more. If that's you, I'd suggest using a second bookmark to allow you to flip back and forth at will. I've endeavoured to keep the average quality of the notes high to make them worth your time (they are rarely just a bare citation). I've tried to be disciplined in keeping the main text on a straight path to its destination, so the scenic detours are all hidden in the notes. You may also be interested in the appendices, the list of further reading (p. 245), or the book's website, theprecipice.com, for even more information and discussion.

8 Of course even after extensive fact-checking, it would be naïve to think no bias or error has slipped through, so I hope readers will help catch and correct such weaknesses as remain.

1 STANDING AT THE PRECIPICE

1 Sagan (1994), pp. 305–6.

2 Many of the dates in this chapter are only roughly known or apply to transitions that unfolded over a range of years. Rather than littering the main text with ranges of estimates or frequent use of 'around' and 'about', I'll just use the scientific convention of numbers that have been rounded off to reflect their degree of certainty.

There is a wide range of uncertainty about when *Homo sapiens* began. There are remains from 200,000 years ago that are generally considered to be anatomically modern humans. There is an active debate over whether more recent fossil discoveries, dated to around 300,000 years ago, should be classified as *Homo sapiens* (Galway-Witham & Stringer, 2018). More importantly, it is unclear who to count as 'human' or even what we might mean by that. Our genus, *Homo*, is between 2 and 3 million years old, and if we include all of our tool-making ancestors, that would include the Australopithecines, more than 3 million years ago (Antón, Potts

& Aiello, 2014). I focus on fossil evidence rather than molecular phylogeny (which generally gives longer lifespan estimates), as the former methodology is more widely accepted.

3 At the time, the modern continent of Australia was joined with the modern island of New Guinea into a larger continent, sometimes called Sahul. It was separated from Asia by at least 100 miles of open sea—a great voyage for the time (Christian, 2004, p. 191). So when humans first set foot into this new world with its unique animals and plants, they would have actually done so in part of what is now New Guinea.

4 We also adapted the environments to us. Even before the advent of agriculture, we used fire to change the face of continents, creating many of the wild grasslands we now think of as timeless (Kaplan et al., 2016).

5 Figures 1.1 and 1.2 are adapted, with permission, from Christian (2004), pp. 193, 213.

6 *Homo sapiens* and our close relatives may have some unique physical attributes, such as our dextrous hands, upright walking and resonant voices. However, these on their own cannot explain our success. They went together with our intelligence to improve our ability to communicate and to harness and create technology.

7 It may seem naïvely idealistic to talk of all humans across increasingly large scales cooperating together, for the word is sometimes used to suggest working together out of altruistic motivations. However, I'm using it in a wider sense, of humans coordinating their behaviours to achieve things they desire that couldn't be achieved alone. Some of this is altruistic (and even unilateral), but it can also be driven by various forms of exchange.

For an explanation of just how important social learning was to our success see, for example, Henrich (2015).

8 It is, of course, somewhat arbitrary how many to include and how to individuate each one. If I were to count just two, I'd say the Agricultural Revolution and the Industrial Revolution (perhaps defining the latter more broadly to include the time of the Scientific Revolution). If I were to include four, I'd probably break the Agricultural Revolution into two parts: the beginning of farming and the rise of civilisation (which I'd then date to the appearance of cities, about 5,000 years ago).

For a rigorous exploration of the big picture of humanity's development, I recommend *Maps of Time* by David Christian (2004).

9   In some ways the term 'Revolution' is misleading: it was neither rapid nor universal. The transition from foraging to farming unfolded over thousands of years, and it was thousands more years before the rise of cities and the development of writing and other things we think of as characteristic of civilisation. And some peoples continued foraging throughout that period. (Similar remarks could be made regarding the Industrial Revolution, with the timescale sped up by a factor of ten.) But it was rapid by the standards of human development during the prior 200,000 years, and it did usher in an extremely different way of life.

10   Here I am using 'power' in the sense of physics: energy per unit time. Draft animals dramatically increased the harnessable energy per human per day.

11   There is evidence of the emergence of agriculture in the Fertile Crescent (12,000 BP, before present), the Yangtze and Yellow River basins (10,000–6,000 BP), Papua New Guinea (10,000–7,000 BP), Central Mexico (10,000 BP), South America (9,000–5,000 BP), eastern North America (5,000–4,000 BP) and sub-Saharan Africa (4,000–2,000 BP).

12   Foraging typically required about ten square miles of land to support each person (Christian, 2004). So a band had to move frequently to draw sustenance from the hundreds of square miles that sustained it. The required land per person shrank so significantly because so much more of the land's productive capacity was being devoted to human-edible plants.

13   McEvedy & Jones (1978), p. 149.

14   Durand (1960). The Roman Empire reached a similar size shortly after, with most people in the world living in one of these two civilisations.

15   Like agriculture, each was independently developed in multiple places across the world.

16   Other scholars sometimes include this revolution under the name of the Enlightenment, or bundle it together with the Industrial Revolution.

17   Important scholars include Ibn al-Haytham (c. 965–1040 CE), whose use of experimental methods in optics was a major influence

on Robert Grosseteste and Roger Bacon in the thirteenth century. And the roots of the idea of unlimited incremental improvements in our understanding can be seen in Seneca's *Natural Questions*, written in 65 CE (Seneca, 1972) (see p. 49).

18 Francis Bacon's *Novum Organum* (1620) is the canonical exposition of the scientific method, and is a convenient dating for the Scientific Revolution. There is substantial debate over why earlier advances outside Europe did not lead to the sustained knowledge-creation we have seen since the seventeenth century. See, for example, Sivin (1982).

19 Because only a tiny proportion of organisms became fossil fuels, the energy within the entire global supply of fossil fuels is not millions of years' worth of solar energy. It is 'only' about 20 to 200 years' worth of global plant growth, which equals about four to 40 days' worth of all sunlight intercepted by the Earth (author's calculations). Nonetheless, fossil fuels provided vastly more energy than could have been practicably harvested from water wheels or burning timber. Humanity may have been able to eventually reach the modern level of total wealth without fossil fuels, though it is not clear whether the growth in income could have broken free of the growth in population, allowing higher prosperity per capita.

A striking consequence of these numbers is that solar energy has the potential to eventually produce more energy in a year than is contained in all fossil fuels that ever existed.

20 In particular, James Watt's improved design of 1781. Earlier engines were so inefficient that they were only financially viable for a narrow range of tasks. The diesel engine would be another major breakthrough, more than 100 years later.

21 Many other factors are also important, notably political, economic and financial systems.

22 Mummert et al. (2011).

23 Consider, for example, the vivid description of working and living conditions in industrial England by Engels (1892).

24 Van Zanden et al. (2014). See Milanovic (2016) for a book-length study of inequality within and between countries over the last two centuries.

25 Throughout this book, I shall use the word 'history' in its everyday unrestricted sense of everything that has happened (in this case,

to humanity). This is the common dictionary definition and suits the universal scope of this book. Historians, by contrast, typically restrict the term 'history' to refer only to events in times and places where there are written accounts of what happened, thus referring to events in Mesopotamia more than 6,000 years ago or Australia before 1788 CE as 'prehistory'.

26 This is roughly the idea of a Malthusian trap. Note that it is very diffi-cult to compare incomes between different times. The idea of 'subsist-ence' does this by determining how much of your income would be needed to keep you alive. But it does not adequately capture changes in the quality or variety of the food consumed, or in the quality of what you could buy with the modest amount of money remaining, or any other aspect of your quality of life. So it is possible for things to be getting better (or worse) for each person (or society) while staying locked at a subsistence level. All that said, breaking free from this Malthusian dynamic was a very big change to human affairs.

27 These figures for the $2 a day poverty line are from *Our World in Data: Global Extreme Poverty* (Roser & Ortiz-Ospina, 2019a). There is strong improvement at higher poverty lines too. I chose $2 a day not because it is an adequate level of income, but because it shows how almost everyone's income used to be *deeply* inad-equate before the Industrial Revolution. It is not that things are great today, but that they were terrible before.

People in richer countries are sometimes disbelieving of these statistics, on the grounds that they can't see how someone could live at all in their town with just $2 each day. But sadly the statistics are true, and even adjust for the fact that money can go further in poorer countries. The answer is that people below the line have to live with accommodation and food of such a poor standard, that equivalents are not even offered by the market in richer countries.

28 While universal schooling was a major factor, note that the positive trend of improving literacy pre-dates the Industrial Revolution, with literacy already at 50% in the United Kingdom. Significant credit for improvements may also be due to the Scientific Revolution. Literacy figures are from *Our World in Data: Literacy* (Roser & Ortiz-Ospina, 2019b).

29 Note that these historical life expectancy figures are challenging to interpret, as they represent the average (mean) age at death, but

not the typical age at death. There were still people who lived long enough to become grandparents, but there were also a great many who died in infancy or childhood, bringing down the average a lot.

30 Longterm life expectancy figures from Cohen (1989), p. 102. Pre-industrial life expectancy in Iceland from *Gapminder* (2019). Current life expectancy from WHO's *Global Health Observatory* 2016 figures (WHO, 2016).

31 Adapted from Roser (2015).

32 One of the earliest relevant texts is the Code of Hammurabi (eighteenth century BCE). While it is a legal text, it is suggestive of the morals underpinning it, and shows how far our norms have come. See Harari (2014), pp. 117–22, for an insightful discussion.

33 See Pinker's *The Better Angels of Our Nature* (2012) for a wealth of examples. Though note that the evidence for declining violence is weakest when it comes to state-level violence (war and genocide) in the twentieth century.

34 For example, the centuries it took Western Europe to recover from the fall of classical civilisation loom large in most Western historical narratives, but when one takes a global view, looking at China, India, the Islamic world, the Americas and even the Eastern Roman Empire, we see that the overall trend did not reverse nearly so much.

A useful analogy is to the stock market. On a scale of days or months, individual stocks are roughly as likely to go up as to go down. But when we take the stock market as a whole and a time-scale of decades, the upward trend is very clear, and persistent over centuries. Or as Thomas Macaulay put it in 1830: 'A single breaker may recede; but the tide is evidently coming in' (1900, p. 542).

As more and more information comes in, it becomes increasingly clear that to treat talk of progress as off limits in historical analysis would be to bury the lede on the story of human history. If academic historians want to restrict their study and analysis to descriptive matters, that is their choice. But we need not follow them. Some of the most important events in our past have major evaluative and normative consequences that need to be discussed if humanity is to learn from its history.

35 For academics, scepticism can also come from a reluctance to evaluate times at all: because it is often done poorly, because it is

not the job of the historian, or because of philosophical beliefs that it is impossible.

36 Estimating the number of people who have ever lived is made difficult by the fact that we have no population data for the majority of human history. The figure is particularly sensitive to estimates of long-run life expectancy. Haub & Kaneda (2018) give an estimate of 108 billion. Rolling forward older estimates from Goldberg (1983) and Deevey (1960) yields 55 billion and 81 billion respectively (author's calculations). Altogether, 100 billion is a safe central estimate, with a credible range of 50–150 billion.

37 Macaulay (1900), p. 544.

38 Estimates for the average lifespan of mammalian species range from 0.6 million years (Barnosky et al., 2011) to 1.7 million years (Foote & Raup, 1996).

The oldest fossils regarded as *Homo erectus* are the Dmanisi specimens from present-day Georgia, dated to 1.8 million years ago (Lordkipanidze et al., 2006). The most recent fossils are from present-day Indonesia, and have been dated to 0.1 million years ago (Yokoyama et al., 2008).

39 The entire twenty-first century would be just three days in this life—three fateful days for the sake of which humanity's entire life was put at risk.

40 We can be reasonably confident that the runaway and moist greenhouse effects (discussed in Chapter 4) pose an upper bound on how long life can continue to exist on Earth, but we remain uncertain about when they will occur, due to the familiar limitations of our climate models. Wolf & Toon (2015) find a moist greenhouse will occur at around 2 billion years, whereas Leconte et al. (2013) place a lower bound at 1 billion years.

The open question is whether carbon dioxide depletion or temperature increases will render Earth uninhabitable before the runaway or moist greenhouse limits are reached. Rushby et al. (2018) estimate carbon dioxide depletion will occur in around 800 million years for C_3 photosynthesis, and around 500 million years later for C_4 photosynthesis.

Over such long timespans, we cannot ignore the possibility that evolution may lead to new life forms able to exist in climates inhospitable to presently existing life forms. Indeed, the first C_4 plants appeared around 32 million years ago (Kellog, 2013).

41 This quote is attributed to Konstantin Tsiolovsky (Siddiqi, 2010, p. 371).

42 It is difficult to produce a precise date for this. I have chosen the first atomic detonation as that brought with it the possibility of global destruction through igniting the atmosphere (see Chapter 4 for details). One could also choose a later date, when the nuclear arsenals were large enough to make nuclear winter a real possibility. If one broadens the idea from extinction to existential risk (see Chapter 2), then one could perhaps start it a few years earlier with the threat of permanent global totalitarianism that came with the Second World War.

43 This was the US 'Ivy Mike' test of 1952. Its explosion was 10.4 megatons (of TNT-equivalent, the standard unit for explosive yield), compared with approximately 6 megatons for all of World War II (including Hiroshima and Nagasaki) (Pauling, 1962). But it would be several more years before thermonuclear weapons were miniaturised sufficiently to fit on a bomber.

44 Note that many respectable scientists expressed what can now be seen as extreme over-confidence, predicting that humanity would certainly be destroyed within the twentieth century. See Pinker (2018, p. 309).

45 The case is by no means proven. Even now, very little research has been done on what appears to be the greatest mechanism for destruction in a nuclear war.

46 The DEFCON levels were a system of war-readiness levels whose precise meaning changed during the course of the Cold War. Lower numbers meant war was more imminent. The most extreme level ever reached was DEFCON 2, later in the Cuban Missile Crisis, and again during the first Gulf War, in 1991.

47 This forces one to wonder how much more dangerous such a crisis would be if other leaders had been in power at the time. Would our current leaders have been able to navigate to a peaceful resolution?

48 Around 100 of these weapons were operational during the Crisis: the 92 tactical missiles, and 6 to 8 medium-range ballistic missiles. All figures from Norris & Kristensen (2012). The scale of the

Soviet troops also went far beyond expectations: there were not 7,000, but 42,000 stationed in Cuba (Ellsberg, 2017, p. 209).

49 This quote was by Castro at the summit for the fortieth anniversary of the crisis, as reported by Robert McNamara in the documentary *Fog of War* (Morris, 2003). Castro's letter to Khrushchev has since come to light. On the Friday, Castro wrote: 'I believe the imperialists' aggressiveness is extremely dangerous and if they actually carry out the brutal act of invading Cuba in violation of international law and morality, that would be the moment to eliminate such danger forever through an act of clear legitimate defence, however harsh and terrible the solution would be for there is no other.' On the Sunday, just after issuing his statement that the (known) missiles would be removed, Khrushchev replied that 'Naturally, if there's an invasion it will be necessary to repulse it by every means' (Roberts, 2012, pp. 237–8).

50 As Daniel Ellsberg, a military consultant during the missile crisis, puts it: 'The invasion would almost surely trigger a two-sided nuclear exchange that would with near certainty expand to massive U.S. nuclear attacks on the Soviet Union' (Ellsberg, 2017, p. 210).

51 A chief difficulty is that it is unclear what this even means. We can talk clearly of the credences that the principal actors had at the time (10%–50%). And we know whether it happened or not (in this case, not, so 0%). But there is an important sense of probability that is more objective than the former, while not constrained to be 0% or 100% like the latter. For example, we'd like it to be sensitive to later revelations such as the existence of tactical nuclear weapons on Cuba or the events on submarine B-59. We want to know something like: if we had 100 crises like this one, how many of them would lead to nuclear war, but it is hard to interpret the words 'like this one'. I suspect there is a real and useful form of probability here, but I don't think it has yet been properly explicated and we risk confusing ourselves when thinking about it.

52 John F. Kennedy, quoted in Sorenson (1965), p. 705.

53 Ellsberg (2017), p. 199. While not providing a probability estimate, McNamara later said 'I remember leaving the White House at the end of that Saturday. It was a beautiful fall day. And thinking that might well be the last sunset I saw' (Ellsberg, 2017, pp. 200–1).

54 Daniel Ellsberg's estimate in light of all the recent revelations is 'Far greater than one in a hundred, greater that day than Nitze's one in ten' (Ellsberg, 2017, p. 220).

55 I was particularly surprised in January 2018 to see the *Bulletin of the Atomic Scientists* setting their famous Doomsday Clock to '2 minutes to midnight', stating that the world is 'as dangerous as it has been since World War II' (Mecklin, 2018). Their headline reason was the deepening nuclear tensions between the United States and North Korea. But the clock was intended to show how close we are to the end of civilisation, and there was no attempt to show how this bears any threat to human civilisation, nor how we are more at risk than during the Cuban Missile Crisis or other Cold War crises.

56 The United States still has 450 silo-based missiles and hundreds of submarine-based missiles on hair-trigger alert (UCS, n.d.).

57 This is related to the 'pacing problem' considered by those who study the regulation of technology. The pacing problem is that technological innovation is increasingly outpacing the ability of laws and regulations to effectively govern those technologies. As Larry Downes (2009) put it: 'technology changes exponentially, but social, economic, and legal systems change incrementally'.

A key difference between the two framings is that the pacing problem refers to the speed of technological change, rather than to its growing power to change the world.

58 Sagan (1994), pp. 316–17.

59 Barack Obama, Remarks at the Hiroshima Peace Memorial (2016). Consider also the words of John F. Kennedy on the twentieth anniversary of the nuclear chain reaction (just a month after the end of the Cuban Missile Crisis): 'our progress in the use of science has been great, but our progress in ordering our relations small' (Kennedy, 1962).

60 In his quest for peace after the Cuban Missile Crisis, Kennedy (1963) put it so: 'Our problems are manmade—therefore, they can be solved by man. And man can be as big as he wants. No problem of human destiny is beyond human beings.'

Of course one *could* have some human-made problems that have gone past a point of no return, but that is not yet the case for any of those being considered in this book. Indeed, we could prevent them simply through inaction: stopping doing the thing that threatens harm.

61 The problem is especially bad here as the words we use to describe the size of a risk are often affected both by its probability and by the stakes. A 1% chance of losing a game of cards is not a grave risk, but a 1% chance of losing your child may well be. This same issue besets the IPCC approach of using qualitative terms to describe the probability of various climate outcomes (Mastrandrea et al., 2010), which I think is mistaken.

62 There the detailed definition of 'existential risk' can be found on p. 35. It includes the risk of extinction as well as other ways of permanently destroying the potential of humanity, such as an unrecoverable collapse of civilisation.

63 See Chapter 6 for a comparison to the level of natural risk.

64 A point made by Bostrom (2013).

65 There are some ways it could happen, though I don't see them as likely. For example, the risk might top out at about one in six, but we could be lucky enough to survive ten or more such centuries. Or perhaps our efforts to control the risks are only half-successful, lowering the risk back to the twentieth-century level, but no further, and then surviving a hundred or more such centuries.

66 In May 2019 the Anthropocene Working Group of the International Commission on Stratigraphy voted to make the Anthropocene a new epoch with a starting date in the mid-twentieth century. A formal proposal will be made by 2021, which will include a suggested starting point (Subramanian, 2019). In 2017 the Working Group said that markings associated with nuclear arms testing were the most promising proxy (Zalasiewicz et al., 2017).

2 EXISTENTIAL RISK

1 Baier (1981).

2 See Bostrom (2002b; 2013).

3 Bostrom (2013) defined existential risk as 'one that threatens the premature extinction of Earth-originating intelligent life or the permanent and drastic destruction of its potential for desirable future development'. My definition (and clarifications below) are very much in line with the second half of Bostrom's. I didn't echo the first part as it is logically unnecessary (our 'premature extinction'

would itself be a 'permanent and drastic curtailment') and thus draws attention away from the heart of the matter: the destruction of our longterm potential.

Note that on my definitions, an existential risk is simply the risk of an existential catastrophe. I could even have defined it this way directly, but wanted the definition to stand on its own.

4 I'm making a deliberate choice not to define the precise way in which the set of possible futures determines our potential. A simple approach would be to say that the value of our potential is the value of the best future open to us, so that an existential catastrophe occurs when the best remaining future is worth just a small fraction of the best future we could previously reach. Another approach would be to take account of the difficulty of achieving each possible future, for example defining the value of our potential as the expected value of our future assuming we followed the best possible policy. But I leave a resolution of this to future work.

I define existential catastrophes in terms of the destruction of our potential rather than the permanence of the outcome for two key reasons. The first is that it is a more helpful definition for identifying a set of risks with key commonalities in how they work and how we must overcome them. The second reason lies in my optimism about humanity. Given a long enough time with our potential intact, I believe we have a very high chance of fulfilling it: that setbacks won't be permanent unless they destroy our ability to recover. If so, then most of the probability that humanity fails to achieve a great future comes precisely from the destruction of its potential—from existential risk.

5 There are other senses of potential one could discuss, such as a narrower kind of potential that only takes into account what we currently can do, or are likely to be able to do, and thus could be increased by doing work to expand our capabilities. However, in this book I shall only be concerned with our longterm potential (in the sense described in the main text). When I simply say 'potential' (for brevity) it should be treated as 'longterm potential'.

6 By leaving open that there may be a remote chance of recovery, some of my more blunt claims are not literally true. For instance, that existential catastrophe involves 'no way back'. This is unfortunate, but I think it is a cost worth bearing.

The point of leaving open remote chances of recovery is to avoid responses that there is always some chance—perhaps incredibly small—that things recover. So that on a more strict reading of potential, it could never be *completely* destroyed. But taking this extremely strict reading wouldn't be useful. It doesn't really matter for our decision-making whether a scenario would have zero chance of recovery, or merely a 0.1% chance of recovery. Both cases are almost equally bad compared to the current world, and are bad for the same reason: that they are extremely hard to reverse, (almost completely) destroying our longterm potential. Moreover, they warrant similar diagnoses about what we should do, such as needing proactive work, rather than learning from trial and error. A possibility with literally 'no way back' should be avoided for the same reasons as one with 'practically no way back'. Thus it is most useful to include nearly inescapable situations in the definition as well as completely inescapable ones.

7   If our potential greatly exceeds the current state of civilisation, then something that simply locks in the current state would count as an existential catastrophe. An example would be an irrevocable relinquishment of further technological progress.

It may seem strange to call something a catastrophe due to merely being far short of optimal. This is because we usually associate events that destroy potential with those that bring immediate suffering and rarely think of events that could destroy one's potential while leaving one's current value intact. But consider, say, a choice by parents not to educate their child. There is no immediate suffering, yet catastrophic longterm outcomes for the child may have been locked in.

8   In some cases, the ideas and methodology of this book can be applied to these local 'existential threats', as they have a somewhat similar character (in miniature).

9   This is not without its own issues. For instance, we shall sometimes have to say that something used to be a risk (given our past knowledge), but is no longer one. One example would be the possibility that nuclear weapons would ignite the atmosphere (see p. 90). But note that this comes up for all kinds of risk, such as thinking last week that there was a risk of the elevator falling due to a frayed cable, but having this risk drop to zero given our current

knowledge (we inspected the cable and found it to be fine). For more on objective versus evidential probability in defining existential risk see Bostrom & Ćirković (2008) and Bostrom (2013).

10 One often sees lists containing many historical civilisations that have collapsed, such as the Roman Empire or the Mayans. But this is not what I'm talking about in this book when I speak of the (global) collapse of civilisation. The statistics of these smaller collapses have little bearing on whether global civilisation will collapse.

   The particular civilisations that collapsed were highly localised, and were more akin to the collapse of a single country than that of global civilisation. For example, even the Roman Empire at its height was much smaller than Brazil is now in both land area and population. These small civilisations were much more prone to regional climatic effects, a single bad government and attacks from without. Furthermore, the collapses were much less deep than what I'm considering: often entire cities and towns survived the 'collapse', the people weren't reduced to a pre-agricultural way of life, and many aspects of the culture continued.

11 Or some more direct way of preventing civilisation or agriculture, such as extreme environmental damage or a continuing debilitating disease.

   It is possible that due to some form of increased fragility, the world is less resistant to collapse than medieval Europe was, and thus that a loss of less than 50% of the population could cause this. I'm sceptical though, and find it just as likely that even a 90% loss may not cause the complete loss of civilisation.

12 From the perspective of recent history, these agricultural revolutions began far apart in time, the thousands of years' head start for some civilisations playing a major role in their subsequent influence on the world stage. But from the broader perspective, these independent developments of agriculture occurred at remarkably similar times: just a few thousand years apart in a story spanning *hundreds* of thousands of years. This suggests that agriculture was not an unlikely technological breakthrough, but a fairly typical response to a common cause. The most likely trigger was the end of the great 'ice age' that lifted between 17,000 and 10,000 years ago, just as agriculture began. This had dramatic effects on the

environment, making the world less suitable for hunting and more suitable for farming.

13 Overall the trend is towards resources becoming harder to access, since we access the easy ones first. This is true for untouched resources in the ground. But this leads people to neglect the vast amount of resources that are already in the process of being extracted, that are being stored, and that are in the ruins of civilisation. For example, there is a single open-cut coal mine in Wyoming that produces 100 million tons of coal each year and has 1.7 billion tons left (Peabody Energy, 2018). At the time of writing, coal power plants in the US hold 100 million tons of ready-to-use coal in reserve (EIA, 2019). There are about 2 billion barrels of oil in strategic reserves (IEA, 2018, p. 19), and our global civilisation contains about 2,000 kg of iron in use per person (Sverdrup & Olafsdottir, 2019).

14 Though we shall see in Chapter 4 that even extreme nuclear winter or climate change would be unlikely to produce enough environmental damage in every part of the world to do this.

15 The question of minimal viable population also comes up when considering multi-generational space travel. Marin & Beluffi (2018) find a starting population of 98 to be adequate, whereas Smith (2014) argues for a much higher minimum of between 14,000 and 44,000. It might be possible for even smaller populations to survive, depending on the genetic technologies available to minimise risks of inbreeding and genetic drift.

16 I believe this has also led people to think of the possibility of human extinction as trite, rather than profound. Its use as a narrative device in such films is indeed trite. But it would be a great mistake to let that judgement about its use in fiction cloud our understanding of its import for our future.

17 Nor do we often encounter serious emotional explorations, such as Bob Dylan's 'Let Me Die in My Footsteps' (1963) or Barry McGuire's 'Eve of Destruction' (1965).

18 See Slovic (2007).

19 Parfit (1984), pp. 453–4.

20 This qualitative difference is also what makes the difference according to the views we'll see later, regarding our past, virtue and cosmic significance.

21 Schell (1982), p. 182.

22 This warrants some elaboration. Following Parfit (1984), we can think of what would be lost were we to go extinct in two parts.

First is the loss of what we could *be*. The loss of each and every person who could have lived. The children and grandchildren we would never have: millions of generations of humanity, each comprised of billions of people, with lives of a quality far surpassing our own. Gone. A catastrophe would not *kill* these people, but it would foreclose their very existence. It would not *erase* them, but ensure they were never even written. We would lose the value of everything that would make each of these lives good—be it their happiness, freedom, success or virtue. We would lose the very persons themselves. And we would lose any value residing in the relationships between people or in the fabric of their society—their love, camaraderie, harmony, equality and justice.

Second is the loss of what we could *do*. Consider our greatest achievements in the arts and the sciences, and how many of them have been reached in just the last few centuries. If we make it through the next few centuries with our potential intact, we will likely produce greater heights than any we've seen so far. We may reach one of the very peaks of science: the complete description of the fundamental laws governing reality. And we will continue expanding the *breadth* of our progress, reaching new provinces yet to be explored.

Perhaps the most important are potential moral achievements. While we have made substantial progress over the centuries and millennia, it has been much slower than in other domains and more faltering. Humanity contains the potential to forge a truly just world, and realising this dream would be a profound achievement.

There is so much that we could be and do, such a variety of flourishing and achievement ahead, that most conceptions of value will find something to mourn should we fail, should we squander this potential. And because this flourishing and achievement is on such a grand scale, the safeguarding of our potential is of the greatest importance.

23 So the scale of our future is not just important in consequentialist terms. It also fuels arguments for reducing existential risk that are rooted in considerations of fairness or justice.

24 By 'matter just as much' I mean that each good or bad thing in their life matters equally regardless of when they live.

On average people's lives today are better than people's lives a thousand years ago because they contain more good things, and may be more instrumentally important too, because we live at a more pivotal time. So in these other senses, our lives may matter more now, but these other senses are compatible with the kind of temporal neutrality I endorse.

25 This has been suggested by J. J. C. Smart (1984, pp. 64–5) and G. E. Moore (1903, § 93).

26 New generations will have new risks that they can help reduce, but only we can reduce the risks being posed now and in coming decades.

27 The name was coined by William MacAskill and myself. The ideas build on those of our colleagues Nick Beckstead (2013) and Nick Bostrom (2002b, 2003). MacAskill is currently working on a major book exploring these ideas.

28 We will see in Appendix E that as well as safeguarding humanity, there are other general ways our acts could have a sustained influence on the longterm future.

29 On a discount rate of 0.1% per annum (low by economists' standards), the intervening million years make suffering in one million years more than 10^{434} times as important as the same amount of suffering in two million years.

30 One could cash this out in different ways depending on one's theory of value. For some it may literally involve the badness of the death of the group agent or species, humanity. For others it will be the absence of human lives in the future and everything good about them.

31 There are serious challenges in doing so if the other time (or place) involves a very different culture, but that is not relevant to this example.

32 Imagine what we'd think if we found out that our government ignored the risk of nuclear war on the grounds that if we were all dead, that couldn't be bad.

33 And even if after that scepticism we still leaned towards such a theory, we should remain very cautious about following its advice regarding the particular area where it most deviates from our

intuition and from the other theories we find plausible—on the value of our longterm future. See Beckstead (2013, p. 63).

34　Indeed, we'd be stuck in almost exactly the same condition as our earliest human ancestors (and there would be far fewer of us).

35　Burke (1790), para. 165. In her seminal work on the rights of future persons, Annette Baier (1981) makes a related point: 'The crucial role we fill, as moral beings, is as members of a cross-generational community, a community of beings who look before and after, who interpret the past in light of the present, who see the future as growing out of the past, who see themselves as members of enduring families, nations, cultures, traditions.' As does John Rawls in *A Theory of Justice* (1971, § 79): 'The realizations of the powers of human individuals living at any one time takes the cooperation of many generations (or even societies) over a long period of time.'

36　Seneca (1972), pp. 279–91. Sixteen centuries later, in 1704, Isaac Newton made a similar remark (Newton & McGuire, 1970): 'To explain all nature is too difficult a task for any one man or even for any one age. 'Tis much better to do a little with certainty, & leave the rest for others that come after you . . .'

In 1755, Denis Diderot expressed related ideas in his *Encyclopédie* (Diderot, 1755, pp. 635–48A): '. . . the purpose of an encyclopedia is to collect knowledge disseminated around the globe; to set forth its general system to the men with whom we live, and transmit it to those who will come after us, so that the work of preceding centuries will not become useless to the centuries to come; and so that our offspring, becoming better instructed, will at the same time become more virtuous and happy, and that we should not die without having rendered a service to the human race.'

37　Perhaps even more astounding is that some of the mysteries of comets whose depth inspired Seneca to write this passage have only recently been revealed—and contributed directly to our understanding of existential risk: 'Some day there will be a man who will show in what regions comets have their orbit, why they travel so remote from other celestial bodies, how large they are and what sort they are' (Seneca, 1972, p. 281).

The nature of their highly eccentric orbits and their size have been key aspects in our current understanding of the risks comets pose to civilisation and humanity. Further understanding of both these features would be among the most useful progress in reducing the risk posed by impacts from space. See Chapter 3.

38 See Scheffler (2018) for interesting additional discussion of reciprocity-based reasons for concern about future generations, and other potential considerations not covered here.

39 As Sagan (1983, p. 275) put it: 'There are many other possible measures of the potential loss—including culture and science, the evolutionary history of the planet, and the significance of the lives of all of our ancestors who contributed to the future of their descendants. Extinction is the undoing of the human enterprise.'

40 See, for example Cohen (2011), Scheffler (2009), Frick (2017).

41 Nick Bostrom (2013) expanded upon this idea: 'We might also have custodial duties to preserve the inheritance of humanity passed on to us by our ancestors and convey it safely to our descendants. We do not want to be the failing link in the chain of generations, and we ought not to delete or abandon the great epic of human civilisation that humankind has been working on for thousands of years, when it is clear that the narrative is far from having reached a natural terminus.'

42 Stewart Brand (2000) has spoken eloquently of this civilisational patience: 'Ecological problems were thought unsolvable because they could not be solved in a year or two . . . It turns out that environmental problems are solvable. It's just that it takes focused effort over a decade or three to move toward solutions, and the solutions sometimes take centuries. Environmentalism teaches patience. Patience, I believe, is a core competency of a healthy civilization.'

43 While it takes us a bit further afield, we might also consider civilisational virtues related to our relationships with the wider world. For example, mistreatment of our fellow animals and our environment suggests deficiencies in our compassion and stewardship.

And we could also consider how safeguarding our future can be motivated by virtues for individuals such as gratitude (to past generations), compassion and fairness (towards future generations), and unity or solidarity towards the rest of humanity. Jonathan Schell (1982, pp. 174–5) considers love in the sense of a generalisation of parental or procreative love: the love with which we bring others into the world.

44 I've been lucky enough to get to work on this question with my colleagues at the Future of Humanity Institute: Anders Sandberg and Eric Drexler. In our paper, 'Dissolving the Fermi Paradox' (Sandberg, Drexler & Ord, 2018) we quantified the current scientific understanding and uncertainties around the origin of life and intelligence. And we showed that it is a mistake to think that since there are billions of billions of stars there must be alien intelligence out there. For it is entirely plausible (and perhaps even likely) that the chance of life starting on any of them is correspondingly tiny. Our scientific knowledge is just as compatible with being alone as with being in a universe teeming with life. And given this, the lack of any sign of intelligent alien life is not in any way surprising or paradoxical—there is no need to invoke outlandish proposals to explain this; the evidence simply suggests that we are more likely to be alone.

We suggest that previous researchers had been led astray by using 'point estimates' for all the quantities in the Drake equation. When these are replaced by statistical distributions of plausible values, we see that even if the mean or median number of alien civilisations is high, there is also a large chance of none. And we update towards this possibility when we see no sign of their activity.

45 This cosmic significance might be thought of as a way of illuminating the moral value that humanity has for other reasons, or as an additional source of value, or as something important that goes beyond value. Here are some of the leading proponents of our cosmic significance in their own words:

Martin Rees (2003, p. 157): 'The odds could be so heavily stacked against the emergence (and survival) of complex life that Earth is the unique abode of conscious intelligence in our entire Galaxy. Our fate would then have truly cosmic resonance.'

Max Tegmark (2014, p. 397): 'It was the cosmic vastness that made me feel insignificant to start with. Yet those grand galaxies are visible and beautiful to us—and only us. It's only we who give them any meaning, making our small planet the most significant place in our entire observable Universe.'

Carl Sagan (1980, p. 370): 'The Cosmos may be densely populated with intelligent beings. But the Darwinian lesson is clear: There will be no humans elsewhere. Only here. Only on this small planet. We are a rare as well as an endangered species. Every one of us is, in the cosmic perspective, precious.'

Derek Parfit (2017b, p. 437): 'If we are the only rational beings in the Universe, as some recent evidence suggests, it matters even more whether we shall have descendants or successors during the billions of years in which that would be possible. Some of our successors might live lives and create worlds that, though failing to justify past suffering, would have given us all, including those who suffered most, reasons to be glad that the Universe exists.'

James Lovelock (2019, p. 130): 'Then, with the appearance of humans, just 300,000 years ago, this planet, alone in the cosmos, attained the capacity to know itself . . . We are now preparing to hand the gift of knowing on to new forms of intelligent beings. Do not be depressed by this. We have played our part . . . perhaps, we can hope that our contribution will not be entirely forgotten as wisdom and understanding spread outward from the earth to embrace the cosmos.'

One way of understanding cosmic significance in consequentialist terms is to note that the more rare intelligence is, the larger the part of the universe that will be lifeless unless we survive and do something about it—the larger the difference *we* can make.

46 Eventually, we may become significant even in terms of raw scale. Cosmologists believe that the largest coherent structures in the universe are on the scale of about a billion light years across, the width of the largest voids in the cosmic web. With the accelerating expansion of the universe tearing things apart, and only gravity to work with, lifeless matter is unable to organise itself into any larger scales.

However, there is no known physical limit preventing humanity from forming coherent structures or patterns at much larger scales—up to a diameter of about 30 billion light years. We might thus create the largest structures in the universe and be unique even in these terms. By stewarding the galaxies in this region, harvesting and storing their energy, we may also be able to create the most energetic events in the universe or the longest-lasting complex structures.

47 As explained earlier in this chapter, it is not that only humans matter, but that humans are the only moral agents.

48 I think this is a very valuable perspective, which will yield insights that reach far beyond what I am able to explore in this book. I hope that others will adopt it and take it much further than I have been able to.

49 The theory of how to make decisions when we are uncertain about the moral value of outcomes was almost completely neglected in moral philosophy until very recently—despite the fact that it is precisely our uncertainty about moral matters that leads people to ask for moral advice and, indeed, to do research on moral philosophy at all.

Remedying this situation has been one of the major themes of my work so far (Greaves & Ord, 2017; MacAskill & Ord, 2018; MacAskill, Bykvist & Ord, forthcoming).

50 Nick Bostrom (2013, p. 24) put this especially well: 'Our present understanding of axiology might well be confused. We may not now know—at least not in concrete detail—what outcomes would count as a big win for humanity; we might not even yet be able to imagine the best ends of our journey. If we are indeed profoundly uncertain about our ultimate aims, then we should recognise that there is a great option value in preserving—and ideally improving—our ability to recognise value and to steer the future accordingly. Ensuring that there will be a future version of humanity with great powers and a propensity to use them wisely is plausibly the best way available to us to increase the probability that the future will contain a lot of value. To do this, we must prevent any existential catastrophe.'

I think the condition that you find it plausible is important. I'm not suggesting that this argument from uncertainty works even if you are extremely confident that there are no duties to protect our future. It might be possible to make such an argument based on expected value, but I am wary of expected value arguments when the probabilities are extremely small and poorly understood (see Bostrom, 2009).

51 Even if one were committed to the bleak view that the only things of value were of negative value, that could still give reason to

continue on, as humanity might be able to prevent things of negative value elsewhere on the Earth or in other parts of the cosmos where life has arisen.

52   Another way of saying this is that protecting our future has immense *option value*. It is the path that preserves our ability to choose whatever turns out to be best when new information comes in. This new information itself also ends up being extremely valuable: whether it is empirical information about what our future will be like or information about which putative moral considerations stand the test of time. Of course, it only holds this option value to the degree to which we expect our future to be responsive to new information about what is morally best.

    For more on moral option value and the value of moral information from the perspective of humanity see Bostrom (2013, p. 24), MacAskill (2014) and also Williams (2015), who generalises this idea: '. . . we should regard intellectual progress, of the sort that will allow us to find and correct our moral mistakes as soon as possible, as an urgent moral priority rather than as a mere luxury; and we should also consider it important to save resources and cultivate flexibility, so that when the time comes to change our policies we will be able to do so quickly and smoothly.'

53   These ideas are beautifully expressed by Sagan (1994).

54   The 2019 budget was $1.4 million (BWC ISU, 2019). Between 2016 and 2018, McDonald's company-operated restaurants incurred an average of $2.8 million expenses per restaurant per year (McDonald's Corporation, 2018, pp. 14, 20). The company does not report costs for its franchised restaurants.

55   Farquhar (2017) estimated global spending on reducing existential risk from AI at $9 million in 2017. There has been substantial growth in the field since then, perhaps by a factor of 2 or 3. I'm confident that spending in 2020 is between $10 and $50 million.

    IDC (2019) estimate global AI spending will reach $36 billion in 2019, a significant proportion of which will be devoted to improving AI capabilities.

56   The global ice cream market was estimated at $60 billion in 2018 (IMARC Group, 2019), or ~0.07% of gross world product (World Bank, 2019a).

    Precisely determining how much we spend on safeguarding our future is not straightforward. I am interested in the simplest understanding of this: spending that is aimed at reducing existential risk. With that understanding, I estimate that global spending is on the order of $100 million.

Climate change is a good example of the challenges in determining the sort of spending we care about. One estimate of global spending on climate change is around $400 billion (~0.5% of gross world product) (Buchner et al., 2017). This is likely an overestimate of the economic cost, since most of the total is spending on renewable energy generation, much of which would otherwise have been spent on building non-renewable capacity. Moreover, as we will see in Chapter 4, most existential risk from climate change comes from the most extreme warming scenarios. It is not clear, therefore, how much of the average dollar towards climate-change mitigation goes towards reducing existential risk. Risks from engineered pandemics present similar challenges—US federal funding on biosecurity totals $1.6 billion, but only a small proportion of this will be aimed at the very worst pandemics (Watson et al., 2018).

Setting aside climate change, *all* spending on biosecurity, natural risks and risks from AI and nuclear war is still substantially less than we spend on ice cream. And I'm confident that the spending actually focused on existential risk is less than one-tenth of this.

57 Robock, Oman & Stenchikov (2007) and Coupe et al. (2019).

58 King et al. (2015). As we will see in Chapter 4, warming of 6 °C is quite plausible given our current scientific understanding.

59 These features are known as non-excludability (the provider can't limit the benefit to those who pay) and non-rivalry (an individual's consumption of the good doesn't limit anyone else's). As one can see, most goods and services offered by the market are both excludable and rivalrous.

60 See Bostrom (2013, p. 26). Building on the idea of the tragedy of the commons, the economist Jonathan Wiener (2016) has called this situation 'the tragedy of the uncommons'.

61 There are some areas of longterm thinking, such as energy policy, pensions and large infrastructure projects. These typically involve thinking on a timescale of one or two decades, however, which is still quite short term by the standards of this book.

62 For more on heuristics and biases in general, see Kahneman (2011). See Wiener (2016) for a detailed discussion on these and other biases affecting public judgement of rare catastrophic risks.

63 Indeed, we sometimes suffer from a version of this effect called *mass numbing*, in which we are unable to conceptualise harms affecting thousands or more people and treat them as even *less* important than the same harm to a single identifiable person. See Slovic (2007).

64 Here I'm setting aside religious discussions of the end times, which I take to be very different from discussion of naturalistic causes of the end of humanity.

65 Russell (1945). Midway through the essay, Russell notes he has just learned of the bombing of Nagasaki, most likely on the morning of 9 August 1945: 'As I write, I learn that a second bomb has been dropped on Nagasaki. The prospect for the human race is sombre beyond all precedent. Mankind are faced with a clear-cut alternative: either we shall all perish, or we shall have to acquire some slight degree of common sense. A great deal of new political thinking will be necessary if utter disaster is to be averted.'

66 *The Bulletin* exists to this day, and has long been a focal point for discussions about extinction risk. In recent years, they have broadened their focus from nuclear risk to a wider range of threats to humanity's future, including climate change, bioweapons and unaligned artificial intelligence.

67 The 1955 Russell–Einstein Manifesto states (Russell, 2002): 'Here, then, is the problem which we present to you, stark and dreadful and inescapable: Shall we put an end to the human race; or shall mankind renounce war? . . . there lies before you the risk of universal death.'

Signing the manifesto was one of Einstein's last acts before his death in 1955, as described by Russell (2009, p. 547): 'I had, of course, sent the statement to Einstein for his approval, but had not yet heard what he thought of it and whether he would be willing to sign it. As we flew from Rome to Paris, where the World Government Association were to hold further meetings, the pilot announced the news of Einstein's death. I felt shattered, not only for the obvious reasons, but because I saw my plan falling through without his support. But, on my arrival at my Paris hotel, I found a letter from him agreeing to sign. This was one of the last acts of his public life.'

68 In a private letter, Eisenhower (1956) contemplated this possibility and its consequences for grand strategy: 'When we get to the point, as we one day will, that both sides know that in any outbreak of general hostilities, regardless of the element of surprise, destruction will be both reciprocal and complete, possibly we will have sense enough to meet at the conference table with the understanding that the era of armaments has ended and the human race must conform its actions to this truth or die.'

In a speech to the United Nations, Kennedy (1961) said: 'For a nuclear disaster, spread by winds and waters and fear, could well engulf the great and the small, the rich and the poor, the committed and the uncommitted alike. Mankind must put an end to war— or war will put an end to mankind . . . Today, every inhabitant of this planet must contemplate the day when this planet may no longer be habitable. Every man, woman and child lives under a nuclear sword of Damocles, hanging by the slenderest of threads, capable of being cut at any moment by accident, or miscalculation, or by madness. The weapons of war must be abolished before they abolish us.'

And Brezhnev suggested that 'Mankind would be wholly destroyed' (Arnett, 1979, p. 131).

69 Schell (1982) was the first to publish, provoked by the new scientific theory that nuclear weapons could destroy the ozone layer, which might make life impossible for humans. This theory was soon found wanting, but that did not affect the quality of Schell's philosophical analysis about how bad extinction would be (analysis that was especially impressive as he was not a philosopher). Sagan (1983) was compelled to think deeply about extinction after his early results on the possibility of nuclear winter. Parfit's magnum opus, *Reasons and Persons* (1984), ended with his crisp analysis of the badness of extinction, which went on to have great influence in academic philosophy. Sagan cited Schell's work, and Parfit was probably influenced by it.

In that same year, *The Imperative of Responsibility* by Hans Jonas (1984) was released in an English translation. Originally written in 1979, it also raised many of the key questions concerning our ethical duties to maintain a world for future generations.

70 In 1985, Reagan said (Reagan & Weinraub, 1985): 'A great many reputable scientists are telling us that such a war could just end up in no victory for anyone because we would wipe out the earth as we know it. And if you think back to a couple of natural calamities . . . there was snow in July in many temperate countries. And they called it the year in which there was no summer. Now if one volcano can do that, what are we talking about with the whole nuclear exchange, the nuclear winter that scientists have been talking about?'

Speaking in 2000, Mikhail Gorbachev reflected (Gorbachev & Hertsgaard, 2000): 'Models made by Russian and American scientists showed that a nuclear war would result in a nuclear winter that would be extremely destructive to all life on earth; the knowledge of that was a great stimulus to us.'

71 The crowd size has been estimated from 600,000 to 1 million, with 1 million being the most common number reported (Montgomery, 1982; Schell, 2007). There have since been even larger protests on other issues.

3 NATURAL RISKS

1 As recollected by Thomas Medwin (1824).

2 Estimates of the impact energy, the crater dimensions and ejecta are from Collins, Melosh & Marcus (2005). Other details from Schulte et al. (2010).

3 Schulte et al. (2010); Barnosky et al. (2011). One could argue that the reign of the dinosaurs continues through their descendants, the birds.

4 The largest is Ceres, at 945 kilometres. Asteroids actually range in size all the way down to fine dust, but are typically called 'meteoroids' when too small to be observed with our telescopes.

5 The largest known comet is Hale-Bopp, at roughly 60 kilometres, though there may be larger ones that are currently too far away to detect. Astronomers have had difficulty detecting comets smaller than a few hundred metres, suggesting that they may not survive very long at these sizes.

6 Many people must have witnessed them falling to the ground, and the use of black metal of celestial origin appears in several myths.

Indeed, the earliest known iron artefacts are a set of small beads made from meteoric iron 5,200 years ago, before iron ore could be smelted. However, it is only 200 years ago that their origin was established to scientific standards (consider how many other phenomena with eyewitness accounts failed to withstand scrutiny).

7   Strictly speaking their paper gets its key estimate of 10 km (±4) by averaging four methods, of which the iridium method is just one, and suggested a slightly smaller size of 6.6 km (Alvarez et al., 1980).

8   The 'impact winter' hypothesis was introduced by Alvarez et al. (1980) in their initial paper. It was finally confirmed by Vellekoop et al. (2014). The suggestion that impact-generated sulphates were responsible for this effect traces back to the 1990s (Pope et al., 1997).

9   As can be seen from the chapter's epigram, Lord Byron had thought of the threat from comets and even the possibility of planetary defence as early as 1822. The threat from comets began to be discussed in more depth at the turn of the nineteenth century, most famously in H. G. Wells' 'The Star' (1897), but also in George Griffith's 'The Great Crellin Comet' (1897) which featured the Earth being saved by an international project to deflect the comet. Bulfin (2015) contains detailed information on these and other Victorian explorations of the ways humanity could be destroyed. There was also non-impact-related concern in 1910 when it was suggested that the tail of Halley's comet might contain gases that could poison our atmosphere (Bartholomew & Radford, 2011, ch. 16).

The threat from asteroids was first recognised in 1941 (Watson, 1941). In 1959 Isaac Asimov urged the eventual creation of a space programme for detecting and eliminating such threats (Asimov, 1959).

10 Media interest was fuelled by the 1980 Alvarez hypothesis, the 1989 near miss with the asteroid 4581 Asclepius and the 1994 collision of the comet Shoemaker-Levy 9, which left a visible mark on Jupiter comparable in size to the entire Earth.

11 This goal was achieved in 2011, for a total cost of less than $70 million (Mainzer et al., 2011; U.S. House of Representatives, 2013).

12 It is often reported that asteroids are a hundred times as common as comets, suggesting that they make up the overwhelming bulk of

the risk. At one level this is true. At the time of writing, 176 near-Earth comets had been identified, compared with 20,000 asteroids (JPL, 2019b). But while comets are 100 times less frequent, they are often larger, so of the 1–10 km NEOs (near-Earth Objects), comets are only 20 times less frequent. Of the NEOs greater than 10 km, four are asteroids and four are comets. Thus for the purposes of existential risk, the background risk from comets may not be that different from that of asteroids.

13   Note that the mass, and thus the destructive energy, of an asteroid is proportional to the cube of the diameter, such that a 1 km asteroid has only a thousandth the energy of a 10 km asteroid. They can also vary in terms of their density and their speed relative to the Earth—for a given size, a denser or faster asteroid has more kinetic energy and is thus more dangerous.

14   Longrich, Scriberas & Wills (2016).

15   When I wrote the first draft of this chapter, most of the risk from the tracked asteroids was in the 2 km asteroid '2010 GZ$_{60}$'. At the time of writing, the chance of a collision over the next century was put at the low but non-negligible level of one in 200,000. Happily, we now know this will miss the Earth. The most risky tracked asteroid is now the 1.3 km asteroid '2010 GD$_{37}$', whose impact probability in the next century is a mere one in 120,000,000 (JPL, 2019b).

16   The main uncertainty about whether we have found them all stems from those asteroids with orbits close to 1 AU (the distance between the Earth and the Sun) and a period of close to one year, making them undetectable for many years at a time. Fortunately, it is very unlikely that there is such an asteroid. And if there were, it would become visible years before a potential impact (Alan Harris, personal communication).

17   The risk from asteroids between 1 and 10 km in size is even lower than this probability suggests, as the 5% of these remaining untracked are disproportionately at the small end of the scale (the vast majority were near the small end of this range to begin with, and our detection methods have been better at finding the bigger ones).

18   Hazard descriptions are from Alan Harris. Estimates of impact probabilities are from Stokes et al. (2017, p. 25).

    The total number of 1–10 km near-Earth asteroids has most recently been estimated at 921 ± 20 (Tate, 2017). As of April 2019,

895 have been discovered: 95–99% of the total (JPL, 2019a). In order to be conservative, I take the lower bound.

Four near-Earth asteroids over 10 km have been identified (JPL, 2019a): 433 Eros (1898 DQ); 1036 Ganymed (1924 TD); 3552 Don Quixote (1983 SA); 4954 Eric (1990 SQ). NASA (2011) believes this is all of them.

19 The International Asteroid Warning Network was established in 2014 on the recommendation of the UN. The international Spaceguard Foundation was founded in 1996 (UNOOSA, 2018).

20 In 2010 annual funding was $4 million. This was increased to $50 million in 2016, and is understood have remained at similar levels since (Keeter, 2017).

21 Unfortunately comets can be much more difficult to characterise and to divert. Short-period comets (those with orbits of less than 200 years) present some novel problems: they are subject to forces other than gravity, making their trajectories harder to predict, and it is more difficult to rendezvous with them. Things get substantially worse with long-period comets since they are so far from the Earth. We understand neither their overall population, nor the detailed trajectories of those that might threaten us (if they pose a threat next century, it would be on their very first observed approach towards us). Moreover, they would be extremely difficult to deflect as we would have less than a year from first detection (at around the orbit of Jupiter) to the time of impact (Stokes et al., 2017, p. 14).

22 Asteroids are about twenty times as frequent in the 1–10 km size category, but tracking those asteroids has reduced the risk by that same multiple. And comets and asteroids are about equally frequent in the 10 km+ size category (JPL, 2019b).

23 Deflection would be very expensive, but the costs would only need to be paid if a large asteroid on a collision course for Earth were discovered. In such a situation, the people of the Earth would be willing to pay extremely high costs, so it would be more a question of what we can achieve in the time available than one of price. In contrast, the costs of detection and tracking need to be paid whether or not there really is a dangerous asteroid, so even though they are much lower in dollar terms, they may be the greater part of the expected cost, and the part that is harder to get funded.

24 National Research Council (2010), p. 4.

25 See Sagan & Ostro (1994) for an early discussion, and Drmola & Mareš (2015) for a recent survey.

26 One reason it is unlikely is that several of the deflection methods (such as nuclear explosions) are powerful enough to knock the asteroid off course, but not refined enough to target a particular country with it. For this reason, these might be the best methods to pursue.

27 Eruptions are measured using two scales. The volcanic explosivity index (VEI) is an ordinal scale classifying eruptions in terms of their ejecta volume. The magnitude scale is a logarithmic scale of eruption mass, given by $M = \log_{10}$ [erupted mass in kg] $- 7$. The magnitude scale is generally preferred by scientists, due to practical problems estimating eruption volumes, and the usefulness of a continuous scale in analysing relationships between magnitude and other parameters. All VEI 8 eruptions with a deposit density of greater than around 1,000 kg/m^3 (most of them) will have magnitudes of 8 or more.

There is no sharp line between supervolcanic eruptions and regular eruptions. Supervolcanic eruptions are those with VEI 8—ejecta volume greater than 1,000 km^3. It is not clear whether flood basalts should count as supervolcanoes, and they have generally been considered separately. See Mason, Pyle, & Oppenheimer (2004) for a discussion of the scales.

28 Not all calderas are the result of supereruptions, however. For example, Kilauea in Hawaii has a caldera that was produced by lava flows, rather than from an explosive eruption.

29 This was its last supervolcanic eruption. It had a lesser eruption 176,000 years ago (Crosweller et al., 2012).

30 There is significant uncertainty about the magnitude of global cooling, with estimates ranging from 0.8 °C to 18 °C. The key driver of climatic effects is the amount of sulphate injected into the upper atmosphere, estimates of which vary by several orders of magnitude. This is usually expressed as a multiple of the Pinatubo eruption in 1991, for which we have accurate measurements.

Early research (Rampino & Self, 1992) found cooling of 3–5 °C, with sulphate levels of 38x Pinatubo. More recently, Robock et al. (2009) use a central estimate of 300x Pinatubo, and find cooling of up to 14 °C. Recent work by Chesner & Luhr (2010) suggests a sulphate yield of 2–23x Pinatubo—considerably less than early numbers. Yost et al. (2018) offer an extensive review of estimates and methodology, arguing that estimates from Chesner & Luhr (2010) are more robust, and calculating an implied global cooling of 1–2 °C. More research is needed to better constrain these estimates.

31 Raible et al., 2016. This inspired Byron to write his poem 'Darkness', which begins: 'I had a dream, which was not all a dream. / The bright sun was extinguish'd, and the stars / Did wander darkling in the eternal space, / Rayless, and pathless, and the icy earth / Swung blind and blackening in the moonless air; / Morn came and went – and came, and brought no day . . .'

The year without summer also inspired Mary Shelley to write *Frankenstein*, while travelling with Byron and Percy Shelley. In her introduction to the 1831 edition she describes how, forced indoors by the 'wet, ungenial summer and incessant rain', the group entertained themselves by telling ghost stories, one of which became *Frankenstein* (Shelley, 2009).

32 This is known as the 'Toba catastrophe hypothesis' and was popularised by Ambrose (1998). Williams (2012) argues that imprecision in our current archaeological, genetic and paleoclimatological techniques makes it difficult to establish or falsify the hypothesis. See Yost et al. (2018) for a critical review of the evidence. One key uncertainty is that genetic bottlenecks could be caused by founder effects related to population dispersal, as opposed to dramatic population declines.

33 Direct extinction, that is. Such an event would certainly be a major stressor, creating risk of subsequent war. See Chapter 6 on risk factors.

The Toba eruption had a magnitude of 9.1, and is the largest eruption in the geological record (Crosweller et al., 2012). On a uniform prior, it is unlikely (4%) that the largest eruption for 2 million years would have occurred so recently. This raises the possibility that the record is incomplete, or that the estimate of Toba's magnitude is inflated.

34 Barnosky et al. (2011). Though note that *many* things have been suggested as possible causes of the end-Permian mass extinction.

See Erwin, Bowring & Yugan (2002) for a survey of proposed causes.

35 Rougier et al. (2018). Estimating the return period for Toba-sized eruptions (magnitude 9 or more) is difficult, particularly with only one data point. Rougier's model suggests between 60,000 and 6 million years, with a central estimate of 800,000 years (personal communication). This estimate is highly sensitive to the upper limit for eruptions, which Rougier places at 9.3. I've rounded all these numbers to one significant figure to reflect our level of confidence.

36 Wilcox et al. (2017); Denkenberger & Blair (2018).

37 This could involve both climate modelling and analysis of the fossil record to see whether past eruptions caused any global or local extinctions. This latter angle might be easier than with asteroids due to the higher frequency of supervolcanic eruptions.

38 When one of the forces would increase or decrease, the star's size changes in response until they are balanced again. The rapid collapsing and exploding can be seen as failed attempts to rebalance these forces.

39 This can happen when the nuclear fuel of a large star runs low, reducing the pressure, or when a tiny white dwarf star drains off too much mass from a close companion star, increasing the gravitational squeezing. The former is more common and is known as a core-collapse supernova. The latter is known as a thermonuclear supernova (or Type Ia supernova).

40 Baade & Zwicky (1934); Schindewolf (1954); Krasovsky & Shklovsky (1957).

41 Bonnell & Klebesadel (1996).

42 This same effect massively increases the range at which they could be deadly to the Earth, which is sometimes interpreted as a reason to be more concerned about gamma ray bursts relative to supernovae. However, it also makes it possible that the explosion will miss us, due to being pointed in the wrong direction. In a large enough galaxy, these effects would exactly cancel out, with the narrowness of the cone having no effect on the average number of stars that a stellar explosion exposes to dangerous radiation levels. In our own galaxy, the narrowness of the cone actually *reduces* the typical number of stars affected by each blast, since it increases the

chance that much of it is wasted, shooting out beyond the edges of our galaxy.

43 This gamma ray burst (GRB 080319B) occurred about 7.5 billion years ago at a point in space that is now more than 10 billion light years away (due to cosmic expansion—author's calculation). This is 3,000 times further away than the Triangulum Galaxy, which is usually the most distant object visible with the naked eye (Naeye, 2008).

44 When these cosmic rays interact with our atmosphere, they also cause showers of high-energy particles to reach the Earth's surface, including dangerous levels of muon radiation.

An event of this type may have played a role in initiating the Ordovician mass extinction approximately 440 million years ago (Melott et al., 2004).

45 We can eliminate the risk from core collapse supernovae as nearby candidates would be very obvious and are not present. However, about a tenth of the risk is from thermonuclear supernovae (type Ia) which are harder to detect and thus it is harder to be sure they are absent (The et al., 2006). Similarly, it is very hard to find binary neutron stars that might collide in order to rule out the risk from this type of gamma ray burst. Given the difficulties and our more limited of understanding of gamma ray bursts, I have declined to give a numerical estimate of how much lower the risk over the next century is compared to the baseline risk.

46 Melott & Thomas (2011) estimate the frequency of extinction-level supernovae events at one in every 5 million centuries. This uses a distance threshold of 10 parsecs. Wilman & Newman (2018) reach a similar estimate, of one in every 10 million centuries.

Melott & Thomas (2011) estimate the total rate of extinction-level GRB (gamma ray burst) events at one in every 2.5 million centuries. This includes both long and short GRBs. Piran & Jimenez (2014) give probabilistic estimates of extinction-level GRB events having happened in the past. They find a probability of over 90% for such a long GRB having occurred in the last 5 billion years, and 50% of one having occurred in the last 500 million years. For short GRBs, they find much lower probabilities—14% in the last 5 billion years, and 2% in the last 500 million years.

These probability estimates (and particularly those of GRBs) are much rougher than those for asteroids and comets, due to the field being at an earlier stage. For example, the estimates for the energy released by supernovae and gamma ray bursts (and the cone angle for GRBs) are based on individual examples that are thought to be representative, rather than on detailed empirical distributions of known energy levels and cone angles for these events. In addition, while we have a reasonable understanding of which events could cause a 30% depletion of global ozone, more work is needed to determine whether this is the right threshold for catastrophe.

47 Some more detailed examples include: determining if there is a level of fluence that is a plausible extinction event, based on the predicted effects of ozone depletion on humans and crops; incorporating the observed distribution of supernova and gamma ray burst energy levels (and cone angles) into the model instead of relying on a paradigm example for each; taking into account the geometrical issues around cone angles from note 42 to this chapter; and doing a sensitivity analysis on the model to see if there are any plausible combinations of values that could make this risk competitive with asteroid risk (then attempting to rule those out).

I'd also encourage blue-sky thinking to see if there are any possible ways existing models could be underestimating the risk by an order of magnitude or more.

48 This is addressed in detail in Chapter 8.

49 Masson-Delmotte et al. (2013) say it is 'virtually certain' (over 99%) that orbital forcing—slow changes in the Earth's position relative to the Sun—cannot trigger widespread glaciation in the next thousand years. They note that climate models simulate no glaciation in the next 50,000 years, provided atmospheric carbon dioxide concentrations stay above 300 ppm. We also know that glacial periods are common enough that if they posed a high risk of extinction, we should see it in the fossil record (indeed most of *Homo sapiens'* history has been during glacial periods).

However, since we are considering many other risks that are known to have low probabilities, it would be good if we could improve our understanding of just how low the probability of entering a glacial period is, and just how much risk of global civilisation collapse it might pose. Notably, the Agricultural Revolution happened just as the last glacial period was ending, suggesting that even if they pose very little extinction risk, they may make agricultural civilisation substantially harder.

50 Adams & Laughlin (1999). Disruption by a passing black hole is less likely still, since stars vastly outnumber black holes.

51 This probability is difficult to estimate from observation as it would destroy any observers, censoring out any positive examples from our data set. However, Tegmark & Bostrom (2005) present an ingenious argument that rules out vacuum collapse being more likely than one in a billion per year with 99.9% confidence. Buttazzo et al. (2013) suggest that the true chance is less than one in 10^{600} per year, while many others think that our vacuum actually is the true vacuum so the chance of collapse is exactly zero.

It might also be possible to trigger a vacuum collapse through our own actions, such as through high-energy physics experiments. Risks from such experiments are discussed on pp. 160–161.

52 The geological record suggests a chance of roughly one in 2,000 per century. It is still debated whether the process is random or periodic. See Buffett, Ziegler & Constable (2013) for a summary of recent developments.

53 Lingam (2019).

54 The idea that our species' longevity is evidence that natural risk is low is briefly mentioned by Leslie (1996, p. 141) and Bostrom (2002b). To my knowledge, the first attempt to quantify evidence from the fossil record is Ord & Beckstead (2014). My colleagues and I have further developed this line of reasoning in Snyder-Beattie, Ord & Bonsall (2019).

55 All these techniques give estimates that are based on averages. If we know we are not in an average time, they might no longer apply. It is extremely unlikely that we will detect a 10 km asteroid on a collision course for Earth. But if we did, we would no longer be able to help ourselves to laws of averages. I do not know of any natural threats where our current knowledge suggests that our risk of extinction is significantly elevated (and we should be surprised to find ourselves in such a situation, since they must be rare). Indeed, it is much more common for the knowledge we acquire to show that the near-term risk from a threat—such as asteroids or supernovae—is actually lower than the longterm average.

56 There are some indirect applications to unrecoverable collapse. The major hazards that have been identified appear to pose a broadly similar level of risk of unrecoverable collapse and extinction (say, within a factor of ten), so there is some reason to think that finding the extinction risk is very small would also have bearing on the collapse risk. What would be needed to break this is a natural hazard that is especially good at posing collapse risk relative to extinction risk. This is harder to find than one might think, especially when you consider that a lot of the fossil evidence we use is for species that are much less robust to extinction than we are, due to smaller geographical ranges and dependence on a small number of food sources—it would have to be able to permanently destroy civilisation across the globe without causing such species to go extinct.

Another approach would be to run the first of my methods using the lifetime of civilisation instead of the lifetime of *Homo sapiens*. As this is about 100 centuries, we'd get a best-guess estimate between 0% and 1% per century. Which is something, but definitely not as comforting. More research on how to strengthen these estimates and bounds for irrevocable civilisation collapse would be very valuable.

57 There is considerable uncertainty about the 200,000-year estimate for the origin of *Homo sapiens* and for many related dates we will consider later. My 200,000 estimate refers to the 'Omo' fossil remains, which are widely regarded as being *Homo sapiens*. More recently discovered fossils from Jebel Irhoud in Morocco are dated to roughly 300,000 years ago, but it is still debated whether they should be considered *Homo sapiens* (see note 2 to Chapter 1). But all these dates are known to be correct to within a factor of two and that is sufficient accuracy to draw the qualitative conclusions later on. Feel free to substitute them with any other estimates you prefer, and see how the quantitative estimates change.

58 This is not strictly true, as in some mathematical contexts probability zero events can happen. But they are infinitely unlikely, such as flipping a coin forever and never getting tails. Of course, we also don't have enough evidence to suggest that human extinction is infinitely unlikely in this sense.

59 This area of research is ongoing and clearly very important for the study of existential risk, which is all about unprecedented events.

One way to state the question is: what probability should we assign to failure if something has succeeded on each of the n trials so far? This is sometimes known as the problem of zero-failure data. Estimators that have been suggested include:

0	Maximum likelihood estimator.
$\frac{1}{3n}$	'One-third' estimator (Bailey, 1997).
$\sim\frac{1}{2.5n}$	Approximation of method from Quigley & Revie (2011).
$\frac{1}{2n+2}$	Bayesian updating with a maximum entropy prior.
$\sim\frac{1}{1.5n}$	50% confidence level (Bailey, 1997).
$\frac{1}{n+2}$	Bayesian updating with a uniform prior.
$\frac{1}{n}$	'Upper bound' estimator.

Note that the widespread 'rule of three' (Hanley, 1983) is not attempting to answer the same question: it suggests using $\frac{3}{n}$, but as an upper bound (at 95% confidence) rather than as a best guess. We will use a more direct approach to estimate such bounds and ask for a higher confidence level.

I think the arguments are strongest for the Bayesian updating with a maximum entropy prior, which ends up giving an estimate right in the middle of the reasonable range (after a lot of trials, or when the possibility of failure is spread continuously through time).

60 The general formula for the upper bound is $1-(1-c)^{100/t}$, where c is the level of confidence (e.g. 0.999) and t is the age of humanity in years (e.g. 200,000).

61 This is not quite the same as saying that there is a 99.9% chance that the risk is below 0.34%. It just means that if the risk were higher than 0.34%, a 1 in 1,000 event must have occurred. This should be enough to make us very sceptical of such an estimate of the risk without substantial independent reason to think it so high. For example, if all other observed species had natural-cause extinction rates of 1% per century, we might think it is more likely that we do too—and that we got very lucky, rather than that we are exceptional. However, we shall soon see that related species have lower extinction risk than this, so this does not offer a way out.

62 On the basis of fossil evidence, we are confident that divergence occurred at least 430,000 years ago (Arsuaga et al., 2014). See White, Gowlett & Grove (2014) for a survey of estimates

from genomic evidence, which range from around 400,000 to 800,000 years ago.

63 This special form of survivorship bias is sometimes known as anthropic bias or an observation selection effect.

64 My colleagues and I have shown how we can address these possibilities when it comes to estimating natural existential risk via how long humanity has survived so far (Snyder-Beattie, Ord & Bonsall, 2019). We found that the most biologically plausible models for how anthropic bias could affect the situation would cause only a small change to the estimated probabilities of natural risk.

65 All species' age estimates in this chapter are from the fossil evidence.
It is difficult to get the complete data for all species in the genus *Homo* as new species are still being discovered, and there are several species which have only been found in a few sites. Therefore, using the dates between the earliest and latest known fossils for each species will probably greatly underestimate their lifespan. One response is to restrict our attention to the species found in more than a few sites and this is what I've done. Unfortunately, this increases a type of bias in the data where we are less likely to know about short-lived species as there will typically be fewer fossilised remains (causing this to underestimate natural risk). However, since *Homo sapiens* is known to have a longevity greater than 200,000 years there is evidence that its extinction chance is not very similar to those of any extremely short-lived species that may have been missed out.

66 One might wonder whether a constant hazard rate is a reasonable model of how species go extinct. For example, it assumes that their objective chance of going extinct in the next century is independent of how long they have lived so far, but perhaps species are more like organisms, in that older species are less fit and at increased risk. Such a systematic change in extinction risk over time would affect my analysis. However, it appears that species lifespans within each family are indeed fairly well approximated by a constant hazard rate (Van Valen, 1973; Alroy, 1996; Foote & Raup, 1996).

67 This could also be said of the previous method, as *Homo sapiens* is arguably a successful continuation of the species before it.

68 All dates given are for when the event ended. Extinction rates are from Barnosky et al. (2011).

It has recently been suggested that the Devonian and Triassic events may have reduced species numbers more from lowering the origination rate of new species than raising the extinction rate of existing species (Bambach, 2006). If so, this would only strengthen the arguments herein, by reducing the frequency of the type of extinction events relevant to us to a 'Big Three'.

Note also that there is a lot of scientific uncertainty around what caused most of these, including the biggest. But for our purposes this doesn't matter too much, since we still know that these events are extremely rare and that is all we use in the argument.

69 Even in the case of an asteroid impact, where technology and geographical distribution are very helpful, we could be at increased risk due to our reliance on technology and the farming of very few types of crop. It is conceivable that a smaller society of hunter-gatherers would be more resilient to this, since they would have skills that are rare now but which might become essential (Hanson, 2008). However, I very much doubt that this risk has increased overall, especially considering the fact that our world still contains people who are relatively isolated and live in relatively untouched tribal groups.

4 ANTHROPOGENIC RISKS

1 Toynbee (1963).

2 The contribution of fusion to an atomic bomb goes far beyond this higher efficiency. A fission bomb has a natural size limit set by the critical mass of its fuel (some tricks allow this to be exceeded, but only by a small multiple). In contrast, the fusion fuel has no such constraints and much larger bombs could be built. Moreover, the neutrons emitted by the fusion can cause fission in the bomb's massive uranium tamper. This is known as a fission-fusion-fission bomb and this final stage of fission can produce most of the energy.

3 Compton (1956), pp. 127–8.

4 Albert Speer, the German minister of armaments, gave a chilling account (Speer, 1970, p. 227): 'Professor Heisenberg had not given any final answer to my question whether a successful nuclear fission could be kept under control with absolute certainty or might continue as a chain reaction. Hitler was plainly not delighted with the possibility that the earth under his rule might be transformed into a

glowing star. Occasionally, however, he joked that the scientists in their unworldly urge to lay bare all the secrets under heaven might some day set the globe on fire. But undoubtedly a good deal of time would pass before that came about, Hitler said; he would certainly not live to see it.'

One cannot be sure from this whether it was exactly the same concern (a thermonuclear reaction spreading through the atmosphere) or a related kind of uncontrolled explosion.

5 Teller had made very 'optimistic' assumptions about the parameters involved in getting the fusion reaction going, and had not taken account of the rate at which the heat of the explosion would radiate away, cooling it faster than the new fusion heated it up (Rhodes, 1986, p. 419).

6 This report has subsequently been declassified and is available as Konopinski, Marvin & Teller (1946).

7 The report ends: 'One may conclude that the arguments of this paper make it unreasonable to expect that the N + N reaction could propagate. An unlimited propagation is even less likely. However, the complexity of the argument and the absence of satisfactory experimental foundation makes further work on the subject highly desirable' (Konopinski, Marvin & Teller, 1946).

In contemporary discussion, the probability of 'three in a million' is often given, either as the estimate of the chance of ignition or as a safety threshold that the chance needed to be below. This number does not occur in the report and appears to have entered the public sphere through an article by Pearl S. Buck (1959). While intriguing, there is no convincing evidence that such a probability was used by the atomic scientists in either manner.

8 David Hawkins, the official historian of the Manhattan Project, has said that the possibility kept being rediscovered by younger scientists and that the leadership at Los Alamos had to keep 'batting it down', telling them that it had been taken care of. In the end, Hawkins did 'more interviews with the participants on this particular subject, both before and after the Trinity test, than on any other subject'. (Ellsberg, 2017, pp. 279–80).

9 Peter Goodchild (2004, pp. 103–4): 'In the final weeks leading up to the test Teller's group were drawn into the immediate preparations when the possibility of atmospheric ignition was revived by Enrico

Fermi. His team went to work on the calculation, but, as with all such projects before the introduction of computers, these involved simplifying assumptions. Time after time they came up with negative results, but Fermi remained unhappy about their assumptions. He also worried whether there were undiscovered phenomena that, under the novel conditions of extreme heat, might lead to an unexpected disaster.'

10 From his private notes written the next day (Hershberg, 1995, p. 759). The full quotation is: 'Then came a burst of white light that seemed to fill the sky and seemed to last for seconds. I had expected a relatively quick and light flash. The enormity of the light quite stunned me. My instantaneous reaction was that something had gone wrong and that the thermal nuclear [sic] transformation of the atmosphere, once discussed as a possibility and jokingly referred to a few minutes earlier, had actually occurred.'

Perhaps from staring into this abyss, Conant was one of the first people to take the destruction of civilisation due to nuclear war seriously. When the war ended, he returned to Harvard and summoned its chief librarian, Keyes Metcalf, for a private meeting. Metcalf later recalled his shock at Conant's request (Hershberg, 1995, pp. 241–2): 'We are living in a very different world since the explosion of the A-bomb. We have no way of knowing what the results will be, but there is the danger that much of our present civilization will come to an end . . . It might be advisable to select the printed material that would preserve the record of our civilization for the one we can hope will follow, microfilming it and making perhaps 10 copies and burying those in different places throughout the country. In that way we could ensure against the destruction that resulted from the fall of the Roman Empire.'

Metcalf looked into what this would require, and prepared a rough plan for microfilming the most important 500,000 volumes, or a total of 250 million pages. But in the end they did not pursue this, reasoning both that its becoming public would cause significant panic, and that written records would probably survive in the libraries of university towns that would not suffer direct hits from atomic weapons. However, when Metcalf resigned from Harvard, he began a project of ensuring vast holdings of important works in major universities in the southern hemisphere, perhaps inspired by the conversation with Conant and fear of nuclear catastrophe (Hershberg & Kelly, 2017).

11 Weaver & Wood (1979).

12 If a group cares deeply about the accuracy of their own internal work, they can set up a 'red team' of researchers tasked with proving the work wrong. This team should be given ample time, until they overcome their initial loyalties to the work so far, beginning to hope it is wrong rather than right. They also should be given ample resources, praise and incentives for finding flaws.

13 Part of any proper risk analysis is a measure of the stakes and a comparison with the benefits. These benefits became *much* smaller once Hitler was defeated, necessitating a much lower probability threshold for the disaster, but it seems that the risk wasn't re-evaluated.

14 At least a few people in government do appear to have found out about it. Serber (1992, p. xxxi): 'Compton didn't have enough sense to shut up about it. It somehow got into a document that went to Washington. So every once in a while after that, someone happened to notice it, and then back down the ladder came the question, and the thing never was laid to rest.'

15 One of the best methods for eliciting someone's subjective probability for an event is to offer a series of small bets and see how extreme the odds need to be before they take them. As it happened, Fermi did exactly this, the evening before the Trinity test, taking bets on whether the test would destroy the world. However, since it was obvious that no one could collect on their bets if the atmosphere *had* ignited, this must have been at least partly in jest. History does not relate who took him up on this nor what odds they took.

16 The fuel was a compound of lithium with deuterium (an isotope of hydrogen that is conducive to fusion reactions). The purpose of the lithium was to react with a neutron, producing the extremely rare hydrogen isotope, tritium. This tritium would then fuse with the deuterium, releasing a lot of energy.

17 15 Mt fell well outside their uncertainty range of 4 to 8 Mt (Dodson & Rabi, 1954, p. 15).

18 The Japanese were understandably upset at being hit by a US nuclear weapon again, just nine years after Nagasaki, and this caused a diplomatic incident. Even the scientific results were a

disaster: they collected relatively little useful data as the larger blast destroyed much of their test equipment.

19 The lithium-7 was reacting in an unanticipated way, producing both more tritium and more neutrons, which drove the fusion and fission reactions much further than expected. It is difficult to make any precise claims about the relative contributions of the two lithium isotopes as the weapon involved several interacting stages, but I believe it is roughly right to say that the contributions of the two kinds of lithium were similar, for the weapon had an amount of lithium-7 equal to 150% of the amount of lithium-6 and got an additional 150% energy release.

20 One reason is that they appear to have taken more caution with the first of these calculations. For the atmosphere to ignite, they would not only need to be wrong in their calculations, but wrong by an amount in excess of their safety factors.

There is also the fact that their first calculation was a yes/no question, while the second wasn't. So there were more ways the second could go wrong. I've demonstrated that it contained a major mistake, but the fuel they recommended *did* still explode, so I suppose a more coarse-grained assessment might still count that calculation as a success.

And finally, there is a question of priors. Even if their method of answering questions was completely unreliable (e.g. flipping a coin), all that means is that it doesn't provide a useful update to your prior probability estimate for the event. It is hard to say what that should have been in the case of igniting the atmosphere, but it is reasonable for it to have been well below 50%, perhaps below 1%.

21 The report on Soviet bombing targets was delivered on 30 August 1945 (Rhodes, 1995, p. 23).

22 Another key technical development was multiple independently targetable re-entry vehicles (MIRVs). These enabled a single ICBM to split and hit several locations. This shifted the strategic balance towards first strike, as the power that struck first could potentially take out several enemy ICBMs with each of its own. This in turn increased the reliance on hair-trigger alert as the retaliating power would need to launch its missiles while the attacking missiles were still on their way in.

23 There were often further checks that may have prevented these incidents escalating all the way to nuclear war. For a sceptical take on how close these close calls were, see Tertrais (2017).

24 There are far more close calls and accidents than I could do justice to in this book. For example, NORAD reported that false alarms led to six Threat Assessment Conferences and 956 Missile Display Conferences even just in the five years from January 1978 to May 1983 (Wallace, Crissey & Sennott, 1986).

25 Brezhnev (1979); Gates (2011); Schlosser (2013).

26 Reports differ on whether there were five missiles shown, or just a single missile (with four more appearing in a second event later that night).

27 Lebedev (2004); Schlosser (2013); Chan (2017).

28 Forden, Podvig & Postol (2000); Schlosser (2013).

29 Within five months 140,000 people died in Hiroshima from a yield of about 15 kilotons. The world's arsenal is about 200,000 times this, so the naïve extrapolation would suggest about 30 billion deaths, or about four times the world's population. But such a calculation makes two major mistakes.

First, it ignores the fact that many people do not live in big dense cities: there are nowhere near enough nuclear warheads to hit all towns and villages. And second, it ignores the fact that bigger nuclear weapons become less efficient at killing, per kiloton. This is because the blast energy is spread out in a three-dimensional ball, while the city is in a two-dimensional disc, which occupies a smaller and smaller fraction of the ball as the energy increases. Thus an increasing fraction of the blast energy is wasted as the weapon is scaled up. Mathematically, the blast damage scales as the two-thirds power of the energy.

30 Ball (2006) estimates 250 million direct deaths from an all-out nuclear war.

The Office of Technology Assessment (1979) describes US government estimates of direct death tolls ranging from 20 to 165 million in the US, and 50 to 100 million in the Soviet Union. Note that these estimates should be adjusted for a present-day case—the population of US cities has increased substantially since the 1970s, and the collapse of the Soviet Union has presumably restricted the targets of a US attack to Russia. Ellsberg (2017, pp. 1–3) describes a classified report prepared for President

Kennedy by the Joint Chiefs of Staff, which estimated the immediate deaths from a nuclear attack on the Soviet Union and China at 275 million, rising to 325 million after six months, numbers that would also have to be scaled up for present populations.

31   Feld (1976) estimates that a one-megaton warhead can irradiate an area of roughly 2,500 km^2 with a lethal dose, implying one would need at least 60,000 such weapons to irradiate the Earth's land area. This significantly exceeds current stockpiles of around 9,000 deployed warheads, which have an average yield considerably below one megaton.

32   Such a 'doomsday device' was first suggested by Leo Szilard in 1950 and its strategic implications were more fully developed by Herman Kahn (Bethe et al., 1950). A cobalt bomb (or similar salted weapon) plays a major role in the plots of *On the Beach* and *Dr Strangelove*, taking nuclear war in both cases from a global catastrophe to an extinction threat.

The greatest obstacle to destroying all of humanity with such a weapon is ensuring that lethal radiation is distributed evenly across the Earth, particularly when taking into account shelters, weather and oceans.

Russia's *Poseidon* nuclear torpedo—currently being developed—is allegedly equipped with a cobalt warhead. Information about the weapon was ostensibly leaked by accident, but is suspected to have been deliberately released by the Russian government, so should be viewed with some scepticism (BBC, 2015).

33   It is sometimes said that the burning of the oil wells in Kuwait refuted nuclear winter. But this isn't right. Carl Sagan thought the burning of the oil wells would cause detectable global cooling because the soot would reach the stratosphere. But the oil-well fires were too small to loft it high enough. This puts a small amount of pressure on the part of the model about how high soot from firestorms would be lofted, but doesn't affect anything that comes after that. There are examples of forest fires elevating smoke as high as nine kilometres (Toon et al., 2007).

34   Robock, Oman & Stenchikov (2007).

35   Though there wouldn't be enough time for great ice sheets to build up over Europe and North America. The Last Glacial Maximum saw global mean temperatures around 6 °C cooler than pre-industrial levels (Schneider von Deimling et al., 2006).

36 Cropper & Harwell (1986); Helfand (2013); Xia et al. (2015).

37 Baum et al. (2015); Denkenberger & Pearce (2016).

38 While Sagan (1983) and Ehrlich et al. (1983), who previously worked on nuclear winter, did suggest extinction was possible, those in the field now do not.

Luke Oman (Oman & Shulman, 2012): 'The probability I would estimate for the global human population of zero resulting from the 150 Tg of black carbon scenario in our 2007 paper would be in the range of one in 10,000 to one in 100,000. I tried to base this estimate on the closest rapid climate change impact analogue that I know of, the Toba supervolcanic eruption approximately 70,000 years ago. There is some suggestion that around the time of Toba there was a population bottleneck in which the global population was severely reduced. Climate anomalies could be similar in magnitude and duration. Biggest population impacts would likely be Northern Hemisphere interior continental regions with relatively smaller impacts possible over Southern Hemisphere island nations like New Zealand . . . I don't know offhand anyone that would estimate higher but I am sure there might be people who would. [I asked two colleagues] who did respond back to me, saying in general terms "very close to 0" and "very low probability".'

Richard Turco (Browne, 1990): 'my personal opinion is that the human race wouldn't become extinct, but civilization as we know it certainly would'.

Alan Robock (Conn, Toon & Robock, 2016): 'Carl [Sagan] used to talk about extinction of the human species, but I think that was an exaggeration. It's hard to think of a scenario that would produce that. If you live in the Southern Hemisphere, it's a nuclear free zone, so there wouldn't be any bombs dropped there presumably. If you lived in New Zealand and there wasn't that much temperature change because you're surrounded by an ocean, there's lots of fish and dead sheep around, then you probably would survive. But you wouldn't have any modern medicine . . . you'd be back to caveman days. You wouldn't have any civilization, so it's a horrible thing to contemplate, but we probably couldn't make ourselves extinct that way.'

Mark Harwell and Christine Harwell (1986): 'It seems possible that several hundred millions of humans could die from the direct effects of nuclear war. The indirect effects could result in the loss of one to several billions of humans. How close the latter projection would come to loss of all humans is problematical, but the current best estimation is that this result would not follow from the physical societal perturbations currently projected to occur after a large-scale nuclear war.'

39   There would be serious issues with advanced electronic technology as these locations wouldn't always have the factories or knowledge to make replacement parts. But things look a lot better for the thousands of technologies humans invented prior to the last hundred years. For example, I can't see why they would be reduced to a pre-industrial level, nor why they wouldn't be able to eventually recover current technology levels.

40   For example, a recent paper from the US Department of Energy argued that much less soot would reach the upper atmosphere compared with the main nuclear winter models (Reisner et al., 2018).

41   In some cases, additional uncertainty can make things better. In particular, it can increase the amount of regression to the mean (or regression to one's prior). So if the estimated outcome seemed unlikely initially, residual uncertainty provides a reason to fall back towards this initial guess. But in this case I don't see any good reasons to assign substantially greater prior probability to a small amount of cooling rather than a large amount, or to a small famine rather than a large one. Moreover, if you count existential catastrophe as much worse than the deaths alone and think that the median case is very unlikely to cause this, then uncertainty makes things a lot worse.

42   Presumably the risk of accidental war is also substantially lower than during the Cold War—as the probability of deliberate war goes down, the probability of interpreting a false alarm as a deliberate strike should also go down, at least insofar as there are humans in the loop. This fits with the track record of so many serious false alarms occurring during times of extreme tension, such as the Cuban Missile Crisis.

43   Both the 70,000 and 14,000 figures include retired warheads. There are around 9,000 'active' warheads today (Kristensen & Korda, 2019d).

44   Adapted from Kristensen & Korda (2019d). Total yield from author's calculations, using data from Kristensen & Korda (2018,

2019a–e), Kristensen & Norris (2018), Kristensen, Norris & Diamond (2018).

45   Robock et al. (2007). Modelling by Reisner et al. (2018), mentioned in note 40, finds a much smaller effect from a similar exchange.

46   The collapse of the Intermediate-Range Nuclear Forces (INF) treaty, which saw the US and Soviet Union agree to eliminate short- and medium-range land-based missiles, is particularly concerning.

   Russian President Vladimir Putin's 2018 speech to the Federal Assembly painted a worrying picture of mistrust between the US and Russia, and the ongoing efforts to modernise and strengthen Russia's nuclear capacity (Putin, 2018).

47   Horowitz (2018).

48   While the greenhouse effect is real, it turned out not to be the reason garden greenhouses work. Most of the warming in a greenhouse is actually due to the fact that the glass physically traps the warm daytime air, preventing it from floating away at night via convection. Greenhouses made of substances that are transparent to both visible and infrared light still work, while those with a small hole at the top to let the warm air out don't.

49   This is why I don't consider climate change to be the first anthropogenic risk. While the mechanism of CO_2 production from burning fossil fuels pre-dates nuclear weapons, emissions have only recently become high enough to start to pose a threat to humanity.

   Between 1751 and 1980 cumulative global carbon emissions from fossil fuels were around 160 Gt, compared with over 260 Gt from 1981 to 2017 (Ritchie & Roser, 2019).

50   Pre-industrial figure from Lindsey (2018); 2019 figure from NOAA (2019).

51   Allen et al. (2018), p. 59. This compares the period 1850–1900 with a 30-year period centred on 2017, assuming the recent rate of warming continues.

52   From 1880 to 2015 (CSIRO, 2015, LSA, 2014). The one standard deviation confidence interval is 19 cm to 26 cm.

53   From 1750 to 2011 (IPCC, 2014, p. 4).

54   Also known as a positive feedback. Unfortunately this causes some confusion as positive climate feedbacks are bad and negative feedbacks are good. I shall thus use the clearer terms 'amplifying feedbacks' and 'stabilising feedbacks' instead.

55 To see how this is possible, suppose that the background sound starts at a level of 100 W/m^2 and that the contribution from the speaker is 10%. In that case the first amplification adds 10 W/m^2 to the sound level near the microphone. When this additional sound is amplified, it adds 1 W/m^2, then this adds 0.1 W/m^2, and so forth. Even though sound continually creates more sound, the total effect would be modest, summing to 111.11 . . . W/m^2. If the speaker's volume was turned up (or the microphone brought closer) such that the speaker added 100% (or more) to what was there already, the sum would diverge (100 + 100 + 100 + . . .) and the sound would quickly grow louder until it hit the physical limits of what the microphone can register or what the speaker can produce.

56 Gordon et al. (2013) find an amplifying effect of 2.2 W/m^2/K in their observation window of 2002 to 2009, and estimate the longterm feedback strength between 1.9 and 2.8 W/m^2/K.

57 The greenhouse effect makes Venus far hotter than Mercury despite being almost twice as far from the Sun. We will return to the longterm fate of the Earth in Chapter 8.

58 Goldblatt et al. (2013) find no runaway greenhouse effect with atmospheric concentrations of 5,000 ppm CO_2. Tokarska et al. (2016) find that the burning of 5,000 Gt C, a low estimate of the total fossil fuel reserves, results in atmospheric concentrations just under 2,000 ppm CO_2, suggesting that even if we burned all the fossil fuels, we would not trigger a runaway greenhouse effect.

59 Both moist and runaway greenhouse effects can be understood in terms of the equilibrium on Earth, between incoming solar radiation and outgoing radiation in heat and reflected light. In our current stable regime, increases in surface temperature are matched by increases in the radiation escaping Earth, keeping our climate relatively stable. But there are limits to the amount of radiation that can escape the atmosphere, which are determined, in part, by its water vapour content.

In a *runaway greenhouse*, the Earth's temperature exceeds one of these limits, beyond which its surface and atmosphere can warm, but no more thermal radiation can escape. This results in runaway warming, with the Earth's surface warming until it reaches a new equilibrium, hundreds of degrees warmer, by which point the oceans have boiled off entirely. A *moist greenhouse* is a stable intermediate state, much warmer than our own, and with much more water vapour in the atmosphere. Over geological timescales, a *moist greenhouse* will also result in the complete loss of Earth's water, due to vapour loss from the upper atmosphere into space.

60 This required a very large amount of greenhouse gas to trigger: about 1,550 ppm of carbon dioxide. This is higher than the amount of carbon dioxide in the atmosphere by 2100 in the IPCC's most pessimistic scenario (Collins et al., 2013, p. 1096). When the simplifications are accounted for, it may require much more than this, or be completely impossible without additional solar radiation (Popp, Schmidt & Marotzke, 2016). The model did not produce a useful estimate of the timeframe for this warming (due to its simplifications), but the author suggests it would probably take many thousands of years, which might provide time for mitigation (Popp, personal communication).

61 The planet was modelled as entirely ocean, this ocean was only 50 metres deep, and there were no seasons. The paper's authors are well aware that these simplifications mean these results may not apply to the actual Earth, and do not claim otherwise.

62 McInerney & Wing (2011).

63 The *permafrost region* occupies 23 million square kilometres—24% of the land area of the Northern Hemisphere—but the extent of actual permafrost is estimated to cover 12 to 17 million square kilometres (Zhang et al., 2000).

64 There is an estimated 1,672 Gt C in Arctic permafrost (Tarnocai et al., 2009). Emissions from 1750 to 2017 are estimated at 660 ± 95 Gt C (Le Quéré et al., 2018).

65 The IPCC says: 'Overall, there is *high confidence* that reductions in permafrost extent due to warming will cause thawing of some currently frozen carbon. However, there is *low confidence* on the magnitude of carbon losses through CO_2 and CH_4 emissions to the atmosphere' (Ciais et al., 2013, p. 526).

66 The estimate is 0.29 ± 0.21 °C (Schaefer et al., 2014).

67 1,500 to 7,000 Gt C (Ciais et al., 2013, p. 473).

68 The IPCC says it is 'very unlikely that methane from clathrates will undergo catastrophic release (high confidence)' (Collins et al., 2013, p. 1,115). This sounds reassuring, but in the official language of the IPCC, 'very unlikely' translates into a 1% to 10% chance,

which then sounds extremely alarming. I don't know what to make of this as the context suggests it was meant to reassure.

69 These are cumulative emissions from 2012–2100 on the RCP 6.0 and RCP 8.5 pathways, consistent with 'baseline scenarios' without additional efforts to constrain emissions. Emissions compatible with the Paris Agreement, which commits countries to keep warming below 2 °C, are much lower. The IPCC (2014, p. 27) estimates that 2018–2100 emissions must be kept below ~340 Gt C for a 66% chance of keeping warming below 2 °C.

70 Assuming that the emission rate continues to grow at 3% per year (Pierrehumbert, 2013).

71 The more recent estimates are towards the upper end of this range. This includes fuel which it is not currently economical to recover, as well as undiscovered fuel. Of course it is possible that even more than this could be found, extracted and burnt. Over time, more and more types of fossil fuel deposit have become economically viable (witness the rise of fracking). While new types of deposit may become cheaper by historical standards, solar power is also becoming cheaper at a very fast rate and is already competitive on price with fossil fuels in some places. I therefore doubt that new types of deposit will become economically viable, when solar energy is considered as an alternative. Global reserves—known, economically viable deposits—contain ~1,000–2,000 Gt C (Bruckner et al., 2014, p. 525).

72 Tokarska et al. (2016), p. 852. The few existing studies of the consequences of burning all the fossil fuels use the lower bound estimate of 5,000 Gt C. Research into even more extreme scenarios, where we burned 10,000 Gt C or more, would be valuable.

73 It is very difficult to estimate the probability that emissions will exceed a given amount of carbon. The IPCC doesn't attempt to do so at all. It prefers to treat the pathways as a menu of policy options: things you choose from, not things that happen to you. There is value in this approach, but it would also be very useful to have some idea of the likelihood of different pathways—especially as there is no top-level agent who can actually make a choice from this menu.

74 Bar-On, Phillips & Milo (2018). This is overwhelmingly in plants and bacteria, which together contain 96% of biomass carbon.

An estimated further 1,200 Gt C is in dead biomass, usually called necromass (Kondratyev, Krapivin & Varotsos, 2003, p. 88). This is organic matter in soils and can also be emitted, primarily through deforestation and forest fires. Peat (carbon-dense soil used as fuel) is an example of necromass.

75 From a combination of agriculture and industry. It is important to note that not all of this has stayed in the atmosphere.

76 Emissions from 1750 to 2017 are estimated at 660 ± 95 Gt C. Of this, ~430 Gt C is from fossil fuel and industry, and ~235 Gt C is from land-use change (Le Quéré et al., 2018).

77 Ciais et al. (2013), p. 526.

78 There are actually several measures of climate sensitivity; this one is called the Equilibrium Climate Sensitivity. It is technically a measure of warming resulting from a given amount of 'radiative forcing', which includes the effects of greenhouse gases as well as other changes to the Earth that move the balance between how much energy is received in sunlight and how much is radiated back out. The official unit for radiative forcing is watts per square metre, but it is commonly understood in terms of the degrees of warming produced by a doubling of atmospheric carbon dioxide.

79 Beade et al. (2001), p. 93.

80 In the language of IPCC, 'likely' means that there is at least a two-thirds chance the true sensitivity lies in this range (IPCC, 2014, p. 16). For uncertainty about cloud feedbacks see Stevens & Bony (2013).

81 Their term 'likely' officially means 66% to 100% chance, though one would expect them to have used 'very likely' if they thought it was greater than 90%. In keeping with the roughness of the confidence level for this interval, the IPCC makes it clear that these probabilities are not based on statistical measures of the scientific uncertainty, but represent expert judgement (Cubasch et al., 2013, pp. 138–42). When looking in more detail, we can see that the literature contains some climate models with very broad probability distributions over the climate sensitivity, allowing non-trivial chance of a sensitivity greater than 6 °C or even 10 °C. However, the fat right-hand tail of these distributions is very dependent on the choice of prior (Annan & Hargreaves, 2011). This means that the data doesn't rule out high sensitivities, but doesn't support them

either. It is thus hard to say anything precise about the chance that climate sensitivity exceeds 4.5 °C, or about it exceeding higher thresholds such as 6 °C.

See Weitzman (2009) for one attempt at accounting for some of this uncertainty, and its implications for policy-making. He estimates a 5% chance of 'generalized climate sensitivity' (accounting for a wider range of feedback mechanisms) exceeding 10 °C warming within two centuries, and a 1% chance of it exceeding 20 °C, with one doubling of emissions.

82 On top of this, the usefulness of this logarithmic relationship itself has also been questioned. Some scientists have found that when climate feedbacks and the changing properties of carbon sinks are taken into account, their models produce a nearly linear relationship between cumulative emissions (in Gt C) and warming. This gives similar predictions for medium emissions scenarios, but suggests much more warming if the emissions are high. For example, Tokarska et al., 2006, p. 853.

83 In July 1979—the very month I was born (Charney et al., 1979).

84 Rogelj et al. (2016).

85 Tai, Martin & Heald (2014) find that under the IPCC's most pessimistic scenario, global food production would decrease by 16% by 2050 relative to 2000. But this takes into account neither adaptation nor the impact of carbon dioxide on crop yields, both of which are expected to have significant, albeit uncertain, offsetting effects. A recent meta-analysis found that crop-level adaptations alone could increase yields by 7–15% (Challinor et al., 2014).

Such a reduction in food supply would have disastrous consequences for millions of people, but would pose little risk to civilisation.

86 IPCC (2014), pp. 14–15.

87 We don't see such biodiversity loss in the 12 °C warmer climate of the early Eocene, nor the rapid global change of the PETM, nor in rapid regional changes of climate. Willis et al. (2010) state: 'We argue that although the underlying mechanisms responsible for these past changes in climate were very different (i.e. natural processes rather than anthropogenic), the rates and magnitude of climate change are similar to those predicted for the future and therefore potentially relevant to understanding future

biotic response. What emerges from these past records is evidence for rapid community turnover, migrations, development of novel ecosystems and thresholds from one stable ecosystem state to another, but there is very little evidence for broad-scale extinctions due to a warming world.'

There are similar conclusions in Botkin et al. (2007), Dawson et al. (2011), Hof et al. (2011) and Willis & MacDonald (2011). The best evidence of warming causing extinction may be from the end-Permian mass extinction, which may have been associated with large-scale warming (see note 91 to this chapter).

88 The measure is known as the 'wet-bulb temperature' and may become lethal at around 35 °C (Sherwood & Huber, 2010).

89 Author's calculation based on the information in Sherwood & Huber (2010).

90 Indeed the extinction effects of the PETM seem surprisingly mild. For example, McInerney & Wing (2011) state: '[in the PETM] Terrestrial and marine organisms experienced large shifts in geographic ranges, rapid evolution, and changes in trophic ecology, but few groups suffered major extinctions with the exception of benthic foraminifera [a type of micro-organism]'.

91 A recent paper suggests that ocean temperatures may have risen by as much as 8 °C during the end-Permian extinction, possibly driven by a huge injection of atmospheric CO_2 (Cui & Kump, 2015). The precise levels of warming and CO_2 concentration remain uncertain, due to relatively sparse geological evidence over this period. While this is just one of many putative causes for the end-Permian extinction, these uncertainties, and our inability to rule out that the biggest mass extinction was caused by rapid warming, are ultimately bad news.

92 Using geoengineering as a last resort could lower overall existential risk even if the technique is *more* risky than climate change itself. This is because we could adopt the strategy of only deploying it in the unlikely case where climate change is much worse than currently expected, giving us a second roll of the dice.

Here is a simplified numerical example. Suppose climate change has a 0.1% chance of being extremely severe, in which case it has a 50% chance of directly causing our extinction, for an overall extinction risk of 0.05%. And suppose that geoengineering fixes the climate, but produces its own 1% risk of extinction. Starting geoengineering now would be a bad idea, since its 1% risk is higher than the overall 0.05% risk. But if we only commence geoengineering if climate change turns out to be extremely severe, then geoengineering will reduce the overall risk: for we only face its 1% risk of extinction in cases where we were otherwise facing a 50% chance. This conditional geoengineering strategy would thus lower overall extinction risk from 0.05% to 0.001%. This can happen in more realistic models too. The key is waiting for a situation when the risk of using geoengineering is appreciably lower than the risk of not using it. A similar strategy may be applicable for other kinds of existential risk too.

93 Ehrlich (1969).

94 From a speech in 1969, quoted in Mann (2018).

95 16.6 million people in the 1960s, 3.4 million in the 1970s, and an average of ~1.5 million per decade since then (Hasell & Roser, 2019). Note that these are only the deaths from identifiable famines and are not complete accounts of all deaths related to food scarcity.

96 These improvements in productivity came with significant environmental costs.

97 He is often credited with saving roughly a billion lives. There are many challenges with making such an estimate, including the fact that it is very hard to estimate how many lives the green revolution actually saved, that many other people were essential to the green revolution, and that if he hadn't produced these varieties someone else may well have done so. Perhaps the best way to think of it is simply that he played a central role in one of the greatest humanitarian success stories of the twentieth century. But I think there is something to the attempts to quantify the impact, even if the result is only very rough, as it helps us better understand the scale of what an individual can achieve for the world. In the case of Borlaug, my rough guess is that the real number is in the tens to hundreds of millions—still a fantastic achievement, and possibly more than anyone else in human history. A great resource on individuals whose work saved millions of lives is *Scientists Greater than Einstein* (Woodward, Shurkin & Gordon, 2009) and its website www.scienceheroes.com, which estimates that Borlaug saved about 260 million lives.

98   UN DESA (2019). Much of the remaining growth is thus from demographic inertia (the disproportionate fraction of people of child-bearing age due to higher past fertility) rather than because people are having many children.

99   Adapted from Roser, Ritchie & Ortiz-Ospina (2019).

100   Wise (2013); Gietel-Basten (2016); Bricker & Ibitson (2019).

101   It is not clear that further increases in power and prosperity will continue to mean more (adverse) impact on the environment. For one thing, humans value a flourishing environment so they use some of their new-found wealth and power to heal it and to switch their consumption to more expensive but less harmful products. This has led to a theory of the *environmental Kuznets curve*, which posits that adverse environmental impacts increase with per capita income during industrialisation, but eventually start to fall back down as societies become richer still. There is some evidence to support this, but it is mixed. The theory probably applies to some types of environmental damage but not others, and offers no guarantees when the turning point will come. There is also the issue that the poorer countries are still near the start of their curves, so things might get worse before getting better, regardless. See Stern (2004).

It is often suggested that economic growth will have to stop (or consumption plateau) if we are to avoid destruction of our finite planet. But this is not as obvious as it first seems. The way economists define consumption and growth includes goods and services such as education, software, art, research and medicine, where additional value can be created without concomitant environmental costs. We could also have growth in green technologies that displace damaging consumption. While environmentally costly consumption would need to plateau, there is nothing in theory preventing us from focusing on growth in these other areas, creating a world with beauty, knowledge and health that far surpass what we have today, while *reducing* our environmental impact. This seems to me to be a superior goal to that of limiting all growth.

102   Fossil fuels (Roberts, 2004); phosphorous (Cordell, Drangert & White, 2009); topsoil (Arsenault, 2014); fresh water (Gleick & Palaniappan, 2010); metals (Desjardins, 2014).

103 Accessible fresh water is estimated as ~2% of the world's ground-water (Boberg, 2005).

104 As these will then be in people's narrow self-interest. There was a famous bet between the business professor Julian Simon and Paul Ehrlich about this, operationalised in terms of whether a representative basket of raw materials would increase in price (representing scarcity) or decrease (representing abundance) over time. Simon won the bet, though this was quite dependent on the exact choices of resource and time period, so little should be read into it.

105 Kolbert (2014); Ceballos et al. (2015).

106 Neither set of data is representative of all species: the fossil record is biased towards species that fossilise easily, while the modern statistics are biased towards species of interest and species that we already have reason to think might be under threat. There are also special statistical issues to do with the fact that the modern record samples extinction rates over very short time periods, where we should expect much more natural variability than over the million-year periods of the fossil record (Barnosky et al., 2011, pp. 51–2).

107 Ceballos et al. (2015).

108 Barnosky et al. (2011), p. 54. If all currently threatened species went extinct, the fraction would rise to about 30%. But it is unclear how to interpret this. It shows how we might be able to get roughly halfway to the mass extinction level, but it is not at all clear that such species *will* go extinct or that extinction would climb the rest of the way to 75%.

109 While one can never know for certain everything someone said or did not say, there is no reason to believe he ever made any pronouncements on bees. See O'Toole (2013).

110 Aizen et al. (2009).

5 FUTURE RISKS

1 Churchill (1946).

2 Rutherford's comments were made on 11 September 1933 (Kaempffert, 1933). His prediction was in fact partly self-defeating, as its confident pessimism grated on Szilard, inspiring him to search for a way to achieve what was said to be impossible

(Szilard & Feld, 1972, p. 529). There is some debate over the exact timing of Szilard's discovery and exactly how much of the puzzle he had solved (Wellerstein, 2014). Rutherford remained sceptical of atomic power until his death in 1937. There is a fascinating possibility that he was not wrong, but deliberately obscuring what he saw as a potential weapon of mass destruction
(Jenkin, 2011). But the point would still stand that the confident public assertions of the leading authorities were not to be trusted.

This conversation with Fermi was in 1939, just after nuclear fission in uranium had been discovered. Fermi was asked to clarify the 'remote possibility' and ventured 'ten percent'. Isidor Rabi, who was also present, replied, 'Ten percent is not a remote possibility if it means that we may die of it. If I have pneumonia and the doctor tells me that there is a remote possibility that I might die, and it's ten percent, I get excited about it' (Rhodes, 1986, p. 280).

Wilbur Wright explained to the Aéro-club de France in 1908: 'Scarcely ten years ago, all hope of flying had almost been abandoned; even the most convinced had become doubtful, and I confess that in 1901 I said to my brother Orville that men would not fly for 50 years. Two years later, we ourselves were making flight' (Holmes, 2008, p. 91).

3   It is a shame that scientists ruined their reputation for this. One can imagine a world in which scientists (at least at the very top) were more circumspect and only baldly asserted that something was impossible when it really was. When it would merely require a change of paradigm for the thing to be true—or when it would be the biggest surprise of the decade—they could simply say that. Having such a reputation for well-calibrated forecasts would have been a real asset for the scientific community, as well as being valuable for policymakers and the community at large.

But no matter how high a bar you draw for scientific eminence or credentials particularly related to the very claim in question, there are people who made claims of certainty or ridicule, yet turned out quite wrong.

4   The qualifier 'on humans' is necessary. The track record of technological progress and the environment is at best mixed, with many technologies inflicting severe harms. Sometimes cleaner technologies have been able to displace the harmful ones, and I believe

this trend will continue, making continued technological progress ultimately positive for the environment. But the evidence for this is mixed and time may prove me wrong. See Stern (2004) for a discussion of this.

5 As the Swedish mathematician, Olle Häggström, puts it in his excellent treatise on future technology (Häggström, 2016): 'the currently dominant attitude towards scientific and technological advances is tantamount to running blindfold and at full speed into a minefield'. While we don't know exactly how many mines are in the field, or whether we might survive treading on one, running through without looking is not the optimal policy. One disanalogy is that while there are few upsides to running into a minefield, technological and scientific progress holds enormous potential benefits. Perhaps I would alter the image to humanity running blindfolded through a minefield, in order to reach a safer and more desirable location.

6 Christakos et al. (2005), p. 107.

7 Christakos et al. (2005), pp. 108–9. While there are several diseases that could fit with the symptoms, there is substantial evidence that it was the bacterium *Yersinia pestis*, carried in the fleas of black rats. Most damningly, its DNA has been recovered from the bones of those who died in the Black Death (Haensch et al., 2010).

8 The mortality rate of the plague varied with the region and demographics. This makes it extremely difficult to extrapolate out the death toll from the limited historical data. Considering death tolls in England, Ziegler gives a credible range of 23–45%, and suggests that one-third is a reasonable estimate, which can be extended to Europe as a whole (Ziegler, 1969). More recently, Benedictow (2004) gave an estimate of 60%—much higher than traditional estimates and meeting with scepticism from many of his colleagues. My guess is that such a high toll is unlikely, but cannot yet be ruled out. From an estimated historical population for Europe at the time of 88 million, Ziegler's 23–45% range corresponds to 20–40 million and Benedictow's 60% corresponds to 53 million European deaths.

One often sees much higher death tolls for the Black Death in popular articles, including a startling figure of 200 million—far in excess of the 80 million people living in Europe at the time. Luke

Muehlhauser (2017) heroically tracked down the source of this statistic to a popular 1988 article in *National Geographic* (Duplaix, 1988), which was clearly not referring to the Black Death in particular, but was an attempt at a total figure for pandemics of the plague during the Middle Ages. It wasn't until the sixteenth century that the European population recovered to where it was before the Black Death (Livi-Bacci, 2017, p. 25).

9   This estimate is based largely on Muehlhauser (2017). My lower bound assumes 25% mortality in Europe, on a starting population of 88 million; 25% mortality in the Middle East, on a starting population of 5.4 million in Egypt and Syria, and 2 million elsewhere in the region; and no mortality in Asia, for a total of 24 million deaths. My upper bound assumes 50% mortality in Europe, on a starting population of 88 million; 25% mortality in the Middle East on a higher starting population of 9 million in Egypt and Syria, and 6 million elsewhere in the region; and 15 million deaths in China, for a total of 63 million deaths. World population in 1340 is taken as 442 million in both cases (Livi-Bacci, 2017, p. 25), giving global mortality estimates of 5.4% and 14.2%.

This is considerably more than some of the worst catastrophes in recent history, in terms of proportional death tolls: First World War (0.8%); Second World War (2.9%); 1918 influenza (3–6%). See Muehlhauser (2017) for an extensive discussion.

10   Early suggestions of extinction-level pandemics appear in Mary Shelley's novel *The Last Man* (1826) and H. G. Wells' (non-fiction) essay 'The Extinction of Man' (1894). As Wells put it: 'for all we know even now we may be quite unwittingly evolving some new and more terrible plague – a plague that will not take ten or twenty or thirty per cent, as plagues have done in the past, but the entire hundred.'

Joshua Lederberg's article 'Biological Warfare and the Extinction of Man' (1969) is the first I know of to seriously discuss the possibility of human extinction from engineered pathogens.

11   The Plague of Justinian may have altered the course of history by sending the Byzantine Empire into decline and allowing the ascent of Islam in the region.

In the absence of estimates for global death toll in the scholarship, Muehlhauser (2017) consulted an expert who suggested applying the 20% mortality rate in Constantinople (Stathakopoulos, 2004) to the population of the empire of 28 million (Stathakopoulos, 2008), for a total death toll of roughly 5.6 million. World population in 451 CE is estimated at 210 million (Roser, Ritchie & Ortiz-Ospina, 2019).

As with the Black Death, the Plague of Justinian reappeared several times over the following centuries. It took an even greater human toll over such a time frame, but these aggregate numbers don't really represent a single catastrophe and cannot be used to determine the fraction of the world's people who were killed.

12 See Nunn & Qian (2010) for a summary of estimates. Snow & Lanphear (1988) give estimates of mortality for various tribes in the north-eastern US ranging from 67 to 95%.

The island of Hispaniola provides a striking example of the difficulties in determining death toll. While estimates of the local population in the decades *after* colonisation are bounded in the tens of thousands, estimates of the 1492 population range from 60,000 to 8 million (Cook, 1998).

13 For this upper bound, I use the estimate of 60.5 million for the pre-Colombian population of the Americas from Koch et al. (2019), who propose 90% mortality over the next century. I take an estimate of rest-of-world population as 419 million, from Livi-Bacci (2017). Together this gives 54.5 million deaths on a world population of 479.5 million, or 11%, which I round to 10%.

This isn't strictly comparable with the other global mortality estimates in this section as it is a population *reduction*, not a count of deaths. It may undercount the total deaths (since new people were being born) or it may overcount them if it is partly a result of lowered birth rate rather than increased death rate. There is also a large mismatch in timespans. The Columbian exchange figures are estimated over a 100-year period, which is vastly longer than that of the Black Death, Plague of Justinian or 1918 flu.

Even with new immigrants arriving from the rest of the world, it took about three and a half centuries for the population of the Americas to recover to its pre-Columbian levels (1492–1840) (Roser, Ritchie, & Ortiz-Ospina, 2019).

14 Such global reach was made possible by the recent increases in travel speed of motorised transport, as well as the increased use of such transport for trade and troop movements. Estimates of the death toll are from Taubenberger & Morens (2006).

15 In addition to this historical evidence, there are some deeper bio-logical observations and theories suggesting that pathogens are unlikely to lead to the extinction of their hosts. These include the empirical anti-correlation between infectiousness and lethality, the extreme rarity of diseases that kill more than 75% of those infected, the observed tendency of pandemics to become less viru-lent as they progress and the theory of optimal virulence. However, there is no watertight case against pathogens leading to the extinc-tion of their hosts.

16 This is, of course, a near impossible claim to assess. The outbreak is estimated to have killed over 100,000 troops in the final stages of the war, so may have had some influence on the outcome (Wever & van Bergen, 2014). One could also point to the far-reaching impacts of single cases: US President Woodrow Wilson fought an infection in the months leading up to the 1919 Peace Conference, which may have played a role in his failure to secure his vision for peace (Honigsbaum, 2018).

17 At the time of writing, the world population is estimated to be 7.7 billion (UN DESA, 2019).

Estimates of world population immediately before the Agricultural Revolution are generally in the order of several million. Weeks (2015) gives a figure of 4 million. Coale (1974) and Durand (1977) put it at 5–10 million. Livi-Bacci (2017, p. 26) suggests population before 35,000 BCE was no more than several hundred thousand, growing very slowly to several million around the dawn of agriculture. Altogether, for most of human history, our population was 1,000–10,000 times smaller than it is today.

18 HIV (Keele, 2006); Ebola (Leroy et al., 2005); SARS (Cyranoski, 2017); influenza (Ma, Kahn & Richt, 2008). H5N1 flu, the high-mortality strain known as 'bird flu', originated in commercial poultry populations in Asia (Sims et al., 2005).

19 Jones et al. (2008). However, it is hard to be sure, as there are rival explanations for the observed increases, such as our increasing ability to detect and classify pathogens.

20 This may initially seem a profound change, but it probably doesn't weigh too heavily. Much of the history of our species (and related species) did not have the benefit of isolated populations anyway, yet these unified populations still had very low extinction rates.

21 Though of course a few scattered populations surviving the initial devastation would not guarantee humanity ultimately survives. We would still need a minimal viable population to repopulate our species and to successfully restore civilisation. We would also be exceptionally vulnerable to other risks during our reduced state.

22 CRISPR-Cas9 is a tool enabling researchers to edit genomes by removing, adding and changing portions of the DNA sequence. It was a significant breakthrough, allowing for much easier, cheaper and more accurate gene editing than was previously available.

A gene drive is a technique for propagating particular genes within a population, by increasing the probability that a given trait is inherited to over 50%. This is an incredibly powerful tool, allowing researchers to pursue genetic modification on a population level.

23 Cost per genome was $9 million in early 2007 (Wetterstrand, 2019). At the time of writing, Dante Labs offers whole genome sequencing for €599 (~$670) (Dante Labs, 2019).

One study finds publications in synthetic biology increasing by 660% when comparing 2012–2017 to 2000–2006 (Shapira & Kwon, 2018). VC funding for biotechnology increased from ~$3 billion in 2012 to ~$7 billion in 2016 (Lightbown, 2017).

24 Herfst et al. (2012).

25 Taubenberger & Morens (2006) estimate the mortality rate for the 1918 flu at upwards of 2.5%.

26 The experiment by Yoshihiro Kawaoka also involved producing a mammal-to-mammal transmissible strain of H5N1 using ferrets, though Kawaoka's experiment started with a hybrid of H5N1 and H1N1 viruses. He was initially asked to remove some details by the US National Science Advisory Board for Biosecurity, but his full paper ended up published in *Nature* (Imai et al., 2012; Butler & Ledford, 2012).

27 This lack of transparency seems to be driven by fear of embarrassment, which would clearly not be a sufficient reason for stifling this critical information. Stakeholders require these rates to assess whether labs are living up to their claimed standards and how much risk they are posing to the public. Getting BSL-3 and BSL-4 labs to provide this information (whether by appealing to

their conscience or through regulation) would seem to be a clear win for biosecurity.

28 One BSL-4 lab, the Galveston National Laboratory, must be commended for already voluntarily reporting its incidents (GNL, 2019). Others need to follow this lead: either voluntarily, through self-regulation or through government regulation.

 A promising avenue is to follow the Federal Aviation Administration's approach where after each plane crash it looks for lessons from what went wrong and ways that practices could be improved.

29 And there are many other worrying examples too. For instance in 2014, GlaxoSmithKline accidentally released 45 litres of concentrated polio virus into a river in Belgium (ECDC, 2014). In 2004, SARS escaped from the National Institute of Virology in Beijing. They didn't realise some of the workers had been infected until a worker's mother came down with it too. And in 2005 at the University of Medicine and Dentistry in New Jersey, three mice infected with bubonic plague went missing from the lab and were never found (U.S. Department of Homeland Security, 2008).

30 Details from Anderson (2008) and Spratt (2007). The maximum fine for such a breach was £5,000, and it is not clear whether it was even levied. Foot-and-mouth virus is a SAPO Category 4 pathogen, equivalent to BSL-4 but for animal pathogens.

31 The research of both Fouchier and Kawaoka was conducted in enhanced BSL-3 labs. This is the standard level for non-human-transmissible H5N1 (Chosewood & Wilson, 2009). But because the entire point of the experiment was to make it transmissible for a model animal that stands in for human, some of the experts believe that BSL-4 should have been required (Imperiale & Hanna, 2012). Others disagree, saying that enhanced BSL-3 is still appropriate (García-Sastre, 2012).

32 Shoham & Wolfson (2004); Zelicoff & Bellomo (2005).

33 Janet Parker, a medical photographer working at the hospital, may have been the last ever death from smallpox. Depressingly, just twelve years earlier, there had been an outbreak from the same building, which saw 73 people infected with a milder strain of the virus. The source was a medical photographer working in the same

studio in which Janet Parker was subsequently infected (Shooter et al., 1980).

34 Hilts (1994); Alibek (2008). The cause was remarkably mundane, according to Alibek's account of the incident. A technician had removed a clogged air filter for cleaning. He left a note, but it didn't get entered in the main logbook. So they turned the anthrax drying machines on at the start of the next shift and blew anthrax out over the city for several hours before someone noticed.

In a report on the accident, US microbiologist Raymond Zilinskas (1983) remarked: 'No nation would be so stupid as to locate a biological warfare facility within an approachable distance from a major population center.'

Note that while extremely lethal, anthrax does not spread from human to human, so there was no pandemic risk. Instead, the example is important as an example of catastrophic safety failure with known lethal agents.

35 Mutze, Cooke & Alexander (1998); Fenner & Fantini (1999).

36 Sosin (2015). There were no known infections.

37 Trevisanato (2007).

38 While Gabriel de Mussis gave a contemporary account of this taking place during the siege of Caffa, it is probably not a first-hand account and could easily have been an embellishment (Kelly, 2006). Even if true, there may well have been other ways for the Black Death to reach Europe (Wheelis, 2002).

39 The entire episode is extremely well documented, giving an idea of the level of acceptance and motivation behind the attack. Colonel Bouquet, who was to take charge of the fort, responded to Amherst's letter: 'I will try to inoculate [sic] the Indians by means of Blankets that may fall in their hands, taking care however not to get the disease myself.' Amherst wrote back, 'You will Do well to try to Innoculate [sic] the Indians by means of Blankets, as well as to try Every other method that can serve to Extirpate this Execrable Race.'

Even before Amherst's first request, Captain William Trent had documented '[We] gave them two Blankets and an Handkerchief out of the Small Pox Hospital. I hope it will have the desired effect.' Military records confirm that these were 'taken from people in the Hospital to Convey the Smallpox to the Indians'. The fort commander reimbursed them for the 'sundries got to replace in kind those which were taken from people in the hospital to convey smallpox to the Indians' (D'Errico, 2001).

40　The confirmed cases are Canada (1940–58), Egypt (1960s–?), France (1915–66?), Germany (1915–18), Iraq (1974–91), Israel (1948–?), Italy (1934–40), Japan (1934–45), Poland (?), Rhodesia (1977), South Africa (1981–93), Soviet Union (1928–91), Syria (1970s?–?), United Kingdom (1940–57), United States (1941–71) (Carus, 2017).

41　Leitenberg (2001), Cook & Woolf (2002). A great many details about the programme were published by the defector Ken Alibek in his book *Biohazard* (2008). However, it is not clear how reliable his account is, so I have only included details that have been independently substantiated.

42　The CDC (Centers for Disease Control and Prevention) notes 29 deaths and 31 injuries from criminal biological attacks between 1960 and 1999 (Tucker, 1999). During the Second World War 200,000 Chinese civilians are estimated to have been killed through a combination of biological and chemical attacks, perpetrated by the Japanese army. Some of these attacks were primitive, such as the release of thousands of plague-ridden rats, and it remains unclear how many of these deaths are attributable to what we would now consider biological weapons (Harris, 2002). It has also been suggested that the Rhodesian government used biological warfare against its own population in the late 1970s (Wilson et al., 2016).

43　While this must surely be a strike against bioweapons posing a large existential risk, it is unclear how strong the evidence really is. Pandemic deaths are dominated by extremely rare outlier events, which occur less than once a century. So a century of bioweapon development without major catastrophe provides little statistical evidence. To really get all we can out of the raw data, one would want to model the underlying distributions and see whether the biowarfare and bioterror distributions have heavier tails than that of natural pandemics.

44　Technical mastery of the techniques required is not something that can be learned from academic materials alone. Even experienced scientists can find it very difficult to learn new techniques without in-person training. Successfully operating such a large project in complete secrecy is fraught with difficulties, and is perhaps the greatest operational barrier. See Danzig et al. (2011) for a discussion of this in relation to the case of Aum Shinrikyo.

45 Acemoglu (2013); RAND (n.d.).

46 In a power law, the probability of an event of size x is proportional to x^α, where α is a parameter lower than -1. The closer α is to -1, the more slowly the probability of extreme events drops off, and the more extreme the statistical behaviour.

The power laws I mentioned as having especially heavy tails are those with α between -2 and -1. These distributions are so extreme that they don't even have well-defined means: the probability of larger and larger events drops off so slowly that the sum corresponding to their expected size fails to converge. The value of α for war is estimated to be -1.4 (Cederman, 2003) and -1.8 for terrorism with biological or chemical agents (Clauset & Young, 2005).

There is much debate about whether the distributions of various disasters are *really* power laws. For example, log-normal distributions have right-hand tails that approximate a power law, so could be mistaken for them, but have a lower probability of small events than in a true power law. For our purpose, we don't really need to distinguish between different heavy-tailed distributions. We are really just interested in whether the right-hand tail (the distribution of large events) behaves as a power law ($\sim x^\alpha$), what its exponent is, and over what domain the power law relationship actually holds.

Any actual distribution will only be well fitted by a power law up to some level. Beyond some point other limits (such as the total population who could be affected) will kick in and the real probability will usually be less than that given by the power law. This makes the use of power laws to model the chances of events outside the observed domain quite speculative (though this is less problematic if using it as an upper bound to the true probability). This also implies that the actual distributions probably *do* have means, though these could easily be higher than the mean of the historical record, or higher even than the highest observed event so far.

47 The cost to create the first human genome sequence is estimated at \$0.5–1 billion (~\$0.7–1.4 billion when adjusted for inflation)

(NHGRI, 2018). At the time of writing, Dante Labs offers whole genome sequencing for €599 (see note 23 to this chapter).

48 Carlson (2016).

49 The first gene drives were performed in 2015 (DiCarlo et al., 2015), and one team planned on using these techniques in the 2016 competition (iGEM Minnesota Team, 2016). The landmark paper on gene editing using CRISP-Cas9 was published in August 2012 (Jinek et al., 2012), and several teams used this method in the 2013 competition (iGEM, 2013).

50 Danzig et al. (2011). Eight people were killed, and 200 injured, in the group's first sarin attack in 1994, which targeted judges involved in cases against the group. Soon after, they murdered a suspected police informant using VX nerve agent. The following year, they killed 13 people and injured 6,000 in their Tokyo subway attack.

51 In a letter to Einstein regarding the contents of their Manifesto (dated 11 February 1955) Russell wrote: 'although the H-bomb at the moment occupies the centre of attention, it does not exhaust the destructive possibilities of science, and it is probable that the dangers of bacteriological warfare may before long become just as great' (Russell, 2012).

52 From Lederberg's article 'Biological Warfare and the Extinction of Man' (1969).

53 For example, Pinker writes 'Bioterrorism may be another phantom menace. Biological weapons, renounced in a 1972 international convention by virtually every nation, have played no role in modern warfare. The ban was driven by a widespread revulsion at the very idea, but the world's militaries needed little convincing, because tiny living things make lousy weapons' (Pinker, 2018, p. 306).

54 See note 54 to Chapter 2.

55 Tucker (2001). The Nuclear Non-Proliferation Treaty verifies the compliance through the International Atomic Energy Agency, which has 2,500 staff. The Chemical Weapons Convention verifies compliance through the Organisation for the Prohibition of Chemical Weapons, which has 500 staff.

56 The Soviet Union signed the BWC in 1972 and ratified it in 1975 (Davenport, 2018). Their bioweapons programme ran from 1928 to at least 1991 (Carus, 2017).

57 South Africa signed the BWC in 1972 and ratified it in 1975 (Davenport, 2018). Their bioweapons programme ran from 1981 to 1993 (Gould & Folb, 2002).

58 Iraq signed the BWC in 1972 and ratified it in 1991 (Davenport, 2018). Their bioweapons programme ran from circa 1974 to 1991 (Carus, 2017).

59 In 2018, US National Security Advisor John Bolton said 'There are . . . states that are parties to the Biological Weapons Convention that we think are violating it' (Bolton & Azar, 2018).

60 Israel is known to have had a bioweapons programme in the past (Carus, 2017), and is one of just ten states that have neither signed nor ratified (182 states have signed, including all other developed nations) (Davenport, 2018).

61 Note that there are other hurdles remaining. It is not easy to create a functioning virus from its DNA alone (though it has been achieved by a small team). And it is currently difficult to replace the DNA in a bacterium. In addition, there are limits to how long a sequence can be synthesised, with the DNA of many organisms being currently out of reach.

62 DiEuliis, Carter & Gronvall (2017); IGSC (2018).

63 A common objection to mandatory screening is that it would allow rivals access to the intellectual property of the DNA sequences being ordered. But there look to be cryptographic solutions to this problem (Esvelt, 2018).

There is a strong track record of such locks being broken when they are used in consumer products, but they may still be able to provide some useful security by requiring the malicious actors to include a computer expert as well as a bio expert. They may also help to 'keep honest people honest', by removing the temptation for academic researchers to perform experiments on controlled pathogens.

64 The term was coined by Bostrom (2011b). They are also informally known as 'infohazards'.

65 See Bostrom, Douglas & Sandberg (2016) for an introduction to the idea, including a formal analysis and some solutions. Lewis (2018) applies the idea to biotechnology information hazards.

66 This situation is exacerbated by additional uncertainty in the size of the benefit or risk, for that creates a wider distribution of

estimates of the net benefit, such that the most optimistic outlier is even further from the centre of their peers.

I said that it only takes one overly rosy estimate to release the information, but if the scientist requires a journal to publish the information, then it actually requires two—an author and an editor. This hints at a technique for resolving this problem at the level of journal editors, where there are fewer parties to get on board. One approach, suggested by Lewis (2018), is to have the first journal to reject a paper on safety grounds share their decision with others, to avoid the paper's author shopping around until they find one of the journals whose estimate of the danger was especially optimistic.

Esvelt (2018) suggests encouraging pre-registration of potentially dangerous research, so that open and broad discussions about safety can happen before the dangerous information has been generated.

67   In 1999, Ayman al-Zawahiri (now leader of al-Qaeda) wrote of his plan to start researching chemical and biological weapons: 'Despite their extreme danger, we only became aware of them when the enemy drew our attention to them by repeatedly expressing concern that they can be produced simply' (Wright, 2002).

68   The treaty was the Geneva Protocol of 1925, which had a section outlawing first use of bacteriological weapons between its signatory states. The Japanese did not sign it until 1970 but gained information from its existence (Harris, 2002, p. 18).

69   Lewis et al. (2019).

70   For example, I have taken care to use only examples which are sufficiently widely known.

71   McCarthy et al. (1955). AI has some foundations going further back than the Dartmouth conference, but that summer of 1956 is usually considered the beginning of AI as a field of enquiry.

72   This is known as Moravec's Paradox, after AI and robotics pioneer Hans Moravec, who wrote in 1988: 'But as the number of demonstrations has mounted, it has become clear that it is comparatively easy to make computers exhibit adult-level performance in solving problems on intelligence tests or playing checkers, and difficult or impossible to give them the skills of a one-year-old when it comes to perception and mobility.'

73 Major improvements to the structure include convolutional neural networks (CNNs) and recurrent neural networks (RNNs). Major improvements to the training include adaptions to stochastic gradient descent, such as Adam and Nesterov Momentum. Hardware improvements were driven by the switch from CPUs to GPUs, and now to even more specialist hardware such as TPUs. These successes have all built on each other in a virtuous cycle—now that neural networks are so good, it is worth assembling the large datasets to train them and worth making specialist hardware to run them, which makes them all the better, warranting more and more investment.

74 He et al. (2015).

75 Phillips et al. (2011); Ranjan et al. (2018).

76 Translation (Hassan et al., 2018); generating photos (Karras et al., 2017); voice mimicry (Jia et al., 2018); autonomous driving (Kocić, Jovičić & Drndarević, 2019); stacking Lego blocks (Haarnoja et al., 2018).

77 Bernstein & Roberts (1958); IBM (2011). Over the past few decades, chess programs have gained around 50 Elo points per year, roughly half of which came from algorithmic improvements, and half from hardware gains (Grace, 2013).

78 Silver et al. (2018). One needs to be careful with such numbers to also consider the hardware used. AlphaZero was trained using an enormous amount of computing power (5,000 TPUs), meaning that in this four hours it was able to simulate a vast number of games of chess against other versions of itself, using these to work out how best to play. This is an important caveat in comparing AlphaZero's achievement to other breakthroughs as the software improvements are somewhat smaller than they first look. But I think the actual time taken is still the key figure for the relevance to AI risk, showing how quickly in real time an AI system could get out of control.

79 Strogatz (2018).

80 AlphaZero may have even exceeded the level where Go experts had thought perfect play lies. The conventional wisdom was that the best human players would need a three- or four-stone handicap to win if facing perfect play (Wilcox & Wilcox, 1996). After 30 hours, AlphaZero was more than 700 Elo points above the top

professional. While it is difficult to convert between handicap stones and Elo at these extremely high levels of play, this is in the same ballpark as the predictions for perfect play (Labelle, 2017). It would be fascinating to see a version of AlphaZero play against the best humans with increasing handicaps to see how many stones ahead it really is.

81 Technically Ke Jie was referring to the 'Master' version of AlphaGo Zero which preceded AlphaZero (*Wall Street Journal*, 2017).

82 The breakthrough result was the DQN algorithm (Mnih et al., 2015) which successfully married deep learning and reinforcement learning. DQN gave human-level performance on 29 out of 49 Atari games. But it was wasn't fully general: like AlphaZero, it needed a different copy of the network to be trained for each game. Subsequent work has trained a single network that can play all games at human level or above, achieving an average score of 60% human level (Espeholt et al., 2018).

83 Attendance at one of the leading conferences, NeurIPS, increased by a factor of 4.8 between 2012 and 2018. AI VC funding increased by a factor of 4.5 between 2013 and 2018 (Shoham et al., 2018).

84 Adapted from Brundage et al. (2018), Coles (1994) and Shoham et al. (2018). Images from Goodfellow et al. (2014), Radford, Metz & Chintala (2015), Liu & Tuzel (2016) and Karras et al. (2017).

85 The survey was given to all researchers who published at two of the top machine-learning conferences in 2015 (NeurIPS and ICML). The data comes from the 352 researchers who responded (Grace et al., 2018).

86 Interestingly there was a large and statistically significant difference between the timelines of researchers from different continents. Researchers from North America thought the chance would reach 50% in 74 years' time, while those from Asia thought it would reach 50% in just 30 years (Europeans were about halfway between).

Note also that this estimate may be quite unstable. A subset of the participants were asked a slightly different question instead (emphasising the employment consequences by talking of all *occupations* instead of all *tasks*). Their time by which there would be a 50% chance of this standard being met was 2138, with a 10% chance of it happening as early as 2036. I don't know how to interpret this discrepancy, but it suggests we take these estimates cautiously.

87 Taking anything else as the starting point would require believing you could systematically do better than the relevant technical community at predicting their success. Perhaps one ground for thinking this would be that the experts are biased towards optimism about achieving their goals—but note that the public expect AGI even sooner (Zhang & Dafoe, 2019).

88 This analogy isn't perfect. AI researchers are not trying to build a new species to let loose in the wild, but to create new entities that can solve problems. However a growing number of them are trying to do this via general purpose intelligence, which involves agency and initiative. As we shall see, the current paradigm of AGI would naturally acquire sub-goals of controlling the world to protect itself and secure its ends.

In theory, there could be multiple species each in control of their own destiny, if they had limited ambitions and lacked the power to substantially interfere with each other.

89 It is certainly conceivable that our values ultimately boil down to something simple, such as the classical utilitarian doctrine of increasing the sum of positive experience. But even here there are two major challenges. First, even positive experience is too complex and poorly understood for us to be currently able to specify to the agent. Maybe in the future when we understand the nature of experience there will be a simple formulation, but there isn't yet. Second, the question of whether classical utilitarianism is the best moral principle is (fiercely) disputed. If we implemented that and were wrong—perhaps missing other key features of what is good in life, or about how it should ideally be distributed—we could lock in a greatly inferior world. I'm more sympathetic to classical utilitarianism than most philosophers, but wouldn't want to risk this. I think we should all take such moral uncertainty seriously.

90 My own view on this is that human values actually have a lot in common. For good reasons, we devote most of our attention to the differences, rather than to the facts that we are almost all in favour of longer, healthier, more prosperous lives, control over our life paths, a flourishing environment and so forth. I'd propose something like having the AI systems promote lives agreed

values while being cautious about the disputed or uncertain values—making sure to leave humans in control of resolving these disputes or uncertainties in the future, through our own reflection and discussion.

91 There are technical ways of understanding what is going on here. Stuart Russell (2014) likens it to a common issue in optimisation: 'A system that is optimizing a function of n variables, where the objective depends on a subset of size $k<n$, will often set the remaining unconstrained variables to extreme values; if one of those unconstrained variables is actually something we care about, the solution found may be highly undesirable.'

Alignment researchers liken the situation to *Goodhart's Law* (Goodhart, 1975): 'Any observed statistical regularity will tend to collapse once pressure is placed upon it for control purposes.' This law was originally proposed to think about the problems of setting targets that correlate with what we really want. While the targets may get met, they often cease to correspond with what we ultimately cared about in the process.

92 This could come up in one of two ways. Model-based systems would plan out the consequences of being turned off and see that it would greatly restrict the space of all future trajectories, typically cutting off many of the best ones. Actions leading to being turned off would therefore be assigned very low values.

Model-free systems could also learn to avoid being turned off. Orseau & Armstrong (2016) show how if the agent is sometimes shut down while learning, this can lead to biases in its learnt behaviour (they also suggest a potential solution to the problem).

For most of what follows, I'll assume the advanced AI is model-based. Or at least that it is capable of using its background knowledge of the world to succeed at some complex and difficult tasks on the first try, rather than always requiring many thousands of failed attempts before groping its way to success. While this takes us somewhat beyond current systems at the time of writing, it is compatible with the current paradigm and is a prerequisite to being counted as generally intelligent. I won't need to assume that it is better at succeeding on its first attempts than a human.

93 Omohundro (2008); Bostrom (2012). For a detailed explanation of how these instrumental goals could lead to very bad outcomes for humanity, see Nick Bostrom's *Superintelligence* (2014).

94 Learning algorithms rarely deal with the possibility of changes to the reward function at future times. So it is ambiguous whether they would assess future states by the current reward function or the future reward function. Researchers have begun to explore these possibilities (Everitt et al., 2016), and they each come with challenges. Using the future reward function helps with the problem of agents resisting human efforts to bring their reward function into better alignment, but it exacerbates the problem of agents being incentivised to 'wire-head'—changing their own reward function into one that is more easily satisfied.

95 Several of these instrumental goals are examples of 'distribution shifts'—situations where the agent faces importantly different situations during deployment that lead to it taking actions that were never exhibited during training or testing. In this case, the agent may never have opportunities to become more powerful than its human controllers during testing, and thus never have a need to exhibit its behaviours involving deception or seizing control of resources.

96 For example, in *Enlightenment Now*, Steven Pinker (2018, pp. 299–300) says of AI risk scenarios that they: 'depend on the premises that . . . (2) the AI would be so brilliant that it could figure out how to transmute elements and rewire brains, yet so imbecilic that it would wreak havoc based on elementary blunders of misunderstanding'.

97 Also, note that an agent may well be able to notice the general issue that its values are likely to be misaligned with ours (warranting an adversarial approach to humanity) even without having a perfect understanding of our values. In that case, even if it was designed to try to replace its existing values with our own, it could still be misaligned, albeit less dangerously so.

There are several promising lines of alignment research related to allowing agents to update their reward functions to better align with ours. One is the broad set of ideas around 'corrigibility'—how to make agents that don't resist changes to their goals. Another is the uncertainty-based approach to reward learning, where rather than acting as if it is certain of its current guess as to the human's values, the agent acts as if it is in a state of moral uncertainty, with degrees of belief in various human values based on the evidence it has seen so far (Russell, 2019). That incentivises the agent to defer to humans (who know more about their values) and to ask for guidance when it needs it. Given my own work on philosophical issues around moral uncertainty, I find this approach particularly promising (MacAskill & Ord, 2018; MacAskill, Bykvist & Ord, forthcoming). And getting it right will require further engagement with this part of philosophy.

98　Indeed, humans are likely to be cheaper and more effective at general-purpose physical action in the world for some time, making robots very much a second-best choice.

99　Even if 99% were erased, it would still have tens of copies remaining, ready to repopulate onto any new computers that were built.

100　There are several known cases in which criminals have taken over more than a million computers. The largest known botnet was Bredolab, which contained more than 30 million computers. It was created by a hacker network and made money by leasing out time on the hijacked computers to other criminal organisations. At its height it was able to send out more than 3 billion infected emails every day.

101　It is interesting to consider just how much power they acquired. In 1942 the Axis powers (excluding Japan) controlled ~\$1.3 trillion in GDP (1990-dollars) (Harrison, 1998), representing ~30% of the world's GDP (Maddison, 2010).

The Soviet Union covered 22.4 million square kilometres, 16% of the world's land area. At its height, during the reign of Genghis Khan's grandson, Kublai Khan, the Mongol Empire had a population of 100 million (Lee, 2009), representing ~25% of world population (Roser et al., 2019).

Given how much their nations were focused on war, their fractions of world military power would probably be even higher than these numbers indicate, but this is harder to objectively measure.

102　One way this could be a good outcome is if the AI system itself was a worthy successor to humanity, creating as good a future as humanity could have hoped to achieve. One sometimes hears this offered as a reason not to worry about the risks of unaligned AI.

While I think there is something to this idea, it is by no means a panacea. Once we take seriously the idea that our best future may involve replacing ourselves, it would seem unlikely that all such replacements are equally good. And effectively licensing any group of programmers to unilaterally trigger such a whole-sale replacement of humanity would be an appalling process for deciding how we should pass on the torch. Furthermore, if we think that the AI system may itself be a subject of moral value, this raises serious possibilities that it could in fact suffer, or otherwise produce a world of negative value—especially if designed by humanity at a time when we know so little about the nature of conscious experience.

103 Metz (2018).

104 Stuart Russell (2015): 'As Steve Omohundro, Nick Bostrom, and others have explained, the combination of value misalignment with increasingly capable decision-making systems can lead to problems—perhaps even species-ending problems if the machines are more capable than humans. Some have argued that there is no conceivable risk to humanity for centuries to come, perhaps for-getting that the interval of time between Rutherford's confident assertion that atomic energy would never be feasibly extracted and Szilárd's invention of the neutron-induced nuclear chain reac-tion was less than twenty-four hours.'

   Russell's Center for Human-Compatible AI is one of the leading research centres for AI alignment. His book, *Human Compatible* (2019), is a ground-breaking and readable introduction to the problem of building safe artificial general intelligence.

105 Shane Legg, co-founder and Chief Scientist of DeepMind, also leads their research on AI safety. When he was asked in an interview about the chance of human extinction within a year of developing AGI, he said (Legg & Kruel, 2011): 'I don't know. Maybe 5%, maybe 50%. I don't think anybody has a good estimate of this . . . It's my number 1 risk for this century, with an engineered biological pathogen coming a close second (though I know little about the latter).'

106 Alan Turing (1951), co-inventor of the computer and one of the founders of the field of artificial intelligence: '. . . it seems prob-able that once the machine thinking method had started, it would

not take long to outstrip our feeble powers. There would be no question of the machines dying, and they would be able to converse with each other to sharpen their wits. At some stage therefore we should have to expect the machines to take control, in the way that is mentioned in Samuel Butler's "Erewhon".'

This novel by Samuel Butler (1872) was developed from his 1863 essay 'Darwin among the machines' (Butler, 1863), which was arguably the first consideration of existential risk from intelligent machines.

Turing's concerns were echoed by his fellow code-breaker, the eminent statistician and computing pioneer I. J. Good (1959): 'Once a machine is designed that is good enough . . . it can be put to work designing an even better machine. At this point an "explosion" will clearly occur; all the problems of science and technology will be handed over to machines and it will no longer be necessary for people to work. Whether this will lead to a Utopia or to the extermination of the human race will depend on how the problem is handled by the machines. The important thing will be to give them the aim of serving human beings.'

The AI pioneer Norbert Wiener discussed the problem of retaining human oversight over advanced AI systems (Wiener, 1960): 'Though machines are theoretically subject to human criticism, such criticism may be ineffective until long after it is relevant. To be effective in warding off disastrous consequences, our understanding of our man-made machines should in general develop *pari passu* with the performance of the machine. By the very slowness of our human actions, our effective control of our machines may be nullified.'

Marvin Minsky (1984) warned of the risks of creating powerful AI servants who may misunderstand our true goals: 'The first risk is that it is always dangerous to try to relieve ourselves of the responsibility of understanding exactly how our wishes will be realized . . . the greater the range of possible methods we leave to those servants, the more we expose ourselves to accidents and incidents. When we delegate those responsibilities, then we may not realize, before it is too late to turn back, that our goals have been misinterpreted, perhaps even maliciously. We see this in such classic tales of fate as *Faust*, the *Sorcerer's Apprentice*, or the *Monkey's Paw* by W.W. Jacobs.'

Richard Sutton, a pioneer of reinforcement learning and co-author of the most widely used textbook on the subject, states there is 'certainly a significant chance within all of our expected lifetimes' that human-level AI will be created, then goes on to say the AI agents 'will not be under our control', 'will compete and cooperate with us', and that 'if we make superintelligent slaves, then we will have superintelligent adversaries'. He concludes that 'We need to set up mechanisms (social, legal, political, cultural) to ensure that this works out well' but that 'inevitably, conventional humans will be less important' (Sutton, 2015), (Alexander, 2015).

Other top researchers who have spoken publicly about risks from AI include Jeff Clune (2019): '. . . even if there is a small chance that we create dangerous AI or untold suffering, the costs are so great that we should discuss that possibility. As an analogy, if there were a 1% chance that a civilization-ending asteroid could hit Earth in a decade or ten, we would be foolish not to begin discussing how to track it and prevent that catastrophe.'

Ian Goodfellow (OSTP, 2016): 'Over the very long term, it will be important to build AI systems which understand and are aligned with their users' values ... Researchers are beginning to investigate this challenge; public funding could help the community address the challenge early rather than trying to react to serious problems after they occur.'

107   Szilard & Winsor (1968), pp. 107–8.

108   While the open letter spoke mainly in generalities (Future of Life Institute, 2015), its attached research agenda explicitly spoke of the need to investigate risks to humanity of the type discussed in this section (Russell, Dewey & Tegmark, 2015).

109   Future of Life Institute (2017).

110   This was an excerpt from Russell (2014), and is similar to the arguments given here.

111   This is a shockingly high estimate. The only other community of researchers I can imagine who may have estimated such a high probability of their work leading to catastrophically bad outcomes for humanity are the atomic scientists in the lead-up to the bomb. And yet I'm grateful to the researchers for their frank honesty about this.

112   See, for example, Pinker in *Enlightenment Now* (2018, p. 300).

113 Demis Hassabis has addressed these issues explicitly (Bengio et al., 2017): 'The coordination problem is one thing we should focus on now. We want to avoid this harmful race to the finish where corner-cutting starts happening and safety gets cut. That's going to be a big issue on global scale.'

114 If systems could improve their own intelligence, there is a chance this would lead to a cascade called an 'intelligence explosion'. This could happen if improvements to a system's intelligence made it all the more capable of making further improvements. This possibility was first noted by I. J. Good (1959) and is a plausible mechanism whereby progress could rapidly spiral out of control. In the survey mentioned earlier (Grace et al., 2018), 29% of respondents thought it likely that the argument for why we should expect an intelligence explosion is broadly correct.

But it is by no means guaranteed. For one thing, it wouldn't just require an AI that is more intellectually capable of AI research than a human—it would need to be more capable than the entire AI research community (at a comparable cost). So there may be time to detect that a system is at a roughly human level before it is capable of an intelligence explosion.

Moreover, it is possible that the difficulty of making successive improvements to a system's intelligence could increase faster than its intelligence does, making the 'explosion' quickly peter out. Indeed, there presumably must be *some* point at which it peters out like this, so the real question is whether there is any part near the beginning of the process during which improvements beget greater improvements. For some exploration of this, see Bostrom (2014, p. 66).

115 Metz (2018). Stuart Russell concurs (Flatow, Russell & Koch, 2014): 'What I'm finding is that senior people in the field who have never publicly evinced any concern before are privately thinking that we do need to take this issue very seriously, and the sooner we take it seriously the better.'

The advice of Hassabis and Russell echoes a prescient warning by I. J. Good (1970): 'Even if the chance that the ultraintelligent machine will be available is small, the repercussions would be so enormous, good or bad, that it is not too early to entertain the possibility. In any case by 1980 I hope that the implications and the safeguards will have been thoroughly discussed, and this is my main reason for airing the matter: an association for considering it should be started.'

If we had taken Good's advice, we might have been decades ahead of where we are now, facing only a small and well-managed risk.

116 In the worst cases, perhaps even negative value—an outcome worse than extinction.

117 Orwell (1949, p. 121) is explicit on this point: 'There are only four ways in which a ruling group can fall from power. Either it is conquered from without, or it governs so inefficiently that the masses are stirred to revolt, or it allows a strong and discontented Middle Group to come into being, or it loses its own self-confidence and willingness to govern. These causes do not operate singly, and as a rule all four of them are present in some degree. A ruling class which could guard against all of them would remain in power permanently.'

118 One could even make a case that this was the first major existential risk. On this view, the first wave of anthropogenic existential risk came from these dangerous ideologies that had recently been discovered, which in a globalised world might be able to become unshakeable. The technological risks (starting with nuclear weapons) would be the second wave.

In support of this idea, one could point to Nazi Germany's rhetoric of a 'thousand-year Reich' as evidence of their ambitions for designing a truly lasting regime. And note that it would not have needed to be a very high probability for it to count—perhaps a one in a thousand chance would be enough (as that is more than a century worth of natural risk).

While a work of fiction, *Nineteen Eighty-Four* shows that Orwell's concerns were about a truly existential catastrophe. He painted a vision of a world that was in perpetual conflict between three totalitarian superpowers, each of which used insidious social and technological means to make rebellion impossible. A key aspect of the vision was that it might represent a *permanent* dystopia: 'If you want a picture of the future, imagine a boot stamping on a human face – forever.' Foreshadowing the idea of existential risk, he likened the outcome to extinction: 'If you are a man, Winston, you are the last man. Your kind is extinct; we are the inheritors. Do you understand that you are *alone*? You are outside history, you are non-existent.'

While the book was published in 1949, his letters during the war attest that he was genuinely worried about these possibilities (it wasn't mere speculation) and that many of the ideas pre-dated the development of atomic weapons (Orwell, 2013). In Orwell's example, almost no one—not even the elites—had anything of real value in their lives. That is one possibility, but clearly outcomes short of that could also be dystopian, as the immiseration of the many could easily outweigh the enrichment of the few.

While I think this is a credible case, overall I'm inclined to still treat nuclear weapons (either the threat of igniting the atmosphere or of global nuclear war) as the first major (and anthropogenic) existential risk.

119 Though note that even if they didn't initially have aims of global expansion, things may still have gone in that direction once their ability to do so was assured. And even if they only started with the means to create a regime that lasted decades, that may have bought them enough time to develop the technological and social methods to expand that to centuries, which would buy the time to expand the reach much further again.

120 Scott Alexander's *Meditations on Moloch* (2014) is a powerful exploration of such possibilities.

121 C. S. Lewis (1943) alludes to this possibility when considering the increases in humanity's power over nature, particularly through genetic technologies. He notes that this power may well increase to a point where a single generation (or part of one) can effectively control the subsequent direction of humanity and all successive generations: 'The real picture is that of one dominant age – let us suppose the hundredth century AD – which resists all previous ages most successfully and dominates all subsequent ages most irresistibly, and thus is the real master of the human species. but then within this master generation (itself an infinitesimal minority of the species) the power will be exercised by a minority smaller still. Man's conquest of nature, if the dreams of some scientific planners are realized, means the rule of a few hundreds of men over billions upon billions of men.'

122 Most normative views have this counter-intuitive property because the behaviours the view recommends don't factor in the chance that the view itself could be false. Since there would be little cost to locking in a true view and much to gain, they would often recommend this. Notable exceptions would include views that pay no heed at all to consequences (such that they don't see much concern about an action that would lead to a different view taking over forever) and views that have strong liberal presumptions built in (such that even if concerned about that prospect, the normative theory explicitly allows the freedom to reject it).

While I'm concerned about the possibility of prematurely converging to a normative theory we would all have reason to regret, I don't see the fact a theory recommends locking itself in as a reason to reject it. Rather, I think that the correct solution lies at the level of moral uncertainty, which I've written on extensively elsewhere (MacAskill, Bykvist & Ord, forthcoming). People should retain some uncertainty about which normative theories are true and this gives us strong moral uncertainty-driven reasons to resist locking in a theory, even if we think there is a high chance it is true—we need to hedge against the possibility that we are mistaken.

123 Bostrom (2013) warns against this, while Butler (1863, 1872) flirts with the prospect. It would all but guarantee that humanity's lifespan was limited by the background rate of natural risks (though of course anthropogenic risks could do us in even sooner). And it would ensure we never achieve the possibilities of longevity or settlement beyond the Earth which we would otherwise be capable of, and which may be an important part of our potential.

124 Worlds where we renounce any attempts to leave the Earth (perhaps in order to leave the heavens pristine) may be another plausible example. See Chapter 8 for some thoughts on why most of the value we could achieve may lie beyond the Earth.

125 The text was: 'No amendment shall be made to the Constitution which will authorize or give to Congress the power to abolish or interfere, within any State, with the domestic institutions thereof, including that of persons held to labor or service by the laws of said State.'

It passed Congress and was even endorsed by Lincoln, but was never ratified. It is unclear whether it would have actually worked, as it may have still been possible to repeal it first, and then abolish slavery. In addition, it wouldn't have *guaranteed* the perpetuation of slavery, as each slave-owning state could still abolish slavery within its own borders (Bryant, 2003).

126 At least not without being precipitated by an early development of advanced AGI. Such a catastrophe may well be better thought of as an AI risk.

127 Note that even on this view options can be instrumentally bad if they would close off many other options. So there would be instrumental value to closing off such options (for example, the option of deliberately causing our own extinction). One might thus conclude that the only thing we should lock in is the minimisation of lock-in.

This is an elegant and reasonable principle, but could probably be improved upon by simply delaying our ability to choose such options, or making them require a large supermajority (techniques that are often used when setting up binding multi-party agreements such as constitutions and contracts). That way we help avoid going extinct by accident (a clear failing of wisdom in any society), while still allowing for the unlikely possibility that we later come to realise our extinction would be for the best.

128 The main reason for this disparity in likelihood is that self-replicating machines appear to be a much more difficult techno-logical pathway compared to fabricators, taking much longer to become economically self-sustaining and thus receiving much less development work (Phoenix & Drexler, 2004). However, scientific curiosity and niche applications would eventually tempt us to try it, so safeguards would need to be found.

Atomically precise manufacturing could enable cheap production of nuclear weapons. One might have thought this impossible, since the technology only rearranges atoms, and uranium or plutonium atoms are not part of the feedstock. But uranium is actually present in seawater at a concentration that is just short of being economically viable to extract with existing technology (Schneider & Sachde, 2013). By reducing the costs of making the equipment needed to extract it (and reducing the energy costs of running it by allowing cheap solar power), atomically precise manufacturing may make uranium much more accessible. A bad actor would still need to enrich the concentration of uranium-235 to make weapons-grade material, but this too is likely to become cheaper and easier in the wake of such a powerful general-purpose manufacturing technology.

129 We can see that the chance of back-contamination from Mars causing an existential catastrophe must be very small by considering the hurdles it would have to overcome. There would have to *be* life on the Martian surface, despite its extreme inhospitability and the fact that all prior attempts to find evidence of life there have come up empty. And even if there was, it would need to be able to flourish in the very different environment of Earth, and to be able to produce existential damage. Official studies of this risk have generally concluded that it is very low, but have advised extreme caution nonetheless (Ammann et al., 2012).

Even though the chance looks small, the need for measures to prevent back-contamination from other planets or moons is an agreed part of the 1967 Outer Space Treaty, which is the basic legal framework for international space law. And the existential risk from back-contamination was one of the first risks mentioned in the philosophical literature (Smart, 1973, p. 65): 'Similar long-term catastrophic consequences must be envisaged in planning flight to other planets, if there is any probability, even quite a small one, that these planets possess viruses or bacteria, to which terrestrial organisms would have no immunity.'

The risk of back-contamination first came up at the time of the Moon landing. It hadn't yet been confirmed that the Moon was sterile, so the US Surgeon General pushed NASA to institute serious protections, saying that it was not outlandish to set aside 1% of the budget to guard against great catastrophe on Earth (Atkinson, 2009). In the end, they spent around $8 million, less than a twentieth of a percent of the cost of the Apollo Program, on its quarantine system (Mangus & Larsen, 2004).

From our increased knowledge of microbiology, we can now see that this system was inadequate (in particular, we didn't yet know about the possibility of ultramicrobacteria or gene transfer agents). This should give us pause about the validity of our current knowledge.

But worse, there was little attempt to get scrutiny of the techniques, and insufficient will to actually use the quarantine system when it conflicted with other mission objectives. For example, after the original plan to lift the command module from the sea was thwarted by the lack of an appropriate crane, it was decided that the astronauts would have to leave the module while it was floating in the sea. This almost certainly let some of the fine moon dust into the ocean, making many of the other aspects of the quarantine system moot (National Research Council, 2002). We thus failed to properly handle this (very small) existential risk.

From what I've seen of the discussion around back-contamination from Mars, these issues are now being taken much more seriously (and won't run into conflict with the needs of a human crew).

130 The chance of an environmental catastrophe from an unprotected sample is widely considered to be small. We can think of these protections as attempting to reduce this chance by a further factor of a million (Ammann et al., 2012).

I am quite sceptical that the probability can really be reduced by this amount, and they don't seem to be taking into account that even BSL-4 facilities have pathogen escapes (as seen earlier in this chapter). But maybe they will be able to reduce the chance of escape to something like one in a thousand, which would be a significant advance on the quarantine from the Apollo Program.

131 Though I would recommend simply waiting a few more decades, at which point we could presumably perform most of the same tests *in situ* on Mars. The risk may be small and well managed, but do we need to run it at all?

132 One reason it is unlikely is that there is no sign that there is any other intelligent life out there. See note 46 to Chapter 2 for an outline of my own thinking on the 'Fermi paradox' and whether we are indeed alone.

Another is that there are millions of other centuries in which it could have arrived, so the chance it first arrives now is very low. This would be offset if it was lying in wait, but then it would be vastly more advanced than us, leaving us at its mercy. This fact that it is very unlikely we could do anything meaningful to resist does not alter the total probability of existential risk via a hostile alien civilisation, but it does suggest that the risk may be moot, as there may be nothing we can realistically do to substantially increase or decrease it.

133 The dangers from passive SETI are analogous to those of opening email attachments from untrusted third parties. For example, if they have developed superintelligent AI, they could send an algorithm for an advanced hostile AI.

None of this is likely, but since these activities only matter if there really are nearby alien civilisations, the main relevant question is the ratio of peaceful to hostile civilisations. We have very little evidence about whether this is high or low, and there is no scientific consensus. Given the downside could be much bigger than the upside, this doesn't sound to me like a good situation in which to take active steps towards contact.

134 Of course in some sense, almost everything is unprecedented, being different from prior events in the small details. I'm not meaning to suggest that this suffices to produce a risk. Instead I'm only interested in cases where there is a major change to a prima facie important parameter, taking it outside the historical range. The exact kinds of thing that undermine the comfort of having a long track record of events like that with no catastrophes. Sometimes this could involve the conditions being unprecedented in the Universe, on Earth, since *Homo sapiens*, or since civilisation began—it depends on whether the concern is that it would collapse civilisation, cause our extinction, destroy our planet or do damage on a cosmic scale.

Kurt Vonnegut's 1963 novel *Cat's Cradle* is a powerful early exploration of this kind of existential risk. He imagines an artificial form of ice crystal ('ice-nine') which is solid at room temperature and which causes a chain reaction in liquid water, turning it all into ice-nine. Since this form of ice is not naturally occurring in Earth's history and has such dramatic properties, it would count as an unprecedented condition in my sense. In the book, this causes an existential catastrophe when it comes into contact with the Earth's oceans, turning all water into this strange solid state.

135 Consider that there are millions of scientists, so even if the experiments they try have just a one in a million subjective chance of extinction, that may still be too high. (Especially when there are biases and selection effects leading even conscientious scientists to systematically underestimate the risks.)

136   There are several key governance problems. First, the scientists typically apply their epistemic standard of high confidence that there is no catastrophe *conditional upon their base scientific theories and models being correct*, rather than considering the stakes or the track record of such theories and models (Ord, Hillerbrand & Sandberg, 2010).

   Second, the cases tend to be decided solely by scientists. While they are the appropriate experts on much at issue, they lack important expertise on risk analysis or on evaluating the stakes. It also creates various biases and conflicts of interest, where the very people whose jobs (or whose colleagues' jobs) depend on a verdict are responsible for deciding that verdict. And it goes against widely held norms where the people who would be exposed to a (disputed) threat should have some say or representation in determining whether it is allowed to go ahead.

   One possible improvement to this situation would be to have an international body that these issues could be brought to (perhaps part of the UN), with individuals responsible for assembling the best arguments for and against, and a scientifically literate judge to make a ruling on whether the experiment should proceed at this time, or whether it should be delayed until a better case can be made for it. I would envisage very few experiments being denied. (This model may work even if the judgements had no formal power.)

137   Nuclear weapons would not have made the list, as fission was only discovered in 1938. Nor would engineered pandemics, as genetic engineering was first demonstrated in the 1960s. The computer hadn't yet been invented, and it wasn't until the 1950s that the idea of artificial intelligence, and its associated risks, received serious discussion from scientists. The possibility of anthropogenic global warming can be traced back to 1896, but the hypothesis only began to receive support in the 1960s, and was only widely recognised as a risk in the 1980s.

   A 1937 US government report on the future of technology provides an excellent example of the difficulties in forecasting (Ogburn, 1937; Thomas, 2001). It did not include nuclear energy, antibiotics, jet aircraft, transistors, computers or anything regarding space.

138   Bostrom (2013; 2018).

6 The Risk Landscape

1 Einstein & *New York Times* (1946).
2 From an online survey by GitHub user 'Zonination' (Zonination, 2017). These results are very similar to those found in a CIA experiment of intelligence officers (Heuer, 1999).
3 This makes it hard to use the phrase 'it is highly unlikely that X' as part of an argument that more attention needs to be paid to X. This wouldn't be true if 'highly unlikely' corresponded to a probability range such as one in ten to one in 100, as these numbers could easily be used to suggest that the risk needed to be taken seriously. This causes problems for the IPCC, which uses such phrases instead of numerical probabilities in its reports.
4 Pinker (2018), p. 295.
5 A further reason some people avoid giving numbers is that they don't want to be pinned down, preferring the cloak of vagueness that comes with natural language. But I'd love to be pinned down, to lay my cards on the table and let others see if improvements can be made. It is only through such clarity and openness to being refuted that we make intellectual progress.
6 My estimate of one in 10,000 per century is equivalent to a life expectancy of 1 million years. While I think that it is plausible the true extinction rate is given by the mass extinction record, at about one in 100 million years, I also think there is a chance that the typical species lifespans are a better guide, which pulls the average up a lot. Recall that the fossil record only provides a reliable guide to natural *extinction* risks, not to other kinds of existential risk. Here I am assuming that they are roughly the same level as the extinction risks, as this seems about right for the risks we've explored in detail, but this is more uncertain.
7 This corresponds to a Bayesian approach of starting with a prior and updating it in light of the evidence. In these terms, what I'm suggesting is that we start with a weak prior set by base rates and other factors that would go into a best-guess estimate before you saw the hard evidence (if any). I don't see good arguments for starting with a prior probability very close to zero.
8 How do my estimates compare with others? Serious estimates of these risks by scholars who have given them careful attention are rare. But thankfully, several pioneers of existential risk have put

down their own numbers. John Leslie (1996, pp. 146–7) estimated a 30% risk over the next five centuries (after which he thought we'd very likely be on track to achieve our potential). Nick Bostrom (2002b) said about the total existential risk over the long term: 'My subjective opinion is that setting this probability lower than 25% would be misguided, and the best estimate may be considerably higher.' Martin Rees (2003) estimated a 50% chance of a global (though perhaps temporary) collapse of civilisation within the twenty-first century (the text of the book is unclear as to what scale of disaster is being estimated, but he has since clarified that it is the collapse of civilisation). My estimates are in line with theirs.

In addition, Carl Sagan provided an illustrative estimate of 60% chance of extinction over the next 100 years (Sagan, 1980, p. 314; Sagan et al., 1980). Interestingly, this was in 1980, prior to the discovery of nuclear winter and his seminal writing about the badness of human extinction. However, it is not clear if the estimate represents his considered view. For it was simply stated without further comment among a set of vital statistics for Earth as they might appear in a hypothetical 'Encyclopedia Galactica'.

9   Another problem with business-as-usual estimates is that it can be hard to define what it even means: what kind of response by humanity are we to use as the default? While the all-things-considered risk can be hard to estimate, it is quite well defined—it is simply one's credence that a catastrophe will occur.

10 Cotton-Barratt, Daniel & Sandberg (forthcoming).

11 The more precise rule is to imagine more fine-grained decrements, such as the factor that is easiest to reduce by 1% of its current level. Then apply this rule many times, working on one factor until another becomes easier to reduce by a percent and switching to that. If there are sufficient diminishing marginal returns to working on each factor and a large enough budget, one may end up working on all three factors.

12 The CSER team proposes three classifications in all: *critical systems*, *global spread mechanism* (which corresponds closely to 'Scaling'), and *prevention and mitigation failure* (which is related to the human elements of 'Prevention' and 'Response'). Their scheme was designed to classify the broader class of global catastrophic risks (which need not be existential). See Avin et al. (2018) for further detail.

13   One might think that this risk is in fact a certainty; that humanity will come to an end *eventually*. But recall that existential catastrophes are defined relative to the best we could achieve—catastrophes that destroy humanity's longterm potential. If humanity (or our descendants) go extinct after fulfilling our longterm potential, that is existential success, not failure. So the total existential risk is roughly the same as the chance that we fail to fulfil our longterm potential. The roughness comes from the possibility that we fail to fulfil our potential for some other reason, such as very gradually degrading our potential, or keeping our potential intact but never acting to fulfil it. These may still be serious threats to our longterm future, warranting detailed study and action. But they are not ways that our generation could destroy humanity's future and are thus not the subject of this book.

14   There are two main ways this assumption could fail: hellish catastrophes, and correlations between the objective probabilities of risks and the value of realising our potential.

Regarding the former, consider that a precise reckoning would have us compare the expected value of each risk—the product of their probability and their stakes. But if their stakes are very similar (say within 1% of each other), then we only lose a tiny amount of accuracy if we compare them by probability alone. And in many cases there is a strong argument for the stakes being within a percent or so of each other.

This is because the difference in value between a world where humanity fulfils its potential and one where we destroy our potential is typically much greater in absolute terms than the difference between the various outcomes in which our potential has been destroyed. For instance, extinction and a permanent collapse of civilisation are two ways that our future could be very small and contain very little value. The difference between these is therefore much smaller than the difference between either one and an expansive future where we have a thousand millennia of breathtaking achievement.

However, there are also existential risks where the value of the future would not come crashing down to something near zero, but would be very large and negative. These are cases where we achieve nearly the maximum in scale (time, space, technological capability), but fill this future with something of negative value.

The difference in value between such a hellish outcome and extinction could rival the difference between extinction and the best achievable future. For such risks one needs to modify the total risk approach. For example, if the risk involved a future as negative as the best future is positive, then you would want to weight this risk twice as much (or even abandon the total risk approach entirely, switching to the more cumbersome approach of expected value). As I think such risks are very low (even adjusting for this increased weighting) I won't explore the details.

The second issue concerns a subtle form of correlation—not between two risks, but between risks and the value of the future. There might be risks that are much more likely to occur in worlds with high potential. For example, if it is possible to create artificial intelligence that far surpasses humanity in every domain, this would increase the risk from misaligned AGI, but would also increase the value we could achieve using AGI that was aligned with human values. By ignoring this correlation, the total risk approach underweights the value of work on this risk.

This can be usefully understood in terms of there being a common cause for the risk and the benefit, producing the correlation. A high ceiling on technological capability might be another common cause between a variety of risks and extremely positive futures. I will set this possibility aside in the rest of the book, but it is an important issue for future work to explore.

15   Note that if several risks are very highly correlated, it may be best to think of them as a single risk: the risk that any of these related catastrophes happen. It could be named for the common cause of the catastrophes, rather than for the proximate causes.

16   Though it is possible for risks to become anticorrelated when they lie on divergent paths the future could take. For example, risks from lack of global coordination and risks from global totalitarianism.

17   In general, I caution very heavily against assuming statistical independence. Such an assumption often leads one to underestimate the chance of extreme events where all variables move the same way. A version of this assumption is at play when people assume variables are normally distributed (since a normal distribution arises from the sum of many independent variables via the central limit theorem). A famous example of this going wrong is in the Black–Scholes option pricing model, which assumes normal

distributions and thus grossly underestimates the chance of large correlated price movements.

However, the case of aggregating existential risks may be a rare situation where it is not too bad to assume independence as we are *less* worried about cases when many events happen together.

18 Today that would include the United States, Russia, China and Europe. By the end of the coming century the list could be quite different.

19 I was not involved in the original Global Burden of Disease study (World Bank, 1993; Jamison et al., 2006), but I played a very minor role in advising on the normative foundation of its more recent reports (GBD, 2012), making the case that the discount rate on health should be set to zero—one of the main changes from the earlier versions.

I was even more inspired by its companion project, *Disease Control Priorities in Developing Countries* (Jamison et al., 1993; Jamison et al., 2006). Rather than looking at how much ill-health was produced by each cause, it assessed how effective different health interventions were at preventing ill-health for each dollar spent. This opened my eyes to the startling differences in cost-effectiveness between different ways of improving health and how donating to the right charity can have hundreds or thousands of times as much impact (Ord, 2013). I went on to be an advisor to the third edition (Jamison, Gelband et al., 2018; Jamison, Alwan et al., 2018).

20 When I say 'increases' I mean to imply that the risk factor causes the increase in existential risk, not just that they are correlated. In particular, it needs to be the case that action on the risk factor will produce a corresponding change in the level of existential risk. This could be reflected in the mathematics by using Judea Pearl's *do* operator (e.g. $\Pr(X|do(f = f_{min}))$) (Pearl, 2000).

21 Economic risk factors could arise from a change in the absolute level of prosperity (poverty), the direction of change from the status quo (decline), or the rate of change (stagnation).

22 Even if they had only direct effects, we could still count them as risk factors (since the strict definition I'm using specifies only that they increase risk, not that they do so indirectly).

23 We could also think of artificial intelligence as a risk factor. An effect such as AI-related unemployment would not be an existential risk,

but since it may threaten massive political upheaval it could still be a risk factor. And one could even approach unaligned AGI as a risk factor rather than as an existential risk. For it isn't a stand-alone mechanism for destroying our potential, but rather a new source of agency; one that may be motivated to use whatever means are required to permanently wrest control of our future. If an artificially intelligent system does cause our downfall, it won't be by killing us with the sheer force of its intellect, but with engineered plagues or some other existential threat.

24 It is possible that there will be no minimum or maximum achievable value (either because the domain is infinite, or because it is an open interval). I've ignored this possibility in the main text for ease of presentation, but it presents no special obstacle. Since the probability of existential risk is bounded above and below, we can replace the expression $Pr(X|f = f_{max})$ with a supremum (the smallest probability higher than $Pr(X|f = f')$ for all achievable f') and replace $Pr(X|f = f_{min})$ with a corresponding infimum.

25 One could also consider the *range* of F. That is, the difference between $Pr(X|f = f_{min})$ and $Pr(X|f = f_{max})$. This corresponds to the sum of F's contribution and its potential, and is a property that doesn't depend on the status quo level.

26 One could also consider the elasticity of existential risk with respect to f near f_{sq}. That is, the proportional change in $Pr(X)$ with respect to a small proportional change in f. This is a unitless measure of sensitivity to the risk factor, allowing it to be compared between different risk factors.

27 This also raises the question of variables that both increase and decrease existential risk over different parts of their domains (i.e. where existential risk is non-monotonic in that variable). I suggest that so long as the effect on existential risk is monotonic within the domain of plausible levels of f, we think of it as a risk factor or security factor. But if it is non-monotonic within even this range, we need to think of it as a more complex kind of factor instead.

For example, if using degrees of warming above pre-industrial levels as a measure of climate change, it is possible that driving the temperature all the way back to pre-industrial levels (or beyond) would eventually become counterproductive to existential risk. However, so long as there is little practical danger of this kind of over-response, it makes sense to treat warming as a risk factor.

28 We can define measurements of a security factor that mirror those of risk factors. We can speak of its current contribution to our safety (to the chance that we *don't* suffer existential catastrophe), and of its potential for further lowering existential risk, were the factor to be increased as much as possible. Finally, we could consider the effect of marginal improvements in the security factor upon the total amount of existential risk.

29 We might generally suspect it to be easier to lower a risk by a percentage point if it starts off at 20% than if it starts at 5%. I used to think a good heuristic was that it should be equally easy to halve the probability of a risk, regardless of its starting point. But things might look different if the risk were extremely high, such as 99.9%. Then the catastrophe might seem over-determined to happen (for otherwise the risk would be lower). So a more subtle heuristic might be that it is equally easy to reduce the odds ratio of the risk by a factor of two. This would mean that risks with a middling probability (say 30% to 70%) would be in a sweet spot for ease of reducing by a percentage point.

But note that not all such risks are equal. In some cases the middling chances we assign to risks like AI might really represent our uncertainty between worlds where it is very hard and worlds where it is easy, so even though the average is middling, it may not be easy to move the probability with our work. (As a toy example, there could be one risk with a genuine 50% chance of occurring, and another that is either extremely unlikely (0.1%) or extremely likely (99.9%) but where we don't know which. This latter risk would be less tractable than its subjective level of risk (50%) would lead us to think.)

30 MacAskill (2015), pp. 180–5.

31 Sometimes it is the short-term neglectedness that counts: how many resources are being spent on it right now. But more often it is the longterm neglectedness: how many resources will be spent on it in total, before it is too late. When the allocation may soon change dramatically (such as when a field is taking off) these can be very different.

32 The way to define the terms is to note that cost-effectiveness is the rate of change of value with respect to the resources spent: d value

/ d resources. Owen Cotton-Barratt has shown that we can then break this into three factors (Wiblin, 2017):

$$\underbrace{\frac{d \text{ value}}{d \text{ resources}}}_{cost\text{-}effectiveness} = \underbrace{\frac{d \text{ value}}{d\% \text{ solved}}}_{importance} \times \underbrace{\frac{d\% \text{ solved}}{d\% \text{ resources}}}_{tractability} \times \underbrace{\frac{d\% \text{ resources}}{d \text{ resources}}}_{neglectedness}$$

33 I owe this point to Owen Cotton-Barratt.

34 Some risks may have a long time between the last moment when action could be usefully taken and the moment of catastrophe. When prioritising based on which risk comes first, we date them by the last moment action could be usefully taken.

   If our work becomes gradually less effective as time goes on, a more complex appraisal is needed. This is the case with climate change, where even though the catastrophic damages would be felt a long time from now, lowering emissions or developing alternative energy sources makes more difference the sooner we do it.

35 The diameter of NEOs fits a power-law distribution with exponent −3.35 (Chapman, 2004). The size of measles epidemics in isolated communities fits a power law with exponent −1.2 (Rhodes & Anderson, 1996). Fatalities from many other natural disasters—tsunamis, volcanoes, floods, hurricanes, tornadoes—also fit power-law distributions. This fit usually fails beyond some large size, where the actual probabilities of extremely large events are typically lower than a power law would predict (e.g. measles outbreaks are eventually limited by the size of the population). However, the warning-shot analysis still works so long as the power law provides an upper bound for the real probability.

36 I owe this point to Andrew Snyder-Beattie.

37 The 'Collingridge dilemma' is a special case of this leverage/near-sightedness trade-off as it relates to the regulation of new technologies. Collingridge (1982) notes that the further away one is from the deployment of some technology, the more power one has to control its trajectory, but the less one knows about its impacts.

38 He makes this point in his dissertation on existential risk and longtermism (Beckstead, 2013), which is one of the best texts on existential risk.

39 World governments spend 4.8% of GDP on education (World Bank, 2019b)—around $4 trillion per year. Targeted spending

on existential risk reduction is on the order of $100 million. See note 56 to Chapter 2 for more details on estimating spending on existential risk.

40　This is for money directed specifically at lowering existential risk, rather than more general money dedicated to areas associated with existential risk (such as climate change and biosecurity). Also, I'm imagining a world that looks like ours today. If an existential risk were to pose a clear emergency (such as a large asteroid bearing down on us), the amount of direct work warranted may be much higher.

7 SAFEGUARDING HUMANITY

1　Asimov (1979), p. 362. I have capitalised the initial 't'.
2　We could also characterise these as:

1. Avoid failing immediately & make it impossible to fail
2. Determine how to succeed
3. Succeed

3　Protecting our potential (and thus existential security more generally) involves locking in a commitment to avoid existential catastrophe. Seen in this light, there is an interesting tension with the idea of minimising lock-in (p. 158). What is happening is that we can best minimise overall lock-in (coming from existential risks) by locking in a small amount of other constraints.

But we should still be extremely careful locking anything in, as we might risk cutting off what would have turned out to be the best option. One option would be to not strictly lock in our commitment to avoid existential risk (e.g. by keeping total risk to a strict budget across all future centuries), but instead to make a slightly softer commitment that is merely very difficult to overturn. Constitutions are a good example, typically allowing for changes at later dates, but setting a very high bar to achieving this.

4　There are many ways one could do this: by avoiding new fires being started, by making sure the buildings are not fire dangers, or by employing a well-resourced fire department. There are analogous options for protecting our potential.

5　A numerical example may help explain this. First, suppose we succeeded in reducing existential risk down to 1% per century and kept it there. This would be an excellent start, but it would have to be supplemented by a commitment to further reduce the risk. Because at 1% per century, we would only have another 100 centuries on average before succumbing to existential catastrophe. This may sound like a long time, but it is just 5% of what we've survived so far and a tiny fraction of what we should be able to achieve.

In contrast, if we could continually reduce the risk in each century, we needn't inevitably face existential catastrophe. For example, if we were to reduce the chance of extinction by a tenth each successive century (1%, 0.9%, 0.81% . . .), there would be a better than 90% chance that we would never suffer an existential catastrophe, no matter how many centuries passed. For the chance we survive all periods is:

$$(100\% - 1\%) \times (100\% - 0.9\%) \times (100\% - 0.81\%) \times \ldots$$
$$\approx 90.4598\%$$

This means there would be a better than 90% chance we survive until we reach some external insurmountable limit—perhaps the death of the last stars, the decay of all matter into energy, or having achieved everything possible with the resources available to us.

Such a continued reduction in risk may be easier than one would think. If the risks of each century were completely separate from those of the next, this would seem to require striving harder and harder to reduce them as time goes on. But there are actions we can take now that reduce risks across many time periods. For example, building understanding of existential risk and the best strategies for dealing with it; or fostering civilisational prudence and patience; or building institutions to investigate and manage existential risk. Because these actions address risks in subsequent time periods as well, they could lead to a diminishing risk per century, even with a constant amount of effort over time. In addition, there may just be a limited stock of novel anthropogenic risks, such that successive centuries don't keep bringing in new risks to manage. For example, we may reach a technological ceiling, such that we are no longer introducing new technological risks.

6 How we prioritise between these parts is a balancing act between getting sustainable longterm protections in place and fighting fires to make sure we last long enough to enjoy those sustainable protections. And this depends on how we think the risk is distributed over time. It is even possible to have situations where

we might be best off with actions that pose their own immediate risk if they make up for it in how much they lower longterm risk. Potential examples include developing advanced artificial intelligence or centralising control of global security.

7   The name was suggested by William MacAskill, who has also explored the need for such a process and how it might work.

Nick Bostrom (2013, p. 24) expressed a closely related idea: 'Our present understanding of axiology might well be confused. We may not now know—at least not in concrete detail—what outcomes would count as a big win for humanity; we might not even yet be able to imagine the best ends of our journey. If we are indeed profoundly uncertain about our ultimate aims, then we should recognise that there is a great option value in preserving—and ideally improving—our ability to recognise value and to steer the future accordingly. Ensuring that there will be a future version of humanity with great powers and a propensity to use them wisely is plausibly the best way available to us to increase the probability that the future will contain a lot of value. To do this, we must prevent any existential catastrophe.'

It is unclear how exactly how long such a period of reflection would need to be. My guess is that it would be worth spending centuries (or more) before embarking on major irreversible changes to our future—committing ourselves to one vision or another. This may sound like a long time from our perspective, but life and progress in most areas would not be put on hold. Something like the Renaissance may be a useful example to bear in mind, with intellectual projects spanning several centuries and many fields of endeavour. If one is thinking about extremely longterm projects, such as whether and how we should settle other galaxies (which would take millions of years to reach), then I think we could stand to spend even longer making sure we are reaching the right decision.

8   I think that some of the best serious reflection about ideal futures has been within science fiction, especially as it has licence to consider worlds that go far beyond the narrow technological limits of our own generation. 'Hard' science fiction has explored societies and achievements whose ambitions are limited only by the most fundamental physical limits. 'Soft' science fiction has explored

what might go wrong if various ideals of our time were taken to extremes or what new ethical issues would come up if new technologies gave us radically new personal or societal options. A good example combining both aspects is *Diaspora* by Greg Egan (1997), in which almost all beings are digital, greatly changing the space of possible utopias.

However, such thought has also been limited by being fiction. This forces a tension between exploring worlds that could genuinely be utopias and making those worlds sufficiently entertaining for the reader, typically by allowing the possibility of fundamental threats to human wellbeing. And it means that the criticism they receive is mainly directed at writing style, characterisation and so forth, rather than as constructive attempts to refine and develop their visions of the future.

9 The difficulty of anticipating the results may actually make it easier to start the process—for it acts as a veil of ignorance. If we were fairly sure of which future the Long Reflection would ultimately endorse, we'd be tempted to judge this by the lights of our current ethical understanding. Those whose current views are far from where we would end up may then be tempted to obstruct the process. But from a position where we are uncertain of the destination, we can all see the benefits in choosing a future based on further reflection, rather than simply having a struggle between our current views. And this veil of ignorance may even overcome the problems of people having irrationally high confidence in their current views. For if different camps think their view is uniquely well supported by argument, they will also think it has a higher than average chance of being the outcome of a careful process of reflection.

10 If compromise futures were considered even less attractive than either of the 'pure views' we started with, then we could ultimately just choose randomly from among the undefeated pure views—perhaps with probabilities related to their degree of support.

But I'm generally optimistic that there are win-win compromises. For example, moral theories can be divided into those whose judgement of the value of outcomes increases roughly in line with the amount of resources, and those where there are steeply diminishing returns. Classical utilitarianism is an example of the former, where two galaxies could support twice as much happiness as one and would thus be twice as good. Common-sense morality is an example of the latter, where most people's untutored intuitions show little interest in creating flourishing beyond the scale of a planet or galaxy. Such differences give rise to opportunities for deals that both views see as highly favourable. I call this phenomenon *moral trade* (Ord, 2015).

In particular, there is room for a 'grand deal' between moral theories where the views with diminishing marginal utility of resources get to decide the entire future of our galaxy, while the views that highly value additional resources get to decide the future of all the other galaxies in the universe (so long as they won't use these for things which are actively disapproved of by the former theories). This should avoid conflict, by giving both groups an outcome they think of as more than 99% as good as is possible—much better than the expected gains of fighting to control the future or flipping a coin for the right to do so.

11 It is possible that a successful Long Reflection would require the improvements to our abilities stemming from one of these radical changes. If so, we would be in a tricky situation and would have to consider the relative risks of making the change without understanding its consequences versus continuing with our unimproved abilities and potentially missing something important.

12 A useful comparison is to the Renaissance. Most people in Europe were not actively involved in this rebirth of culture and learning, yet it is this grand project for which Europe in the fourteenth to seventeenth centuries is best remembered. Notable differences include that the Long Reflection would be a global project and that it should be more open to everyone to participate.

13 Schell (1982), p. 229.

14 The Long Reflection may overlap with the final phase of achieving our potential, perhaps substantially. For it is only irreversible actions that must await our reflection. There may be domains of irreversible action that get resolved and can then be acted on, while others are still under debate.

15 I sometimes hear colleagues suggest that one thing that could be more important than existential risk is the consideration of how to think about infinite value (Bostrom, 2011a; Askell, 2018). This has two parts: one is a challenge to theories like utilitarianism that are difficult to apply in an infinite universe (such as cosmologists believe our universe to be), and the other is a question of whether there

might be prospects of creating something of infinite value (which would seem to trump the large but finite value of safeguarding humanity).

I'm not sure whether these questions are crazy or profound. But either way, they can be left until after we achieve existential security. The case for avoiding existential catastrophe doesn't rely on theories like utilitarianism, and if infinite value really can be created, then existential security will increase the chances we achieve it. So even questions like these with potentially game-changing effects on how we think of the best options for humanity in the future seem to be best left to the Long Reflection.

There may yet be ethical questions about our longterm future which demand even more urgency than existential security, so that they can't be left until later. These would be important to find and should be explored concurrently with achieving existential security.

16 Stephen Hawking (Highfield, 2001): 'I don't think the human race will survive the next thousand years, unless we spread into space. There are too many accidents that can befall life on a single planet. But I'm an optimist. We will reach out to the stars.'

Isaac Asimov (1979, p. 362): 'And if we do that over the next century, we can spread into space and lose our vulnerabilities. We will no longer be dependent on one planet or one star. And then humanity, or its intelligent descendants and allies, can live on past the end of the Earth, past the end of the sun, past (who knows?) even the end of the universe.'

Michael Griffin, NASA Administrator (2008): 'The history of life on Earth is the history of extinction events, and human expansion into the Solar System is, in the end, fundamentally about the survival of the species.'

Derek Parfit (2017b, p. 436): 'What now matters most is how we respond to various risks to the survival of humanity. We are creating some of these risks, and we are discovering how we could respond to these and other risks. If we reduce these risks, and humanity survives the next few centuries, our descendants or successors could end these risks by spreading through this galaxy.'

Elon Musk (2018): '. . . it's important to get a self-sustaining base. Ideally on Mars because Mars is far enough away from Earth that a war on Earth, the Mars base might survive, is more likely to survive than a moon base.'

Carl Sagan (1994, p. 371) implicitly suggested it when saying: 'In the littered field of discredited self-congratulatory chauvinisms, there is only one that seems to hold up, one sense in which we are special: Due to our own actions or inactions, and the misuse of our technology, we live at an extraordinary moment, for the Earth at least—the first time that a species has become able to wipe itself out. But this is also, we may note, the first time that a species has become able to journey to the planets and the stars. The two times, brought about by the same technology, coincide—a few centuries in the history of a 4.5 billion-year-old planet. If you were somehow dropped down on the Earth randomly at any moment in the past (or future), the chance of arriving at this critical moment would be less than 1 in 10 million. Our leverage on the future is high just now.'

But he goes on to suggest something more like the existential security I am proposing (1994, p. 371): 'In a flash, they create world-altering contrivances. Some planetary civilizations see their way through, place limits on what may and what must not be done, and safely pass through the time of perils. Others, not so lucky or so prudent, perish.'

17   My colleagues Anders Sandberg and Stuart Armstrong spell out the logic and mathematics in more detail (Armstrong & Sandberg, 2013). They show that the number of redundant copies only needs to increase logarithmically in order to have a non-zero chance that at least one copy exists at all times.

18   The spread of these risks *is* limited by the speed of light, so cosmic expansion would limit their spread to a finite region (see Chapter 8). This isn't very helpful for us though, as the region of desolation right now would include everywhere we could ever reach. In the distant future, things are a little better: all the galactic groups will be causally isolated, so we may be spread over millions of independent realms. However, these locations cannot repopulate each other, so a 1% risk would on average permanently wipe out 1% of them. On some views about what matters, this would be just as bad as a 1% chance of losing all of them. Even if we only care about having at least one bastion of humanity survive, this isn't very helpful. For without the ability to repopulate, if there were a sustained one in a thousand chance of extinction of each realm per century, they would all be gone within 5 million years. This might seem like a long time, but since we would have to survive a

hundred billion years to reach the point in time when the universe is divided up like this, it doesn't appear to offer any useful protection. The real protection comes not from redundancy, but from taking the risks seriously and working to prevent them.

19 How much it helps is an open question. Firstly, the extent it helps depends on how much of the existential risk is uncorrelated between planets and how best to model the risk. One model is to think of some fraction of the aggregate risk being uncorrelated and to say that spreading to other planets removes that part. Eliminating this part of our aggregate risk may thus make a major change to our chance of achieving our potential. But we could also model our situation with some fraction of the risk *each century* being uncorrelated (e.g. 5 percentage points out of a total of 10). In this example, eliminating uncorrelated risk (without solving the correlated risk) would just double the expected length of time before an existential catastrophe from ten centuries to 20, increasing humanity's lifespan by just a small fraction. I'm not sure which of these models is the better way of looking at it.

There is also a question of cost-effectiveness. Space settlement does not look to me to be one of the most cost-effective ways to reduce risk over this century. As a simple example, building similar sustainable colonies in remote parts of the Earth (Antarctica, under the sea . . .) would be much cheaper and would protect against many of the same risks. But perhaps it is a mistake here to think of the money for space settlement as coming from the same fixed budget as other risk reduction. It is such an inspiring project that the money may well be found from other sources, and the inspiration it produces (along with the knowledge that humanity really is destined for the stars) may ultimately *increase* the total resources forthcoming for existential security.

20 Depending on the details of how we define existential catastrophes, it may be possible to suffer two of them. For example, suppose that we lost 99% of our potential in an existential catastrophe. As I don't require complete destruction of our potential in order to count as an existential risk, this would count. In the years that follow, it may still make sense to preserve the remaining 1% of our potential from further catastrophes. If our potential were truly vast to begin with, this 1% remaining may also be vast compared to the issues of the day

that usually concern us, and the case for focusing on preserving our remaining potential may still be very strong. In this case, it makes sense for people after the time of the first existential catastrophe to adopt all the thinking surrounding existential risk.

We could define existential risk relative to our remaining potential, such that humanity could suffer several successive existential catastrophes, or relative to our initial potential, such that only the first one counts. I'm not sure which definition is best, so I leave this open. But note that it doesn't really change the arguments of this section. Even though these people would have the ability to learn from the first existential catastrophe, it only helps them preserve the scraps of value that remain possible—the first catastrophe was overwhelmingly more important and it had to be faced by people with no precedent.

21 This is explored by Groenewold (1970) and Bostrom (2002b).

22 These actions may be aimed at any stage of a catastrophe—preventing its initiation, responding to its spread, or creating resilience to its effects—but securing the resources, gathering the information and planning the actions still needs to be pre-emptive.

23 I owe this point to Bostrom (2013, p. 27).

24 This situation is sometimes called 'Knightian uncertainty', or just 'uncertainty', which is then distinguished from situations of 'risk', where we have access to the probabilities (Knight, 1921). There are several slightly different ways of making this distinction, such as reserving the term 'uncertainty' for situations where we have absolutely no quantifiable information about whether the event will occur.

I won't adopt this terminology here, and will carry on using 'risk' for existential risks. Note that almost all uses of 'risk' in this book denote these situations where we don't know the objective probabilities, but where we do have at least a small amount of quantifiable knowledge about whether the catastrophe will occur (e.g. that the chance of nuclear war in the next minute is less than 50%).

25 See Rowe and Beard (2018) for an overview of attempts and methodologies for estimating existential risk.

26 Lepore (2017).

27 A fault tree is a schematic representation of the logical relationships between events, particularly those leading up to failure. This allows the user to identify possible sources of failure in terms of the

sequences and combinations of events that must occur, and estimate their likelihood.

28   The issue of anthropic selection effects when estimating the risk of extinction was raised by Leslie (1996, pp. 77, 139–41) and explored in Bostrom (2002a). See Ćirković, Sandberg & Bostrom (2010) for a detailed analysis of 'anthropic shadow': the censoring of the historical record for various events related to extinction risk.

29   This is outlined, with reference to the Large Hadron Collider, in Ord, Hillerbrand & Sandberg (2010). The situation can be easily understood in a Bayesian framework. We have a prior credence over what the objective probability is, as well as a piece of evidence that it is what the scientists have calculated. Our posterior estimate is therefore somewhere between our prior and the scientists' estimate. When the scientists' estimate is extremely low, this posterior estimate will tend to be higher than it.

   This issue concerns all low-probability risks, but only really needs to be addressed in those with high enough stakes to warrant the extra analysis.

30   This is because for low-probability high-stakes risks there is more room for the true probability to be higher than the estimate, than to be lower. For example, if the estimate is one in a million and there is an equal chance of it being ten times higher or a tenth as high, the former exerts a larger effect on the expected size of the true probability, pulling it up. In other words, if you haven't already adjusted for this effect, then your point estimate of the underlying probability is often lower than your expectation of the underlying probability, and it is this latter number that is the decision-relevant probability.

31   One interesting starting point might be to create a body modelled on the IPCC, but aimed at assessing existential risk as a whole. This would be a new international advisory body under the auspices of the United Nations, focused on finding and explaining the current scientific consensus on existential risk.

32   Beck (2009), p. 57.

33   H. G. Wells was an enthusiastic advocate for world government for much of his career (Wells, 1940, pp. 17–18): 'It is the system of nationalist individualism and unco-ordinated enterprise that is the world's disease, and it is the whole system that has to go

. . . The first thing, therefore that has to be done in thinking out the primary problems of world peace is to realise this, that we are living in the end of a definite period of history, the period of the sovereign states. As we used to say in the eighties with ever-increasing truth: "We are in an age of transition". Now we get some measure of the acuteness of the transition. It is a phase of human life which may lead, as I am trying to show, either to a new way of living for our species or else to a longer or briefer dégringolade of violence, misery, destruction, death and the extinction of mankind.'

Bertrand Russell (1951) wrote: 'Before the end of the present century, unless something quite unforeseeable occurs, one of three possibilities will have been realized. These three are:

1. The end of human life, perhaps of all life on our planet.
2. A reversion to barbarism after a catastrophic diminution of the population of the globe.
3. A unification of the world under a single government, possessing a monopoly of all the major weapons of war.'

And more recently Nick Bostrom has advocated for humanity forming what he calls a 'Singleton' (2006). This could be a form of world government, but it doesn't have to be. As I understand his concept, humanity is a singleton if it is in a situation where it behaves roughly like a coherent agent. This involves humanity avoiding outcomes that are Pareto-inferior by the lights of the people of the world (i.e. negative-sum conflicts such as war). But it needn't involve a single political point of control.

34 Einstein (1948), p. 37. Einstein was motivated by extinction risk, but from a somewhat different dynamic from that which worries me most. Where I've stressed the problem of bad actors or defection from moratoria within individual nations, Einstein was chiefly focused on the need to remove the ability for one nation to wage war on another once the methods of waging war posed a risk of extinction.

35 Somewhat inaccurately, as it is neither red nor a telephone. Nor does it sit on the president's desk. In reality it is a secure teletype link (then fax, and now email), located at the Pentagon.

36 The Soviet ambassador Anatoly Dobrynin (1995, p. 100) gives a memorable recollection: 'Nowadays one can hardly imagine just

how primitive were our embassy's communications with Moscow in the dreadful days of the Cuban crisis, when every hour, not just every day, counted for so much. When I wanted to send an urgent cable to Moscow about my important conversation with Robert Kennedy, it was coded at once into columns of numbers (initially this was done by hand and only later by machine). Then we called Western Union. The telegraph agency would send a messenger to collect the cable . . . who came to the embassy on a bicycle. But after he pedaled away with my urgent cable, we at the embassy could only pray that he would take it to the Western Union office without delay and not stop to chat on the way with some girl!'

37 One could think of this in terms of the modern military concept of an OODA loop (the time needed to Observe, Orient, Decide and Act). The OODA loop of diplomacy was far too slow to appropriately manage the unfolding events on the ground.

They did have some faster options, such as immediate radio and television announcements, but these required conducting the diplomacy in front of the entire world, making it much harder for the parties to back down or agree to domestically or internationally unpopular terms.

38 The OPCW's 2019 budget is €70 million, or $79 million (OPCW, 2018). Staff numbers from OPCW (2017).

39 These commendable efforts have been led by the International Gene Synthesis Consortium (IGSC, 2018).

40 U.S. Department of State (n.d.). By most accounts, the treaty was a significant step towards disarmament, successfully eliminating a large class of weapons from the arsenals of the two great nuclear powers and putting in place robust verification procedures (Kühn & Péczeli, 2017).

41 I've suggested extinction risk rather than the broader idea of existential risk because the latter would involve an additional difficulty of assessing which other outcomes count as existential catastrophes.

Catriona McKinnon (2017) has made some helpful suggestions for how a crime of deliberate or reckless imposition of extinction risk could be created in international criminal law.

42 For example, while many states have laws against imposing risks on others (such as drunk driving) international criminal law has no precedent for this (McKinnon, 2017, p. 406). And while it may seem that nothing could be more fitting of the title 'a crime against

humanity', the use of the word 'humanity' is ambiguous and is sometimes interpreted here in the sense of 'essential human dignity', rather than as the totality of all humans. For example, in relation to the crimes of the Holocaust, Altman & Wellman (2004) describe an understanding of the term as 'Harm was done to the humanity of the Jewish victims, but that is not to say that harm was done to humanity itself.'

The issue of finding thresholds for what counts as increasing a risk unnecessarily (and for adjudicating it) is a serious one. For example, we would not want to include trivial increases, such as releasing carbon dioxide with a short car journey, but we don't have access to precise probabilities with which to set a higher threshold. And if the proposal would make heads of state fear that their everyday actions, such as reducing a petrol levy, may open them up to prosecution, then it would be very hard to get their consent to establish the law.

43 For example, to pose a 1% existential risk. It would of course be illegal to actually kill everyone or cause the collapse of civilisation, but since there would be no punishment after the fact, that is not especially relevant. So the law would need to punish the imposition of risk or the development of the systems that may lead to the catastrophe.

Ellsberg (2017, p. 347) eloquently captures the insanity of the current state of affairs: 'Omnicide—threatened, prepared, or carried out—is flatly illegitimate, unacceptable, as an instrument of national policy; indeed, it cannot be regarded as anything less than criminal, immoral, evil. In the light of recent scientific findings, of which the publics of the world and even their leaders are still almost entirely unaware, that risk is implicit in the nuclear planning, posture, readiness, and threats of the two superpowers. That is intolerable. It must be changed, and that change can't come too soon.'

44 UNESCO (1997).

45 These include Finland (1993–), Canada (1995–), Israel (2001–6), Germany (2004–), Scotland (2005–), Hungary (2008–12), Singapore (2009–), Sweden (2011–), Malta (2012–) and Wales (2016–). There are a lot of lessons to be learned from their mixed record of success and failure, with several being weakened or

abolished, especially after changes of political control at the national level. See Nesbit & Illés (2015) and Jones, O'Brien & Ryan (2018).

46 An early critique was that of Hans Jonas (1984, p. 22): 'One other aspect of the required new ethics of responsibility for the distant future is worth mentioning: the doubt it casts on the capacity of representative government, operating by its normal principles and procedures, to meet the new demands . . . the future is not represented . . . the nonexistent has no lobby, and the unborn are powerless.'

47 And even more problematically, the policy may affect who comes to exist at that future time and whether there are any future people at all.

48 Such an institution could also represent children, as they too have interests that predictably diverge from those of adults, but are denied a vote.

49 Smart (1984), p. 140. Bostrom (2014) expands on this and other aspects of what going slow does and doesn't achieve.

50 Slow progress (as opposed to merely shifting all dates back by some number of years) may also give us more time to identify threats before they strike, and more time to deal with them. This is because it would effectively speed up our reaction time.

51 They also come with a more well-known shadow cost on the environment. The arguments there are similar and I also endorse using some fraction of the gains from technology to counteract these costs.

52 We could also think of this as technological progress in our current time not making *humanity* more prosperous at all. Even thought of narrowly in terms of money, it may be merely making our own generation more prosperous at the expense of large reductions in the expected prosperity of a vast number of future generations.

The same can be said in the narrow terms of technology itself. Creating technology at a breakneck pace is a greedy strategy. It optimises the level of technology next year, but reduces the expected technology level across the longterm future.

53 It is hard to find such time, when the academic incentives are set up to push people towards publishing more papers and rewarding

producers of technical papers, not papers on ethics or govern-ance. But academics are ultimately in charge of their own incentive structure and should push to change it, if it is letting them (and humanity) down.

54 Indeed, a good choice for a young scientist or technologist would be to go to work in government. The lack of scientific literacy and expertise in government is most often lamented by precisely the people who have those skills and could be applying them in government. Put another way, it is unreasonable to blame people working on policy for not being great at science, but more reason-able to blame people who are great at science for not working on policy.

55 UN (n.d.).

56 See Grace (2015). There is some debate over how successful the Asilomar Conference was. In the decades after the guidelines were created, some of the risks envisioned by the scientists turned out not to be as great as feared, and many of the regulations were gradually unwound. Some critics of Asilomar have also argued that the model of self-regulation was inadequate, and that there should have been more input from civil society (Wright, 2001).

57 See Bostrom (2002b).

58 This distinction is from Bostrom (2014), and the analysis owes a great deal to his work on the topic.

59 The precise half-life is the natural logarithm of 2 ($\approx$0.69) divided by the annual risk, whereas the mean survival time is simply 1 divided by the annual risk.

But note that it is just the objective probability of survival that is characterised by a decaying exponential. Since we don't know the half-life, our subjective probability of survival is the weighted average of these exponentials and is typically not itself an expo-nential. In particular, it will typically have a fatter tail. (This same effect comes up when we are uncertain of the discount rate; see pp. 247–248.)

60 There may also be risks that fit neither of these categories.

61 The distinction between state and transition risks is not sharp. In particular, for risks with a changing character, it can depend upon the timescale we are looking at. Take the risk of nuclear winter. At some scales, it acted as a state risk. In the years from 1960 to

1990 this was a reasonable model. But if we zoom in, we see a set of transitions that needed to be managed. If we zoom out to include the time up to the present, then it feels like two regimes (during and post-Cold War) with very different hazard rates. If we zoom out to include the entire time humanity remains vulnerable to nuclear weapons, it may look like a state risk again on a scale of centuries, with the ebb and flow of geopolitics on a decadal level being washed out. And if we zoom out even further, such that there is only a small region of nuclear war risk sandwiched between the pre-nuclear and post-nuclear eras, then it may again be best thought of in terms of a transition risk: how should humanity best navigate the transition to a regime in which we harness nuclear energy.

A similar issue may arise with the risk from AGI. Once it is developed, we would enter a state where humanity is vulnerable to AGI accident or misuse by any of the actors who gain access to it. This state will only end when actions are taken to end it. This period (which may be short) will be an AGI state risk, while when zoomed out it is a transition risk.

Despite mixed cases like these, it remains a useful distinction. Indeed, even in these very cases, it is a useful lens through which to understand the changing nature of the risks.

62 This idea was introduced by Bostrom (2013, pp. 24–6).

63 It is, of course, a cliché for a researcher to suggest that more research is needed. I hope the reader will be able to see why more research on existential risk will indeed be especially valuable to humanity. Here, research isn't needed merely for a more definitive answer to an arbitrary academic question. It is required in order to be able to answer a fundamentally important question (which actions would best safeguard the longterm potential of humanity?) that has had hardly any study thus far.

64 For example, at the time of writing there have only been two published studies on the climate effects of full-scale nuclear war since the Cold War ended in 1991 (Robock, Oman & Stenchikov, 2007; and Coupe et al., 2019), and no detailed study of its agricultural effects since 1986 (Harwell & Hutchinson, 1986).

65 For example, the research fellowship which has given me the time needed to write this book was funded by an individual donor.

66 They have also made a large grant supporting the Future of Humanity Institute at Oxford, where I work.

67 Their focus on the research that is actually most helpful (guided by a passion for the cause) is especially valuable, for a general increase in research funding for existential risk might be expected to flow mainly towards established research (perhaps after suitable rebranding), leaving little for more foundational or daring work. We might also expect general funding to go to more well-understood risks (such as asteroids) at the expense of bigger, but less well-understood risks (such as those from advanced artificial intelligence); or to risks that are catastrophic, but not really existential. If mainstream granting bodies were to start funding calls for existential risk, they would need to take care not to distort the priorities of the field. See also Bostrom (2013, p. 26).

68 Other academic institutes include the Future of Life Institute (FLI) and the Global Catastrophic Risk Institute (GCRI).

69 Environmentalism is a useful example. It was much less of a partisan political issue in the early days, when it had many great successes. It was Richard Nixon who established the US Environmental Protection Agency, and Reagan (1984) who stated 'Preservation of our environment is not a liberal or conservative challenge, it's common sense.' I think it may have had even more success if this non-partisan spirit had been kept alive.

70 Consider environmentalism. The chief issues facing early environmentalists were pollution, biodiversity loss, extinction and resource scarcity. But they didn't call themselves 'extinctionists' or 'pollutionists'. They found their identity not in the problems they were fighting, but in the positive value they were fighting to protect.

71 At the time of writing, DeepMind and OpenAI are the most prominent examples. They are in need of great researchers in AI safety, and also great software engineers—especially those who take existential risk seriously.

72 Organisations focused on reducing existential risk include:

The Future of Humanity Institute (FHI)

The Centre for the Study of Existential Risk (CSER)

The Future of Life Institute (FLI)

The Global Catastrophic Risk Institute (GCRI)

The Berkeley Existential Risk Initiative (BERI)

The Open Philanthropy Project (OpenPhil)

The Nuclear Threat Initiative (NTI)

The Bulletin of the Atomic Scientists

The Global Challenges Foundation

The Law and Governance of Existential Risk group (LGER)

Alliance to Feed the Earth in Disasters (ALLFED)

The high-impact careers site 80,000 Hours maintains an up-to-date job board, including such positions:
80000hours.org/job-board
and explanations of the kinds of careers that can really help:
80000hours.org/career-reviews

73   In keeping with this, I have signed over the entire advance and royalties from this book to charities helping protect the longterm future of humanity.

74   Eig (2014). McCormick should share the credit with the birth-control activist, Margaret Sanger (who first secured McCormick's donations), and the scientists Gregory Pincus and John Rock, whose research she funded.

75   Fleishman, Kohler & Schindler (2009), pp. 51–8. See note 97 to Chapter 4.

8 Our Potential

1   Wells (1913), p. 60.

2   I use the term 'civilisation'—as I have done throughout this book—to refer to humanity since the Agricultural Revolution (which I round to 10,000 years, reflecting our imprecise knowledge of its beginning). This is a broader definition than the more commonly used 5,000 years since the time of the first city states. I use this longer time as I believe the Agricultural Revolution was the more important transition and that many of the things we associate with civilisation will have been gradually accumulating during this time of villages and towns preceding the first cities.

3   Using the fossil record, estimates of median species lifetime for mammals range from 0.6 million years (Barnosky et al., 2011) to 1.7 million years (Foote & Raup, 1996). I set aside estimates using

molecular phylogeny, which generally give longer estimates, as that methodology is less widely accepted.

For all species in the fossil record estimates range from 1 million years (Pimm et al., 1995) to 10 million years (De Vos et al., 2015). May (1997, p. 42) concludes: 'if one is to speak of an average, it might be better to offer a range like 1–10 million years'.

4   Most atmospheric carbon has a lifetime of around 300 years, but there is a long tail of carbon that survives many times longer than this. Archer (2005) finds that within 100,000 years, 7% of fossil fuel carbon will remain.

5   After the end-Permian extinction, which saw 96% of species go extinct around 250 million years ago, a full recovery took around 8–9 million years for marine species, and slightly longer for land species (Chen & Benton, 2012).

6   Forey (1990); Zhu et al. (2012); Shu et al. (1999).

7   The oldest cyanobacteria fossils are somewhere between 1.8 and 2.5 billion years old (Schirrmeister, Antonelli & Bagheri, 2011).

Simple life is generally accepted as having emerged at least 3 billion years ago (Brasier et al., 2006).

'Complex life' is not a precisely defined term. I take it as referring to the level of the Cambrian explosion (541 million years ago), or the appearance of Ediacaran biota (about 600 million years ago). The exact boundary makes little difference in what follows.

8   This scenario is from Christopher Scotese (Barry, 2000). It should be understood as speculative.

9   Indeed, this will happen within just 100,000 years. When our early human ancestors gazed at the stars, they too saw shapes unknown to us.

10   The first set of plants to die are those that use C_3 carbon fixation for photosynthesis, which requires a higher level of carbon dioxide. Roughly 3% of plants use C_4 carbon fixation for photosynthesis, which works at carbon dioxide levels far below the critical limits for C_3 (Kellogg, 2013).

There is significant uncertainty in all of these estimates. We can be reasonably confident that the runaway and moist greenhouse effects pose an upper bound on how long life can continue to exist on Earth, but we remain uncertain about when they will occur, due to the familiar limitations of our climate models. Wolf & Toon (2015) find a moist greenhouse will occur at around 2 billion years, whereas Leconte et al. (2013) place a lower bound at 1 billion years.

The open question is whether carbon dioxide depletion or temperature increases will render Earth uninhabitable before the runaway or moist greenhouse limits are reached. Rushby et al. (2018) estimate carbon dioxide depletion will occur in around 800 million years for C_3 photosynthesis, and around 500 million years later for C_4 photosynthesis.

Over such long timespans, we cannot ignore the possibility that evolution may lead to new species able to exist in climates inhospitable to presently existing life forms. Indeed, the first C_4 plants only appeared around 32 million years ago (Kellog, 2013).

11 The Sun is getting brighter at a rate of about 10% per billion years, and will continue to do so for around 5 billion years, when it enters its red giant phase. It is this surprisingly small relative change that would spell the end of complex life absent our intervention. By about 6 billion years' time, we'd need to absorb or deflect about half the incoming light.

12 Schröder & Connon Smith (2008).

13 Conventional star formation will cease in about one to 100 trillion years, but there are many proto-stars (called brown dwarfs) that are too small to ignite on their own. Over these cosmological timescales, their collisions will create a small but steady stream of new stars that will keep going for at least a million times as long as conventional star formation (Adams & Laughlin, 1997; Adams & Laughlin, 1999).

14 The patterns of the stars are reflected in some of the earliest post-agricultural artefacts, and knowledge of the stars is culturally and practically important today for many indigenous peoples who have retained their forager lifestyle. There is even a tantalising possibility that oral traditions have preserved some ancestral mythology about the stars for more than 50,000 years: there are indigenous groups in North America, Siberia and Australia who all call the same constellation 'The Seven Sisters' in their own languages. The constellation is the Pleiades, which was also known as 'The Seven Sisters' to the ancient Greeks (Wilson, 2001).

15 We don't know precisely how many. Extrapolating the count of galaxies visible in the Hubble Ultra Deep Field image shows there are at least 150 billion galaxies visible with present technologies. But

this will undercount since we can't detect them all, and overcount because there were more galaxies in the early universe, many of which have since merged (the Hubble image shows these distant regions as they were long ago when the light left them). Using a recent estimate of 0.0009 galaxies per cubic megalight-year at the present moment (Conselice et al., 2016), I calculate 400 billion galaxies in the observable universe right now.

Galaxies come in a vast range of sizes, from more than a trillion stars down to perhaps just thousands. Most are much smaller than the Milky Way. This wide variety of scales for the galaxies adds to the uncertainty. We may find that there are many more small and faint galaxies than we had anticipated, greatly increasing the number of galaxies in the observable universe, but simultaneously making the average galaxy less impressive.

16 The ratio is a little better if we compare this just to the *land* area of the Earth, but even our ocean (surface or floor) is much easier to settle than are the distant planets and moons.

17 The hourly energy from the Sun is $\sim 3.2 \times 10^{20}$ J (Tsao, Lewis & Crabtree, 2006), compared with our annual energy consumption of $\sim 6 \times 10^{20}$ J (IEA, 2019).

18 When such a feat of astronomical engineering is discussed, people often jump straight to its ultimate realisation: a Dyson sphere, entirely encapsulating the Sun. But such an extreme version introduces other challenges and downsides. Instead, it is best thought of as a scalable approach.

19 One approach is to make individual solar collectors and put them in orbit around the Sun. Arguably we've already started this, with some of our existing spacecraft and satellites. While there are complexities to do with getting rid of the waste heat and sending the captured energy somewhere useful, this approach is relatively simple at first. However, it becomes more complicated once there are enough collectors to capture a significant fraction of the Sun's energy (as then one needs to orchestrate their orbits to avoid collisions).

Another promising approach is to use not satellites, but 'statites'. These are objects that are not in orbit, but which avoid falling into the Sun by having their gravitational pull towards the Sun exactly cancelled by the pressure of light pushing them away.

Balancing these forces requires the collectors to be very light per unit area, but it does look achievable. While there are substantial challenges in engineering each collector, they would require very little building material and scaling the project up is simply a matter of making more of them and dropping them in place. My colleagues Eric Drexler and Anders Sandberg have done feasibility calculations, estimating the mass needed for enough statites to absorb all sunlight to be around 2×10^{20} kg. This is around the mass of Pallas, the third largest asteroid in our Solar System (Sandberg, n.d.).

20 Even if fossil fuels were to remain useful in vehicles (which need to carry their own energy sources), their carbon emissions could be easily reversed by carbon dioxide scrubbers, powered by the abundant solar energy.

21 There may also be limits to the directions spacecraft can be sent with this method, so we may not be able to use it to send craft directly *to* our nearest stars.

Since the spacecraft we have already sent beyond our Solar System are not aimed at the nearest stars, they will proceed past them. Since they weren't launched with enough speed to escape the Milky Way, they are destined to wander through our galaxy for an extremely long time, perhaps swinging past many stars before eventually being destroyed.

22 Overbye (2016).

23 It is typically imagined that such a settlement would be on a planet, but it could instead be on a moon or in a space station constructed from material in asteroids. The latter might make a superior early base of operations as the initial craft wouldn't need to survive a descent onto a planet, or to construct a massive rocket in order to get back into space.

24 See note 46 to Chapter 2 for an outline of my own thinking on the 'Fermi paradox' and whether we are alone.

25 If the life were less advanced than us, it may pose a test to our morality; if it were more advanced, it may pose a threat to our survival. Or if the ceiling on technological ability lies not too many centuries into the future, then there is a good chance we would meet any other intelligent beings after we had all reached the same level: as technological equals.

26 One can understand this by imagining lines connecting each pair of stars in our galaxy that are closer than some distance d to each other. For small values of d, only a small proportion of stars get connected. But there is a critical level for d at which point a giant connected component appears, which connects almost all stars in the galaxy. My colleague, Anders Sandberg, has calculated this to be at about six light years.

The critical distance would be even less if we took advantage of the fact that stars drift past each other. If we waited for these close approaches, we wouldn't have to travel so far in each step.

There is also a complicating factor that perhaps not every star system is sufficiently settleable to enable new journeys from there. This seems less likely now that we know rocky planets are so common, but may still be true. This would effectively thin out the set of stars and increase the critical distance.

27 This is not the fastest or most efficient way to settle the galaxy, especially if we can travel further in each step. I focus on it because it is the *easiest* way—the one that requires the least technology, the least planning and the fewest resources from our own Solar System.

28 Adams & Laughlin (1997).

29 Large groups are known as 'clusters', but occupy the same position in the scale hierarchy.

30 These intersections are sometimes known as 'superclusters', though that term is also sometimes used to refer to a wider region around the intersection, such that each galaxy is considered part of some supercluster. Either way, superclusters are a useful concept in mapping our environment, but not especially relevant to our potential.

31 The discovery that the universe is expanding is generally credited to Edward Hubble and Georges Lemaître, who reached the conclusion independently in 1927 and 1929 respectively (Gibney, 2018). The accelerating expansion of the universe was discovered only in the late 1990s (Riess et al., 1998)—work that won the 2011 Nobel Prize in Physics.

In the paragraphs that follow, I'm describing the limits as they would be under the simplest known account of accelerating expansion, where it is due to a cosmological constant. This is known as the 'concordance cosmology' or ΛCDM. Other explanations of accelerating expansion (including that it is illusory) may produce quite different limits, or even no limits at all.

32　The light from these galaxies has only had 13.8 billion years (the age of our universe) to reach us, but they are currently 46 billion light years away, as the space in between has been expanding in the meantime.

33　The 63 billion light year limit is the sum of the distance we can currently observe (46.4 billion light years) and the distance we can currently affect (16.5 billion light years).

　　If someone travelled away from the Earth, they could see a bit further in that direction. In the extreme case, if they travelled at the speed of light, they could eventually reach a point currently about 16 billion light years away from Earth and then see an entire eventually observable universe centred on that distant point. They wouldn't see *more* than someone here, but would see different parts including some that will never be visible from here. But according to our current best theories, anything more than 79 billion light years from here (the sum of these distances) is absolutely impossible to observe.

34　Surprisingly this affectable region extends slightly beyond the 'Hubble sphere'—the region containing all galaxies that are receding from us at less than the speed of light (currently with a radius of 14.4 billion light years). This is because it is still possible to reach some of the closest galaxies that are receding faster than the speed of light. This may appear impossible as nothing can travel through space faster than the speed of light, but we can use the same trick that those distant galaxies themselves are using. They recede so quickly not because they are travelling quickly through space, but because the space between us is expanding. If you shine a torch into the sky, the very light you emit will also recede from you faster than the speed of light, since the intervening space itself expands. Some of the photons you release will eventually reach about 2 billion light years beyond the Hubble sphere. Indeed, since almost everything you do affects the pattern of photons reflected off the Earth into deep space, it is almost impossible to avoid affecting things 16 billion light years away as you go about your daily life.

Thus, while people often use the term 'Hubble sphere' or 'Hubble volume' as a stand-in for everything we can hope to affect, they should really use 'affectable universe', which is both more descriptive and more accurate.

35 Author's calculation based on a density of 0.009 galaxies per cubic megalight-year, from Conselice et al. (2016). These numbers for how many more galaxies become visible or cease to be affectable each year depend sensitively on the unresolved question of how many galaxies there are (see note 15 to this chapter).

36 A key challenge is very small particles of dust strewn through the intergalactic void. If a spacecraft collides with these at a substantial fraction of the speed of light, then the collision would be devastating. The chance of a spacecraft not meeting any such dust grains on its journey decreases exponentially with distance travelled, so longer distances travelled in a single step could present a big challenge. Some form of shielding is probably required. My colleague, Eric Drexler, has calculated that sending several layers of shielding material in advance of the spacecraft could protect the payload, but of course this is still speculative.

The distances could probably be reduced (perhaps markedly so) by taking advantage of the sparse scattering of stars that lie between the galaxies.

37 The expected value of our future is something like the product of its duration, scale, quality and the chance we achieve it. Because these terms are multiplied together, increasing any of them by a given factor has the same effect on the expected value. Thus our marginal efforts are best spent on the one that is easiest to improve in relative terms.

38 This argument, about how safety beats haste, was first put forward by Nick Bostrom (2003), though with different empirical assumptions. He measured the annual cost of delay by the energy in the starlight of the settleable region of our universe that we are failing to harness. However, I don't think starlight will be the majority of the resources we could harness (less than a thousandth of a star's mass is turned to energy via starlight), and I think the shrinking size of this settleable region is crucial.

My own guess is that the annual proportional loss is roughly equal to the proportion of galaxies that become unreachable due to cosmic expansion. It is entirely possible that this too will soon come to look like the wrong answer: for example, if it was technologically impossible to ever travel between galaxies, or if the evidence for accelerating expansion was overturned. But I think it is likely that the general point will remain: that the annual proportional loss is very small—likely less than one part in a billion. This is because most of the relevant timescales that would set this proportion are themselves measured in billions of years (the age of the universe, the lifespans of most stars, the time for which galaxies will continue to form stars, the age of the Earth and the duration of life on Earth so far).

39 This is from a speech he gave to the Cambridge Apostles in 1925 when he was twenty-one (Mellor, 1995). This is the same Ramsey who developed the economic theory of discounting and argued against discounting based on the mere passage of time alone (see Appendix A).

40 Some of these thoughts were directly inspired by Bostrom (2005, 2008).

Appendices

1 I probably got my dollar in enjoyment of seeing my whimsical thirty-year-long plan come to fruition, but let's set that aside . . .

2 It is not *exactly* the same effect—my income has increased not only due to economic growth, but also due to being further along in my life.

3 Some economists, such as Dasgupta (2008), also incorporate social preferences for equality in the parameter η. This could increase its size beyond what is implied by individual diminishing marginal utility.

4 It derives from the arguments given in Ramsey (1928).

   An alternative model involves computing the net present value of a future monetary benefit by considering that an earlier monetary benefit could (usually) be increased in size by investing it until that later date. So the net present value of a later benefit could be thought of as the amount of money that we would need to invest today in order for it to compound to the given size by the given future date. This line of argument suggests a discount rate that depends on the interest rate rather than the growth rate.

However, this justification for discounting doesn't apply when the option of such investment is not available. This is the case here—for existential catastrophes would presumably cut off such investments or make it impossible for future generations to be benefited by them. This is clearly the case for extinction and seems likely in other cases too.

5   This point has been made well by Ng (2016) and (2005).

6   Indeed, while many great philosophers have argued against pure time preference—see Sidgwick (1907), Parfit (1984) and Broome (2005)—I know of no philosophers who support its inclusion in the social discount rate, so it may be completely unanimous. For those who know philosophy, this is truly remarkable since philosophers disagree about almost every topic, including whether they are the only person in the world and whether there are any true moral claims (Bourget & Chalmers, 2014).

Note that there is a certain amount of talking past each other in the debate between philosophers and economists over discounting. Philosophers often say they are in favour of a zero discount rate, when that isn't the right term for what they mean (i.e. they have no quarrel with monetary benefits to people mattering less if those people are richer in the future). The philosophers are usually talking about a zero rate of pure time preference, or that the ηg term is inapplicable for the topic they are considering (such as for health benefits).

7   A recent survey of 180 economists who publish on the social discount rate found that their most common estimate for the pure rate of time preference was 0%, with a median of 0.5% (Drupp et al., 2018, p. 120).

8   Ramsey (1928), p. 543; Harrod (1948), p. 40. Arthur Pigou (1920, p. 25) suggested that non-zero pure time preference 'implies . . . our telescopic faculty is defective'.

9   I doubt that this is the job of the economist either. Not only does it give an impoverished conception of economics, yielding all the ground on normative matters, but it yields that ground to what amounts to an opinion poll.

Even if economics does not want to be concerned with normative matters, there are others who are so concerned, reflecting deeply on these matters and arriving at novel conclusions backed by strong arguments. Some of these people work in the economics department, some in the philosophy department, and some scattered across other areas within academia or without. They have certainly not reached agreement on all moral problems, but they do have some hard-won knowledge of normative matters. A decision to replace their insight with a broad survey (or with the opinions of the current government) is itself a normative choice, and seems to me a poor one.

10   While an early paper on this topic (Cropper, Aydede & Portney, 1994) showed participants had a strong preference for saving fewer lives sooner rather than more lives later, this didn't provide any real evidence for pure time preference. For in follow-up questions, many of them said they made their choice because the future is uncertain and because it would likely have technology for saving lives—rather than due to pure time preference (see Menzel (2011) for more details).

A follow-up study by Frederick (2003) was better able to isolate time preference from these confounding explanations. For example, he asked people to compare one person dying next year in the United States from exposure to pollutants to one person dying 100 years from now in the same circumstances: 64% said these were 'equally bad', 28% said the earlier deaths were worse, and 8% said the later deaths were worse. Of the 28% who appeared to exhibit pure time preference, they thought a death now was equal to about three deaths in 100 years. So the average time preference across all the participants was less than a quarter of a percent per annum over the next century (and presumably would decrease further in future centuries in line with other experiments).

11   This could be due to the remaining existential risks, to the possibility that even a successful future doesn't last that long, or to failures to achieve a flourishing future which don't qualify as existential risks (for example a gradual slide towards ruin that we could have prevented at any point, but simply failed to).

12   Stern, 2006. Stern's choice of a catastrophe rate of about 10% per century was fairly arbitrary, and was not an attempt to quantify the best evidence about future risks. However, I think it is in the right ballpark for the risk over the next century.

While most of the discussion about his approach to discounting centred on his very small value for δ, he did effectively include the ŋg term. But as the trajectories of growth rates were generated within his model, it is harder to explicitly see how it affected his results (see Dasgupta (2007) for a deep analysis and comparison to the earlier work of Nordhaus).

13 There are good arguments for why the pure rate of time preference should be a constant over time (thus giving an exponential curve). This is because it is the only way to avoid 'time inconsistency' where as time passes you predictably change your mind about which of two benefits is superior. But this argument doesn't apply for the growth rate or the catastrophe rate, which are empirical parameters and should be set according to our best empirical estimates.

14 We saw in Chapter 3 that the natural risk rate is probably lower than one in 200,000 per year (to account for us surviving 200,000 years so far and for related species surviving much longer than this on average). A 50% chance of getting to a period with a catastrophe rate of at most one in 200,000 has a discounted value in excess of 100,000 years.

15 The argument is based on that given by Martin Weitzman (1998), though he doesn't address discounting based on the catastrophe rate. The main trick is to see that we are discounting for the chance that a catastrophe has occurred before the given time, so the effective discount factor at a given time is the probability-weighted average of the discount factors in the worlds we find credible. It is then fairly simple to see that this is not equivalent to discounting at the average rate (or any fixed rate), and that in the long run it tends towards discounting at the lowest rate in which we have nonzero credence. Thus even if you know there is a constant hazard rate, your uncertainty over what this rate is can produce nonexponential discounting.

16 This depends on whether the action just reduces the risk in the near term, or whether it has sustained effects on risk.

17 For example, if you think there is at least a 10% chance we survive more than a million years, this is equivalent to saying that the certainty-equivalent discount rate declines to a value small enough that the discounted value of the future is at least 10% × 1,000,000 = 100,000 times as much as this year.

18 Parfit (1984); Ng (1989); Arrhenius (2000).

19　This is due to a phenomenon that arises when multiple averages are combined. The comedian Will Rogers is said to have quipped: 'When the Okies left Oklahoma and moved to California, they raised the average intelligence level in both states.' This intitially sounds impossible, but further thought reveals it could happen if the people who moved state were below average for Oklahoma (so their leaving increased its average) but above the average of California (so their arriving increased its average). This is now known as the Will Rogers Phenomenon and has important implications in medical statistics (Feinstein, Sosin & Wells, 1985).

It arises here if it is possible for someone to be born in either of two different generations. If their wellbeing would be below the average of the first generation and above the average of the second, moving them would raise both averages (and thus the sum of generational averages) without affecting any individual's wellbeing nor changing who exists. Since the sum of the averages has gone up by some amount it is possible to modify the example, lowering everyone's utility by a smaller amount such that everyone is slightly worse off but the sum of the generational averages has still gone up. A theory that can prefer an alternative with the same population which is worse for everyone is generally regarded as fatally flawed (especially if there is no gain in some other dimension such as equality).

But is it possible for someone to be born in either of two different generations? It seems that it is. For example, present-day medical technology freezes embryos such that couples can choose when to have them implanted. If the embryo could be implanted immediately, or alternatively 30 years later, then it could be born in two different generations. If it were to have the same wellbeing in each case and this was between the average wellbeing levels of each of the generations, we get the Will Rogers Phenomenon. Similar versions of this apply even if you date the person to conception rather than birth, as existing medical technology can be used to have the same sperm fertilise the same ovum now, or in 30 years. The worry here is not that *in vitro* fertilisation will cause practical problems for conducting longterm analysis, but that the principle of the sum of generational averages is theoretically unsound since it can go up even if all individuals' wellbeing goes down.

More generally, many rating systems based on multiple averages have this flaw. For example, it is possible to make the GDP per capita of all countries go up just by moving individuals between countries. We should always be sceptical of such measures.

20 This would lead to a large, but not vast, value of the future. Depending on one's assumptions, the entire future might be worth something like ten times as much as the present generation.

21 Narveson (1973).

22 See Narveson (1967, 1973), Parfit (1984, 2017a) and Frick (2014).

23 Though if each outcome had completely different people, then the world with low wellbeing levels wouldn't be worse either, so this still doesn't completely capture our intuitions on the case.

24 See Narveson (1973) and Heyd (1988).

25 Examples of problems that occur include:
- being morally indifferent between a future where everyone has low wellbeing and a future (comprised of a different set of people) where everyone has high wellbeing;
- having moral rankings change when 'irrelevant alternatives' are introduced (such as preferring A to B when they are the only choices, but then B over A, when an inferior option C is also available);
- creating cyclical preferences across choice situations (leading you to prefer A to B, B to C and C to A);
- having cyclical orderings of value (saying that A is better than B, B is better than C and C is better than A);
- saying that all outcomes that differ even slightly in how many people exist are incomparable with each other.

26 There is some recent work trying to justify the asymmetry through appeal to something more fundamental, such as that of Frick (2014).

27 At least if considered on their own. Many proponents of person-affecting accounts of the value of the wellbeing of future generations combine this with other moral principles that would recognise the badness of extinction. If so, their overall moral views don't suffer from this kind of counter-intuitiveness, but nor do they pose any threat to the claim I am trying to defend: that human extinction would be extremely bad.

28 Beckstead (2013, p. 63) makes this point particularly well. Similarly, someone who found the Total View most plausible should be very cautious about following its advice in a choice like that of the repugnant conclusion.

29 DoD (1981).

30 DoD (1981), p. 8.

31 DoD (1981), p. 12.

32 Oxnard Press-Courier (1958); DoD (1981), p. 8.

33 DoD (1981), p. 20.

34 DoD (1981), p. 21. Most alarmingly, a critical mechanism that prevented one of the bombs from detonating appears to have failed on the other bomb (Burr, 2014). McNamara is quoted in U.S. Department of State (1963).

35 DoD (1981), p. 22.

36 DoD (1981), p. 28; Broder (1989).

37 DoD (1981), p. 29.

38 Accidental nuclear detonation was more likely then than it would be today, as the B-52's weapons did not yet satisfy the 'one-point safety' standard. See Philips (1998).

39 SAC (1969); Risø (1970); DoD (1981), p. 30; Philips (1998); Taagholt, Hansen & Lufkin (2001), pp. 35–43. The early-warning system had three lines of communication to the US: radio relayed by the B-52 bomber on airborne alert, direct radio and a bomb alarm. The plane crash cut the first of these. If a nuclear bomb had detonated, this would have cut the direct radio link too and triggered the bomb alarm, making the accident impossible to distinguish from a Soviet nuclear strike.

40 DoD (1981), p. 31.

41 Bordne first revealed his story to a Japanese newspaper in 2015, and it was subsequently published in the *Bulletin of the Atomic Scientists* (Tovish, 2015). The account has been disputed by other former missileers (Tritten, 2015).

42 So risk 1 is $(p_1/p_2) \times ((1-p_2)/(1-p_1))$ times as important as risk 2. We could rewrite this as $(p_1/(1-p_1)) / (p_2/(1-p_2))$ which is the ratio of their odds ratios. So if you want a single number to express the counterfactual importance of each risk (which doesn't need to be adjusted depending on the risk you are comparing it to) the odds ratio works perfectly.

43 You might wonder whether this brings up a question about how to individuate risks. For example, what makes something a 90% risk rather than two overlapping 50% risks? It turns out that there is no real dependence on how we individuate risks. Instead, it depends what we are asking. If you are considering lowering a set of risks that would constitute a 90% total existential risk were there no other risks at play, then this is just like lowering an individual risk with a probability of 90%.

The fact that the 'big risk' could be a collection of smaller risks that move together opens up some intriguing possibilities. For example, suppose that risk this century is 10% and eliminating it would leave all future risk as 90%. If you were in a position to either reduce all near-term risk by some amount or all later risk, this effect would come up, suggesting that acting on the later risk gets the nine-fold boost. (That said, acting on future risk would also typically get a penalty, which may exceed this, since we should expect additional people to also be in a position to help with it.)

44 The former reduces the total 91% existential risk by 0.9 percentage points, to 90.1%. The latter reduces it by 0.1 percentage points to 90.9%. The effect is surprising because the reduction of the 90% risk to 89% is also a smaller fractional change to that risk. The key to understanding this intuitively is that it is the fractional change to the chance of *not* having the catastrophe that matters.

45 It also happens regardless of whether there are several large risks or a multitude of smaller risks: for example, the same effects happen if there are 100 independent risks of 2% each.

46 This doesn't require us to assume that the centuries after the catastrophe have zero value, just that they have the same amount of value as each other and that this is less than those before the catastrophe.

47 Another way to see this is that the expected value of our future without risk reduction is some number V. Eliminating the probability of existential catastrophe in the first century is equivalent to getting a safe century for free, followed by the original value of the future from that point on. So the value we add is the value of a century for humanity. Or we could ask how many years it would take to build up the amount of risk you eliminated. In this basic model, the value of eliminating the risk is the value of that many years.

Note also that when I say the value of a century of humanity, I am referring to its *intrinsic* value. Since humanity's actions now can greatly affect the intrinsic value of future centuries, much of the all-things-considered value of a century is in its *instrumental value*. For example, the amount by which we improve the stock of knowledge, technology, institutions and environmental resources that we pass on to future generations. And of course, the amount by which we increase or decrease the total existential risk (though in this basic model, the instrumental value of our existential risk reduction is at most as high as the intrinsic value of the century).

48 The rate of increase of value doesn't have to stay higher than the hazard rate forever, in order to have a big impact. This is good, as the evidence we have suggests that the extremely long-run hazard rate will be roughly constant (due to a small amount of uneliminable risk), which would require value to grow exponentially over the extremely long run to offset it. But it seems likely that value we can create with a given amount of raw resources will eventually reach a limit, and the rate at which we acquire new resources is limited by the speed of light to be cubic, or less.

49 One might object to this based on counterfactuals: these insights are useful, but other people in the future would develop them anyway, so there is no counterfactual impact on the future. There is some important truth to this. While there would still be some counterfactual impact of this book's insights, this would be in terms of advancing the conversation such that future people are in a position to make advances that build on these ones. But those further advances may have diminishing marginal value. So on this counterfactual basis, the value of work now upon future risk will be less than it appears, and may diminish over time.

But this doesn't really affect the case I'm making. For if we are assuming that other people in the future will be doing important cumulative work on existential risk, then that suggests the risk won't be constant over future centuries, but will be decreasing. Either way, there is a similar effect, as we are just about to see.

50 Yew-Kwang Ng (2016) also noticed this counter-intuitive effect.

51 The term comes from Bostrom (2013). David Deutsch (2011) has an admirable defence of the contrary view.

We may also reach a form of social or political maturity, where we are no longer flirting with radical new systems such as the totalitarian and communist experiments of the twentieth century, and have developed a stable society that could never again descend into tyranny.

52 The boost to the value depends roughly on the ratio between the risk over the next century and the longer term risk per century. For example, if there is a one in ten chance of existential catastrophe this century, but this declined rapidly to a background natural risk rate of less than one in 200,000 per century, then the value of eliminating risk this century would be boosted by a factor of 20,000 compared to the basic model.

53 We could think about this mathematically as having an asymmetric probability distribution for the value of the future (one that is perhaps symmetric in log-space). The expected value of the future corresponds to the mean of this distribution and this may be substantially higher than the median.

54 Others have tried extending and interpolating the scale, but have done so in ways that involved arbitrary or problematic elements. Several people have tried to add a K4 level for 'the universe' but applied this to the entire universe or the observable universe, neither of which is the right concept. Carl Sagan had an early form of continuous Kardashev scale, but it forced all levels to be exactly 10 orders of magnitude apart. Since they are actually about 9 and 11, this broke the connection between the Kardashev levels and the structure of the cosmos, which was a very big loss. I think it better to just interpolate between the two integer Kardashev levels, so that each order of magnitude takes you roughly a ninth of the way from K0 to K1, but an eleventh of the way from K2 to K3. Sagan also added a K0 level, but his was just an arbitrary location 10 orders of magnitude before K1, such that the fraction of the way we've travelled from K0 to K1 did not bear any real meaning.

55 This is an order of magnitude estimate, with about 1 million people in ancient Mesopotamia at the time writing was discovered, consuming about 100 W in energy from food and expending about 100 W in work/heat. The truth may be a small factor higher than this due to inaccuracies in those numbers or if their livestock consumed substantially more food than the human population.

BIBLIOGRAPHY

Acemoglu, D. (2013). 'The World Our Grandchildren Will Inherit', in I. Palacios-Huerta (ed.), *In 100 Years: Leading Economists Predict the Future* (pp. 1–36). MIT Press.

Adams, F. C., and Laughlin, G. (1997). 'A Dying Universe: The Long-Term Fate and Evolution of Astrophysical Objects', *Reviews of Modern Physics*, 69(2), 337–72.

—(1999). *The Five Ages of the Universe: Inside the Physics of Eternity*. Free Press.

Aizen, M. A., Garibaldi, L. A., Cunningham, S. A., and Klein, A. M. (2009). 'How Much Does Agriculture Depend on Pollinators? Lessons from Long-Term Trends in Crop Production'. *Annals of Botany*, 103(9), 1,579–88.

Alexander, S. (2014). Meditations on Moloch. https://slatestarcodex.com/2014/07/30/meditations-on-moloch/

—(2015). AI Researchers on AI Risk. https://slatestarcodex.com/2015/05/22/ai-researchers-on-ai-risk/

Alibek, K. (2008). *Biohazard*. Random House.

Allen, M., et al. (2018). 'Framing and Context', in V. Masson-Delmotte, et al. (eds), *Global Warming of 1.5°C*. An IPCC Special Report on the impacts of global warming of 1.5°C above pre-industrial levels and related global greenhouse gas emission pathways, in the context of strengthening the global response to the threat of climate change (pp. 49–91), in press.

Alroy, J. (1996). 'Constant Extinction, Constrained Diversification, and Uncoordinated Stasis in North American Mammals'. *Palaeogeography, Palaeoclimatology, Palaeoecology*, 127(1), 285–311.

Altman, A., and Wellman, C. H. (2004). 'A Defense of International Criminal Law'. *Ethics*, 115(1), 35–67.

Alvarez, L. W., Alvarez, W., Asaro, F., and Michel, H. V. (1980). 'Extraterrestrial Cause for the Cretaceous-Tertiary Extinction'. *Science*, 208(4448), 1,095–108.

Ambrose, S. H. (1998). 'Late Pleistocene Human Population Bottlenecks, Volcanic Winter, and Differentiation of Modern Humans'. *Journal of Human Evolution*, 34(6), 623–51.

Ammann, W., et al. (2012). 'Mars Sample Return Backward Contamination – Strategic Advice and Requirements'. *National Aeronautics and Space Administration*.

Anderson, I. (2008). 'Foot and Mouth Disease 2007: A Review and Lessons Learned – Report to the UK Prime Minister and the Secretary of State for Environment Food and Rural Affairs'. London: The Stationery Office.

Annan, J. D., and Hargreaves, J. C. (2011). 'On the Generation and Interpretation of Probabilistic Estimates of Climate Sensitivity'. *Climatic Change*, 104(3–4), 423–36.

Antón, S. C., Potts, R., and Aiello, L. C. (2014). 'Evolution of Early Homo: An Integrated Biological Perspective'. *Science*, 345 (6192), 1236828.

Archer, D. (2005). 'Fate of Fossil Fuel CO_2 in Geologic Time', *Journal of Geophysical Research*, 110(C9).

Armstrong, S., and Sandberg, A. (2013). 'Eternity in Six Hours: Intergalactic Spreading of Intelligent Life and Sharpening the Fermi Paradox'. *Acta Astronautica*, 89, 1–13.

Arnett, R. L. (1979). 'Soviet attitudes towards nuclear war survival (1962–1977): has there been a change?' [PhD thesis]. The Ohio State University.

Arrhenius, G. (2000). 'An Impossibility Theorem for Welfarist Axiologies'. *Economics and Philosophy*, 16, 247–66.

Arsenault, C. (5 December 2014). 'Only 60 Years of Farming Left If Soil Degradation Continues'. *Scientific American*.

Arsuaga, J. L., et al. (2014). 'Neandertal Roots: Cranial and Chronological Evidence from Sima de los Huesos'. *Science*, 344(6190), 1,358–63.

Asimov, Isaac (August 1959). 'Big Game Hunting in Space'. *Space Age*.

—(1979). *A Choice of Catastrophes: The Disasters that Threaten Our World*. Simon and Schuster.

Askell, A. (2018). 'Pareto Principles in Infinite Ethics' [PhD Thesis]. Department of Philosophy, New York University.

Atkinson, N. (2009). How to Handle Moon Rocks and Lunar Bugs: A Personal History of Apollo's Lunar Receiving Lab. https://www.universetoday.com/35229/how-to-handle-moon-rocks-and-lunar-bugs-a-personal-history-of-apollos-lunar-receiving-lab/

Avin, S., et al. (2018). 'Classifying Global Catastrophic Risks'. *Futures*, 102, 20–6.

Baade, W., and Zwicky, F. (1934). 'On Super-Novae'. *Proceedings of the National Academy of Sciences*, 20(5), 254–9.

Bacon, F. (2004). 'Novum Organum', in G. Rees and M. Wakely (eds), *The Oxford Francis Bacon*, vol. 11: *The Instauratio magna Part II: Novum organum and Associated Texts* (pp. 48–586). Oxford University Press (original work published in 1620).

Baier, A. (1981). 'The Rights of Past and Future Persons', in E. Partridge (ed.), *Responsibilities to Future Generations: Environmental Ethics* (pp. 171–83). Prometheus Books.

Bailey, R. T. (1997). 'Estimation from Zero-Failure Data'. *Risk Analysis*, 17(3), 375–80.

Ball, D. (2006). 'The Probabilities of "On the Beach": Assessing "Armageddon Scenarios" in the 21st Century' (Working Studies Paper No . 401), in Strategic and Defence Studies Centre.

Bambach, R. K. (2006). 'Phanerozoic Biodiversity Mass Extinctions'. *Annual Review of Earth and Planetary Sciences*, 34(1), 127–55.

Bar-On, Y. M., Phillips, R., and Milo, R. (2018). 'The Biomass Distribution on Earth'. *Proceedings of the National Academy of Sciences*, 115(25), 6,506–11.

Barnosky, A. D., et al. (2011). 'Has the Earth's Sixth Mass Extinction Already Arrived?' *Nature*, 471 (7336), 51–7.

Barry, P. L. (2000). Continents in Collision: Pangea Ultima. https:// science.nasa.gov/science-news/science-at-nasa/2000/ast06oct_1

Bartholomew, R. E., and Radford, B. (2011). *The Martians have Landed! A History of Media-Driven Panics and Hoaxes*. McFarland.

Baum, S. D., Denkenberger, D. C., Pearce, J. M., Robock, A., and Winkler, R. (2015). 'Resilience to Global Food Supply Catastrophes'. *Environment Systems and Decisions*, 35(2), 301–13.

BBC (12 November 2015). 'Russia Reveals Giant Nuclear Torpedo in State TV "leak"'. BBC News.

Beade, A. P. M., Ahlonsou, E., Ding, Y., and Schimel, D. (2001). 'The Climate System: An Overview', in J. T. Houghton, et al. (eds), *Climate Change 2001: The Scientific Basis. Contribution of Working Group I to the Third Assessment Report of the Intergovernmental Panel on Climate Change*. Cambridge University Press.

Beck, U. (2009). *World at Risk* (trans. C. Cronin). Polity Press.

Beckstead, N. (2013). 'On the Overwhelming Importance of Shaping the Far Future' [PhD Thesis]. Department of Philosophy, Rutgers University.

Benedictow, O. J. (2004). *The Black Death, 1346–1353: The Complete History*. Boydell Press.

Bengio, Y., et al. (2017). Creating Human-level AI: How and When? (Panel from The Beneficial AI 2017 Conference) [Video]. https://www.youtube.com/watch?v=V0aXMTpZTfc

Bernstein, A., and Roberts, M. de V. (1958). 'Computer v Chess-Player'. *Scientific American*, 198(6), 96–105.

Bethe, H., Brown, H., Seitz, F., and Szilard, L. (1950). 'The Facts About the Hydrogen Bomb'. *Bulletin of the Atomic Scientists*, 6(4), 106–9.

Blanton, T., Burr, W., and Savranskaya, S. (2012). *The Underwater Cuban Missile Crisis: Soviet Submarines and the Risk of Nuclear War*. National Security Archive, Electronic Briefing Book No. 399. National Security Archive.

Boberg, J. (2005). 'Freshwater Availability', in J. Boberg (ed.), *How Demographic Changes and Water Management Policies Affect Freshwater Resources* (pp. 15–28). RAND Corporation.

Bolton, J., and Azar, A. (2018). Press Briefing on the National Biodefense Strategy. https://www.whitehouse.gov/briefings-statements/press-briefing-national-biodefense-strategy-091818/

Bonnell, J. T., and Klebesadel, R. W. (1996). 'A Brief History of the Discovery of Cosmic Gamma-Ray Bursts'. *AIP Conference Proceedings*, 384, 977–80.

Bostrom, N. (2002a). *Anthropic Bias: Observation Selection Effects in Science and Philosophy*. Routledge.

—(2002b). 'Existential Risks: Analyzing Human Extinction Scenarios and Related Hazards'. *Journal of Evolution and Technology*, 9.

—(2003). 'Astronomical Waste: The Opportunity Cost of Delayed Technological Development'. *Utilitas*, 15(3), 308–14.

—(2005). A Philosophical Quest for our Biggest Problems (talk at TEDGlobal). https://www.ted.com/talks/nick_bostrom_on_our_biggest_problems

—(2006). 'What Is a Singleton'. *Linguistic and Philosophical Investigations*, 5(2), 48–54.

—(2008). 'Letter from Utopia'. *Studies in Ethics, Law, and Technology*, 2(1).

—(2009). 'Pascal's Mugging'. *Analysis*, 69(3), 443–5.

—(2011a). 'Infinite Ethics'. *Analysis and Metaphysics*, 10, 9–59.

—(2011b). 'Information Hazards: A Typology of Potential Harms from Knowledge'. *Review of Contemporary Philosophy*, (10), 44–79.

—(2012). 'The Superintelligent Will: Motivation and Instrumental Rationality in Advanced Artificial Agents'. *Minds and Machines*, 22(2), 71–85.

—(2013). 'Existential Risk Prevention as Global Priority'. *Global Policy*, 4(1), 15–31.

—(2014). *Superintelligence: Paths, Dangers, Strategies*. Oxford University Press.

—(2018). 'The Vulnerable World Hypothesis' (Working Paper, v. 3.45).

Bostrom, N., and Ćirković, M. M. (2008). 'Introduction', in N. Bostrom and M. Ćirković (eds), *Global Catastrophic Risks* (pp. 1–30). Oxford University Press.

Bostrom, N., Douglas, T., and Sandberg, A. (2016). 'The Unilateralist's Curse and the Case for a Principle of Conformity'. *Social Epistemology*, 30(4), 350–71.

Botkin, D. B., et al. (2007). 'Forecasting the Effects of Global Warming on Biodiversity'. *BioScience*, 57(3), 227–36.

Bourget, D., and Chalmers, D. J. (2014). 'What Do Philosophers Believe?' *Philosophical Studies*, 170(3), 465–500.

Brand, S. (April 2000). 'Taking the Long View'. *Time*.

Brasier, M., McLoughlin, N., Green, O., and Wacey, D. (2006). 'A Fresh Look at the Fossil Evidence for Early Archaean Cellular Life'. *Philosophical Transactions of the Royal Society B: Biological Sciences*, 361(1470), 887–902.

Brezhnev, L. (1979). Brezhnev Message to President on Nuclear False Alarm, Diplomatic Cable (No. 1979STATE295771) from Sec State (D.C.) to Moscow American Embassy. National Security Archive. United States Department of State.

Bricker, D., and Ibitson, J. (2019). *Empty Planet: The Shock of Global Population Decline*. Crown.

Broder, J. (9 May 1989). 'H-Bomb Lost at Sea in '65 off Okinawa, U.S. Admits'. *Los Angeles Times*.

Broome, J. (2005). 'Should We Value Population?' *Journal of Political Philosophy*, 13(4), 399–413.

Browne, M. W. (23 January 1990). 'Nuclear Winter Theorists Pull Back'. *The New York Times*.

Bruckner T., et al. (2014). '2014: Energy Systems', in O. Edenhofer, et al. (eds), *Climate Change 2014: Mitigation of Climate Change. Contribution of Working Group III to the Fifth Assessment Report of the Intergovernmental Panel on Climate Change* (p. 1,465). Cambridge University Press.

Brundage, M., et al. (2018). The Malicious Use of Artificial Intelligence: Forecasting, Prevention, and Mitigation. ArXiv, https://arxiv.org/pdf/1802.07228.

Bryant, C. (2003). 'Stopping Time: The Pro-Slavery and "Irrevocable" Thirteenth Amendment'. *Harvard Journal of Law and Public Policy*, 26(2), 501–49.

Buchner, B. K., et al. (2017). *Global Landscape of Climate Finance 2017*. Climate Policy Initiative.

Buck, P. S. (1959, March). 'The Bomb – The End of the World?' *The American Weekly*.

Buffett, B. A., Ziegler, L., and Constable, C. G. (2013). 'A Stochastic Model for Palaeomagnetic Field Variations'. *Geophysical Journal International*, 195(1), 86–97.

Bulfin, A. (2015). '"To Arms!" Invasion Narratives and Late-Victorian Literature'. *Literature Compass*, 12(9), 482–96.

Burke, E. (1790). *Reflections on the French Revolution*. James Dodsley.

Burr, W. (2014). 'New Details on the 1961 Goldsboro Nuclear Accident'. National Security Archive, Electronic Briefing Book No. 475. National Security Archive.

Butler, D., and Ledford, H. (2012). 'US Biosecurity Board Revises Stance on Mutant-Flu Studies'. *Nature*.

Butler, S. (13 June 1863). *Darwin Among the Machines*. The Press.

—(1872). *Erewhon*. Ballantyne and Co.

Buttazzo, D., et al. (2013). 'Investigating the Near-Criticality of the Higgs Boson'. *Journal of High Energy Physics*, 2013(12), 89.

BWC ISU (2019). Biological Weapons Convention – Budgetary and Financial Matters (21 January 2019 Letter from BWC Implementation Support Unit to BWC Representatives). https://www.unog.ch/80256EDD006B8954/(httpAssets)/1FE92995054B8108C1258394004233AD/$file/2019-0131+2018+MSP+Chair+letter+on+financial+measures.pdf

Carlson, R. (2016). On DNA and Transistors. http://www.synthesis.cc/synthesis/2016/03/on_dna_and_transistors

Carus, W. S. (2017). 'A Century of Biological-Weapons Programs (1915–2015): Reviewing the Evidence'. *The Nonproliferation Review*, 24(1–2), 129–53.

Ceballos, G., et al. (2015). 'Accelerated Modern Human-Induced Species Losses: Entering the Sixth Mass Extinction'. *Science Advances*, 1(5), e1400253.

Cederman, L.-E. (2003). 'Modeling the Size of Wars: From Billiard Balls to Sandpiles'. *The American Political Science Review*, 97(1), 135–50.

Challinor, A. J., et al. (2014). 'A Meta-Analysis of Crop Yield under Climate Change and Adaptation'. *Nature Climate Change*, 4(4), 287–91.

Chan, S. (18 September 2017). 'Stanislav Petrov, Soviet Officer Who Helped Avert Nuclear War, Is Dead at 77'. *The New York Times*.

Chapman, C. R. (2004). 'The Hazard of Near-Earth Asteroid Impacts on Earth'. *Earth And Planetary Science Letters*, 222(1), 1–15.

Charney, J. G., et al. (1979). 'Carbon Dioxide and Climate: A Scientific Assessment'. *National Academy of Sciences*.

Chen, Z.-Q., and Benton, M. J. (2012). 'The Timing and Pattern of Biotic Recovery Following the End-Permian Mass Extinction'. *Nature Geoscience*, 5(6), 375–83.

Chesner, C. A., and Luhr, J. F. (2010). 'A Melt Inclusion Study of the Toba Tuffs, Sumatra, Indonesia'. *Journal of Volcanology and Geothermal Research*, 197(1), 259–78.

Chosewood, L. C., and Wilson, D. E. (eds) (2009). 'Biosafety in Microbiological and Biomedical Laboratories'. HHS Publication No. (CDC) 21–1112 (5th ed.). Centers for Disease Control and Prevention.

Christakos, G., et al. (eds) (2005). 'Black Death: The Background', in *Interdisciplinary Public Health Reasoning and Epidemic Modelling: The Case of Black Death* (pp. 103–52). Springer.

Christian, D. (2004). *Maps of Time*. University of California Press.

Churchill, W. (1946). Speech, 'The Sinews of Peace', 5 March 1946 at Westminster College, Fulton, Missouri, US [Radio Broadcast]. BBC Archives.

Ciais, P., et al. (2013). 'Carbon and Other Biogeochemical Cycles', in T. F. Stocker, et al. (eds), *Climate Change 2013: The Physical Science Basis. Contribution of Working Group I to the Fifth Assessment Report of the Intergovernmental Panel on Climate Change* (pp. 465–570). Cambridge University Press.

Ćirković, M. M., Sandberg, A., and Bostrom, N. (2010). 'Anthropic Shadow: Observation Selection Effects and Human Extinction Risks'. *Risk Analysis*, 30(10), 1,495–506.

Clauset, A., and Young, M. (2005). Scale Invariance in Global Terrorism. ArXiv, https://arxiv.org/abs/physics/0502014.

Clune, J. (2019). AI-GAs: AI-Generating Algorithms, an Alternate Paradigm for Producing General Artificial Intelligence. ArXiv, http://arxiv.org/abs/1905.10985.

Coale, A. J. (1974). 'The History of the Human Population'. *Scientific American*, 231(3), 40–51.

Cohen, G. A. (2011). 'Rescuing Conservatism: A Defense of Existing Value', in *Reasons and Recognition: Essays on the Philosophy of T. M. Scanlon* (pp. 203–26). Oxford University Press.

Cohen, M. N. (1989). *Health and the Rise of Civilization*. Yale University Press.

Coles, L. S. (1994). 'Computer Chess: The Drosophila of AI'. *AI Expert*, 9(4).

Collingridge, D. (1982). *The Social Control of Technology*. St Martin's Press.

Collins, G. S., Melosh, H. J., and Marcus, R. A. (2005). 'Earth Impact Effects Program: A Web-Based Computer Program for Calculating the Regional Environmental Consequences of a Meteoroid Impact on Earth'. *Meteoritics and Planetary Science*, 40(6), 817–40.

Collins, M., et al. (2013). 'Long-Term Climate Change: Projections, Commitments and Irreversibility', in T. F. Stocker, D. et al. (eds), *Climate Change 2013: The Physical Science Basis. Contribution of Working Group I to the Fifth Assessment Report of the Intergovernmental Panel on Climate Change* (pp. 1,029–136). Cambridge University Press.

Compton, A. H. (1956). *Atomic Quest*. Oxford University Press.

Conn, A., Toon, B., and Robock, A. (2016). Transcript: Nuclear Winter Podcast with Alan Robock and Brian Toon. https://futureoflife.org/2016/10/31/transcript-nuclear-winter-podcast-alan-robock-brian-toon/

Conselice, C. J., Wilkinson, A., Duncan, K., and Mortlock, A. (2016). 'The Evolution of Galaxy Number Density at z<8 and its Implications'. *The Astrophysical Journal*, 830(2), 83.

Cook, M., and Woolf, A. (2002). 'Preventing Proliferation of Biological Weapons: U.S. Assistance to the Former Soviet States' (CRS Report for Congress) [Report RL31368]. U.S. Homeland Security Digital Library.

Cook, N. D. (1998). *Born to Die: Disease and New World Conquest, 1492–1650* (vol. 1). Cambridge University Press.

Cordell, D., Drangert, J.-O., and White, S. (2009). 'The Story of Phosphorus: Global Food Security and Food for Thought'. *Global Environmental Change*, 19(2), 292–305.

Cotton-Barratt, O., Daniel, M., and Sandberg, A. (n.d.). 'Defence in Depth against Human Extinction: Prevention, Response, Resilience, and Why they all Matter' [manuscript in preparation].

Coupe, J., Bardeen, C. G., Robock, A., & Toon, O. B. (2019). 'Nuclear Winter Responses to Nuclear War Between the United States and Russia in the Whole Atmosphere Community Climate Model Version 4 and the Goddard Institute for Space Studies ModelE'. *Journal of Geophysical Research: Atmospheres*, 8,522–43.

Cropper, M. L., Aydede, S. K., and Portney, P. R. (1994). 'Preferences for Life-Saving Programs: How the Public Discounts Time and Age'. *Journal of Risk and Uncertainty*, 8(3), 243–65.

Cropper, W. P., and Harwell, M. A. (1986). 'Food Availability after Nuclear War', in M. A. Harwell and T. C. Hutchinson (eds), *The Environmental Consequences of Nuclear War (SCOPE 28)*, vol. 2: *Ecological, Agricultural, and Human Effects*. John Wiley and Sons.

Crosweller, H. S., et al. (2012). 'Global Database on Large Magnitude Explosive Volcanic Eruptions (LaMEVE)'. *Journal of Applied Volcanology*, 1(1), 4.

CSIRO (2015). Sea Level Data – Update of Reconstructed GMSL from 1880 to 2013. http://www.cmar.csiro.au/sealevel/sl_data_cmar.htm.

Cubasch, U., et al. (2013). 'Introduction', in T. F. Stocker, (eds), *Climate Change 2013: The Physical Science Basis. Contribution of Working Group I to the Fifth Assessment Report of the Intergovernmental Panel on Climate Change*. Cambridge University Press.

Cui, Y., and Kump, L. R. (2015). 'Global Warming and the End-Permian Extinction Event: Proxy and Modeling Perspectives'. *Earth-Science Reviews*, 149, 5–22.

Cyranoski, D. (2017). 'Bat Cave Solves Mystery of Deadly SARS Virus – and Suggests New Outbreak Could Occur'. *Nature*, 552(7683), 15–16.

D'Errico, P. (2001). Jeffery Amherst and Smallpox Blankets. https://people.umass.edu/derrico/amherst/lord_jeff.html

Dante Labs (2019). Dante Labs Tests. https://us.dantelabs.com/collections/our-tests

Danzig, R., et al. (2011). 'Aum Shinrikyo: Insights Into How Terrorists Develop Biological and Chemical Weapons'. Center for a New American Security.

Dasgupta, P. (2007). 'Commentary: The Stern Review's Economics of Climate Change'. *National Institute Economic Review*, 199(1), 4–7.

Dasgupta, P. (2008). 'Discounting Climate Change'. *Journal of Risk and Uncertainty*, 37(2–3), 141–69.

Davenport, K. (2018). Biological Weapons Convention Signatories and States-Parties. https://www.armscontrol.org/factsheets/bwcsig

Dawson, T. P., Jackson, S. T., House, J. I., Prentice, I. C., and Mace, G. M. (2011). 'Beyond Predictions: Biodiversity Conservation in a Changing Climate'. *Science*, 332(6025), 53–8.

De Vos, J. M., Joppa, L. N., Gittleman, J. L., Stephens, P. R., and Pimm, S. L. (2015). 'Estimating the Normal Background Rate of Species Extinction'. *Conservation Biology*, 29(2), 452–62.

Deevey, E. S. (1960). 'The Human Population'. *Scientific American*, 203(3), 194–204.

Denkenberger, D. C., and Blair, R. W. (2018). 'Interventions that may Prevent or Mollify Supervolcanic Eruptions'. *Futures*, 102, 51–62.

Denkenberger, D. C., and Pearce, J. M. (2016). 'Cost-Effectiveness of Interventions for Alternate Food to Address Agricultural Catastrophes Globally', *International Journal of Disaster Risk Science*, 7(3), 205–15.

Desjardins, J. (2014). A Forecast of When We'll Run Out of Each Metal. https://www.visualcapitalist.com/forecast-when-well-run-out-of-each-metal/

Deutsch, D. (2011). *The Beginning of Infinity: Explanations that Transform the World*. Viking.

DiCarlo, J. E., Chavez, A., Dietz, S. L., Esvelt, K. M., and Church, G. M. (2015). 'Safeguarding CRISPR-Cas9 Gene Drives in Yeast'. *Nature Biotechnology*, 33(12), 1,250–5.

Diderot, D. (1755). 'Encyclopedia', in P. Stewart (trans.), *Encyclopédie ou Dictionnaire raisonné des sciences, des arts et des métiers*, vol. 5 (pp. 635–648A). Michigan Publishing.

DiEuliis, D., Carter, S. R., and Gronvall, G. K. (2017). 'Options for Synthetic DNA Order Screening, Revisited'. *MSphere*, 2(4).

Dobrynin, A. (1995). *In Confidence: Moscow's Ambassador to Six Cold War Presidents*. Random House.

DoD (1981). 'Narrative Summaries of Accidents Involving US Nuclear Weapons (1950–1980)'. Homeland Security Digital Library. U.S. Department of Defense.

Dodson, R. W., and Rabi, I. I. (1954). *Meeting Minutes of the Forty-First Meeting of the General Advisory Committee to the U.S. Atomic Energy Commission*. United States Atomic Energy Commission.

Downes, L. (2009). *The Laws of Disruption: Harnessing the New Forces that Govern Business and Life in the Digital Age*. Basic Books.

Drmola, J., and Mareš, M. (2015). 'Revisiting the deflection dilemma'. *Astronomy and Geophysics*, 56(5), 5.15–5.18.

Drupp, M. A., Freeman, M. C., Groom, B., and Nesje, F. (2018). 'Discounting Disentangled'. *American Economic Journal: Economic Policy*, 10(4), 109–34.

Duplaix, N. (1988). 'Fleas: The Lethal Leapers'. *National Geographic*, 173(5), 672–94.

Durand, J. D. (1960). 'The Population Statistics of China, A.D. 2–1953'. *Population Studies*, 13(3), 209–56.

—(1977). 'Historical Estimates of World Population: An Evaluation'. *Population and Development Review*, 3(3), 253.

Dylan, B. (1963). 'Let me die in my footsteps' [Lyrics]. *The Freewheelin' Bob Dylan*. Columbia Records.

ECDC (2014). 'Communicable Disease Threats Report, Week 37, 7–13 September 2014'. European Centre for Disease Prevention and Control.

Egan, G. (1997). *Diaspora*. Millennium.

Ehrlich, P., et al. (1983). 'Long-Term Biological Consequences of Nuclear War'. *Science*, 222(4630), 1,293–300.

Ehrlich, P. R. (September 1969). 'Eco-Catastrophe'. *Ramparts*.

EIA (2019). 'Electric Power Monthly with Data for April 2014'. U.S. Energy Information Administration.

Eig, J. (2014). *The Birth of the Pill: How Four Crusaders Reinvented Sex and Launched a Revolution*. W. W. Norton.

Einstein, A. (1948). 'A Reply to the Soviet Scientists'. *Bulletin of the Atomic Scientists*, 4(2), 35–8.

Einstein, A., and *The New York Times* (25 May 1946). 'Atomic Education Urged by Einstein Scientist in Plea for $200,000 to Promote New Type of Thinking'. *The New York Times*.

Eisenhower, D. (1956). Letter, DDE to Richard L. Simon, Simon and Schuster, Inc. DDE's Papers as President, DDE Diaries Series, Box 14, April 1956 Miscellaneous (5).

Ellsberg, D. (2017). *The Doomsday Machine: Confessions of a Nuclear War Planner*. Bloomsbury Publishing.

Engels, F. (1892). *The Condition of the Working Class in England in 1844* (trans. F. K. Wischnewetzky). Swan Sonnenschein and Co.

Erwin, D. H., Bowring, S. A., and Yugan, J. (2002). 'End-Permian Mass Extinctions: A Review', in C. Koeberl and K. G. MacLeod (eds), *Special Paper 356: Catastrophic Events and Mass Extinctions: Impacts and Beyond* (pp. 363–83). Geological Society of America.

Espeholt, L., et al. (2018). '{IMPALA}: Scalable Distributed Deep-{RL} with Importance Weighted Actor-Learner Architectures', in J. Dy and A. Krause (eds), *Proceedings of the 35th International Conference on Machine Learning* (pp. 1,407–16). PMLR.

Esvelt, K. M. (2018). 'Inoculating Science against Potential Pandemics and Information Hazards'. *PLOS Pathogens*, 14(10), e1007286.

Everitt, T., Filan, D., Daswani, M., and Hutter, M. (2016). 'Self-Modification of Policy and Utility Function in Rational Agents'. *Artificial General Intelligence*, LNAI 9782, 1–11.

Farquhar, S. (2017). Changes in Funding in the AI Safety Field. https://www.centreforeffectivealtruism.org/blog/changes-in-funding-in-the-ai-safety-field/

Feinstein, A. R., Sosin, D. M., and Wells, C. K. (1985). 'The Will Rogers Phenomenon. Stage Migration and New Diagnostic Techniques as a Source of Misleading Statistics for Survival in Cancer'. *The New England Journal of Medicine*, 312(25), 1,604–8.

Feld, B. T. (1976). 'The Consequences of Nuclear War'. *Bulletin of the Atomic Scientists*, 32(6), 10–3.

Fenner, F., and Fantini, B. (1999). 'The Use of Rabbit Haemorrhagic Disease Virus for Rabbit Control', in *Biological Control of Vertebrate Pests: The History of Myxomatosis – an Experiment in Evolution*. CABI Publishing.

Flatow, I., Russell, S., and Koch, C. (2014). 'Science Goes to the Movies: "Transcendence"' (I. Flatow, interviewer) [Audio file from 24:33]. *Science Friday*.

Fleishman, J. L., Kohler, J. S., and Schindler, S. (2009). *Casebook for The Foundation: A Great American Secret*. PublicAffairs.

Foote, M., and Raup, D. M. (1996). 'Fossil Preservation and the Stratigraphic Ranges of Taxa'. *Paleobiology*, 22(2), 121–40.

Forden, G., Podvig, P., and Postol, T. A. (2000). 'False Alarm, Nuclear Danger'. *IEEE Spectrum*, 37(3), 31–9.

Forey, P. L. (1990). 'The Coelacanth Fish: Progress and Prospects'. *Science Progress* (1933–), 74(1), 53–67.

Frederick, S. (2003). 'Measuring Intergenerational Time Preference: Are Future Lives Valued Less?' *Journal of Risk and Uncertainty*, 26(1), 39–53.

Frick, J. (2017). 'On the Survival of Humanity'. *Canadian Journal of Philosophy*, 47(2–3), 344–67.

Frick, J. D. (2014). '"Making People Happy, Not Making Happy People": A Defense of the Asymmetry Intuition in Population Ethics' [PhD Thesis]. Department of Philosophy, Harvard University.

Future of Life Institute (2015). Research Priorities for Robust and Beneficial Artificial Intelligence: An Open Letter. https://futureoflife. org/ai-open-letter/

Future of Life Institute (2017). Asilomar AI Principles. https:// futureoflife.org/ai-principles/

Galway-Witham, J., and Stringer, C. (2018). 'How Did Homo Sapiens Evolve?' *Science*, 360(6395), 1,296–8.

Gapminder. (2019). Life Expectancy (years). https://www.gapminder. org/data/

García-Sastre, A. (2012). 'Working Safely with H5N1 Viruses'. *MBio*, 3(2).

Gates, R. M. (2011). *From the Shadows: The Ultimate Insider's Story of Five Presidents and How They Won the Cold War.* Simon and Schuster.

GBD (2012). 'The Global Burden of Disease Study 2010'. *The Lancet*, 380(9859), 2,053–260.

Gibney, E. (2018). 'Belgian Priest Recognized in Hubble-Law Name Change'. *Nature*.

Gietel-Basten, S. (2016). 'Japan is Not the Only Country Worrying About Population Decline – Get Used to a Two-Speed World'. *The Conversation*.

GiveWell (2019). 2019 GiveWell Cost-effectiveness Analysis — Version 3. https://docs.google.com/spreadsheets/d/1McptF0GVGv-QBlhWx_ IoNVstWvt1z-RwVSu16ciypgs/

Giving What We Can (2019). https://www.givingwhatwecan.org/

Gleick, P. H., and Palaniappan, M. (2010). 'Peak Water Limits to Freshwater Withdrawal and Use'. *Proceedings of the National Academy of Sciences*, 107(25), 11,155–62.

GNL. (2019). Laboratory Safety at UTMB. Galveston National Laboratory, University of Texas Medical Branch. https://www.utmb.edu/gnl/about/lab-safety.

Goldberg, S. (1983). 'How Many People Have Ever Lived?' in S. Goldberg (ed.), *Probability in Social Science* (pp. 19–31). Birkhäuser.

Goldblatt, C., Robinson, T. D., Zahnle, K. J., and Crisp, D. (2013). 'Low Simulated Radiation Limit for Runaway Greenhouse Climates'. *Nature Geoscience*, 6(8), 661–7.

Good, I. J. (1959). 'Speculations on Perceptrons and Other Automata'. Research Lecture, RC-115. IBM, Yorktown Heights, New York, 2 June.

—(1970). 'Some Future Social Repercussions of Computers'. *International Journal of Environmental Studies*, 1(1–4), 67–79.

Goodchild, P. (2004). *Edward Teller, the Real Dr. Strangelove*. Harvard University Press.

Goodfellow, I. J., et al. (2014). Generative Adversarial Networks. ArXiv, https://arxiv.org/abs/1406.2661.

Goodhart, C. (1975). 'Problems of Monetary Management: The U.K. Experience', in *Papers in Monetary Economics*. Reserve Bank of Australia.

Gorbachev, M., and Hertsgaard, M. (24 September 2000). 'Mikhail Gorbachev Explains What's Rotten in Russia'. *Salon*.

Gordon, N. D., Jonko, A. K., Forster, P. M., and Shell, K. M. (2013). 'An Observationally Based Constraint on the Water-Vapor Feedback'. *Journal of Geophysical Research: Atmospheres*, 118(22), 12,435–43.

Gould, C., and Folb, P. (2002). *Project Coast: Apartheid's Chemical and Biological Warfare Programme* (R. Berold, ed.). United Nations Publications UNIDIR.

Grace, K. (2013). 'Algorithmic Progress in Six Domains' [Technical report 2013-3]. Machine Intelligence Research Institute.

—(2015). 'The Asilomar Conference: A Case Study in Risk Mitigation' [Technical report 2015–9]. Machine Intelligence Research Institute.

Grace, K., Salvatier, J., Dafoe, A., Zhang, B., and Evans, O. (2018). 'Viewpoint: When Will AI Exceed Human Performance? Evidence from AI Experts'. *Journal of Artificial Intelligence Research*, 62, 729–54.

Greaves, H., and Ord, T. (2017). 'Moral Uncertainty About Population Axiology'. *Journal of Ethics and Social Philosophy*, 12(2), 135–67.

Griffin, M. (2008). 'NASA's Direction, Remarks at the Mars Society Convention', 3 August 2006, in *Leadership in Space: Selected Speeches of NASA Administrator Michael Griffin, May 2005–October 2008* (pp. 133–8). National Aeronautics and Space Administration.

Griffith, G. (1897, November). 'The Great Crellin Comet'. *Pearsons Weekly's Christmas*.

Groenewold, H. J. (1970). 'Modern Science and Social Responsibility', in P. Weingartner and G. Zecha (eds), *Induction, Physics and Ethics*. Synthese Library (Monographs on Epistemology, Logic, Methodology, Philosophy of Science, Sociology of Science and of Knowledge, and on the Mathematical Methods of Social and Behavioral Sciences), vol. 31 (pp. 359–78). Springer.

Haarnoja, T., et al. (2018). 'Composable Deep Reinforcement Learning for Robotic Manipulation'. 2018 IEEE International Conference on Robotics and Automation (ICRA), 6,244–51. IEEE.

Haensch, S., et al. (2010). 'Distinct Clones of Yersinia Pestis Caused the Black Death'. *PLOS Pathogens*, 6(10), e1001134.

Häggström, O. (2016). 'Here Be Dragons: Science, Technology and the Future of Humanity', in *Here Be Dragons*. Oxford University Press.

Hanley, J. A. (1983). 'If Nothing Goes Wrong, Is Everything All Right?' *JAMA*, 249(13), 1743.

Hanson, R. (2008). 'Catastrophe, Social Collapse, and Human Extinction', in N. Bostrom and M. Ćirković (eds), *Global Catastrophic Risk*. Oxford University Press.

Harari, Y. N. (2014). *Sapiens: A Brief History of Humankind*. Random House.

Harris, S. H. (2002). *Factories of Death: Japanese Biological Warfare, 1932–45, and the American Cover-Up*. Psychology Press.

Harrison, M. (1998). 'The Economics of World War II: An Overview', in M. Harrison (ed.), *The Economics of World War II: Six Great Powers in International Comparison* (pp. 1–42). Cambridge University Press.

Harrod, R. F. (1948). *Towards a Dynamic Economics: Some Recent Developments of Economic Theory and Their Application to Policy*. Macmillan and Co.

Harwell, M. A., and Harwell, C. C. (1986). 'Integration of Effects on Human Populations', in M. A. Harwell and T. C. Hutchinson (eds), *The Environmental Consequences of Nuclear War (SCOPE 28)*, vol. 2: *Ecological, Agricultural, and Human Effects* (pp. 469–92). John Wiley and Sons.

Harwell, M. A., and Hutchinson, T. C. (1986). *The Environmental Consequences of Nuclear War (SCOPE 28)*, vol. 2: *Ecological, Agricultural, and Human Effects*. John Wiley and Sons.

Hasell, J., and Roser, M. (2019). Famines. Our World in Data. https://ourworldindata.org/famines.

Hassan, H., et al. (2018). 'Achieving Human Parity on Automatic Chinese to English News Translation'. *ArXiv*, http://arxiv.org/abs/1803.05567.

Haub, C., and Kaneda, T. (2018). How Many People Have Ever Lived on Earth? https://www.prb.org/howmanypeoplehaveeverlivedon earth/

He, K., Zhang, X., Ren, S., and Sun, J. (2015). 'Delving Deep into Rectifiers: Surpassing Human-Level Performance on ImageNet Classification'. 2015 IEEE International Conference on Computer Vision (ICCV), 1,026–34. IEEE.

Helfand, I. (2013). 'Nuclear Famine: Two Billion People at Risk?' *Physicians for Social Responsibility*.

Henrich, J. (2015). *The Secret of Our Success: How Culture Is Driving Human Evolution, Domesticating Our Species, and Making Us Smarter*. Princeton University Press.

Herfst, S., et al. (2012). 'Airborne Transmission of Influenza A/H5N1 Virus Between Ferrets'. *Science*, 336(6088), 1,534–41.

Hershberg, J. G. (1995). *James B. Conant: Harvard to Hiroshima and the Making of the Nuclear Age*. Stanford University Press.

Hershberg, J., and Kelly, C. (2017). James Hershberg's Interview. https://www.manhattanprojectvoices.org/oral-histories/james-hershbergs-interview

Heuer, R. J. (1999). 'Chapter 12: Biases in Estimating Probabilities', in *Psychology of Intelligence Analysis*. Center for the Study of Intelligence.

Heyd, D. (1988). 'Procreation and Value: Can Ethics Deal with Futurity Problems?' *Philosophia*, 18(2–3).

Highfield, R. (16 October 2001). 'Colonies in Space May Be Only Hope, Says Hawking'. *Daily Telegraph*.

Hilts, P. J. (18 November 1994). 'Deaths in 1979 Tied to Soviet Military'. *The New York Times*.

Hof, C., Levinsky, I., Araújo, M. B., and Rahbek, C. (2011). 'Rethinking Species' Ability to Cope with Rapid Climate Change'. *Global Change Biology*, 17(9), 2,987–90.

Holmes, D. B. (2008). *Wilbur's Story*. Lulu.com.

Honigsbaum, M. (2018). 'Spanish Influenza Redux: Revisiting the Mother of all Pandemics'. *The Lancet*, 391(10139), 2,492–5.

Horowitz, M. C. (2018). 'Artificial Intelligence, International Competition, and the Balance of Power'. *Texas National Security Review*, 1(3), 37–57.

IBM (2011). Deep Blue. https://www.ibm.com/ibm/history/ibm100/us/en/icons/deepblue/

IDC (2019). Worldwide Spending on Artificial Intelligence Systems Will Grow to Nearly $35.8 Billion in 2019, According to New IDC Spending Guide. https://www.idc.com/getdoc.jsp?containerId=prUS44911419

IEA (2018). 'Costs and Benefits of Emergency Stockholding', *Insights Series 2018*, International Energy Agency.

—(2019). 'Global Energy and CO2 Status Report: The Latest Trends in Energy and Emissions in 2018 – Data Tables', International Energy Agency.

iGEM(2013).Jamboree/TeamAbstracts.http://2013.igem.org/Jamboree/Team_Abstracts

iGEM Minnesota Team (2016). Shifting Gene Drives Into Reverse: Now Mosquitoes Are the Yeast of Our Worries. http://2016.igem.org/Team:Minnesota

IGSC (2018). International Gene Synthesis Consortium Updates Screening Protocols for Synthetic DNA Products and Services. https://www.prnewswire.com/news-releases/international-gene-synthesis-consortium-updates-screening-protocols-for-synthetic-dna-products-and-services-300576867.html?tc=eml_cleartime

Imai, M., et al. (2012). 'Experimental Adaptation of an Influenza H5 HA Confers Respiratory Droplet Transmission to a Reassortant H5 HA/H1N1 Virus in Ferrets'. *Nature*, 486(7403), 420–8.

IMARC Group (2019). 'Ice Cream Market: Global Industry Trends, Share, Size, Growth, Opportunity and Forecast 2019–2024'. *IMARC*.

Imperiale, M. J., and Hanna, M. G. (2012). 'Biosafety Considerations of Mammalian-Transmissible H5N1 Influenza'. *MBio*, 3(2).

IPCC. (2014). 'Summary for Policymakers', in C. B. Field, et al. (eds), *Climate Change 2014: Impacts, Adaptation, and Vulnerability. Part A: Global and Sectoral Aspects. Contribution of Working Group II to the Fifth Assessment Report of the Intergovernmental Panel on Climate Change* (pp. 1–32). Cambridge University Press.

Jamison, D. T., et al. (2018). 'Universal Health Coverage and Intersectoral Action for Health: Key Messages from Disease Control Priorities', 3rd ed. *The Lancet*, 391(10125), 1,108–20.

Jamison, D. T., et al. (eds) (2006). *Disease Control Priorities in Developing Countries*, 2nd ed. Oxford University Press.

Jamison, D. T., et al. (eds) (2018). *Disease Control Priorities: Improving Health and Reducing Poverty*, vol. 9: *Disease Control Priorities*, 3rd ed. Washington, D.C.: World Bank.

Jamison, D. T., et al. (eds) (2006). *Global Burden of Disease and Risk Factors*. World Bank and Oxford University Press.

Jamison, D. T., Mosley, W. H., Measham, A. R., and Bobadilla, J. L. (eds) (1993). *Disease Control Priorities in Developing Countries*. Oxford University Press.

Jenkin, J. G. (2011). 'Atomic Energy is "Moonshine": What Did Rutherford Really Mean?' *Physics in Perspective*, 13(2), 128–45.

Jia, Y., et al. (2018). 'Transfer Learning from Speaker Verification to Multispeaker Text-to-Speech Synthesis'. *Advances in Neural Information Processing Systems*, 4,480–90.

Jinek, M., et al. (2012). 'A Programmable Dual-RNA-Guided DNA Endonuclease in Adaptive Bacterial Immunity'. *Science*, 337(6096), 816–21.

Jonas, H. (1984 [1979]). *The Imperative of Responsibility*. University of Chicago Press.

Jones, K. E., et al. (2008). 'Global Trends in Emerging Infectious Diseases'. *Nature*, 451(7181), 990–3.

Jones, N., O'Brien, M., and Ryan, T. (2018). 'Representation of Future Generations in United Kingdom Policy-Making'. *Futures*, 102, 153–63.

JPL (2019a). Discovery Statistics – Cumulative Totals. https://cneos.jpl.nasa.gov/stats/totals.html

—(2019b). Small-Body Database. https://ssd.jpl.nasa.gov/sbdb.cgi

Kaempffert, W. (12 September 1933). 'Rutherford Cools Atom Energy Hope'. *The New York Times*.

Kahneman, D. (2011). *Thinking, Fast and Slow*. Macmillan.

Kaplan, J. O., Pfeiffer, M., Kolen, J. C. A., and Davis, B. A. S. (2016). 'Large-Scale Anthropogenic Reduction of Forest Cover in Last Glacial Maximum Europe'. *PLOS ONE*, 11(11), e0166726.

Karras, T., Aila, T., Laine, S., and Lehtinen, J. (2017). Progressive Growing of GANs for Improved Quality, Stability, and Variation. ArXiv, http://arxiv.org/abs/1710.10196.

Keele, B. F. (2006). 'Chimpanzee Reservoirs of Pandemic and Nonpandemic HIV-1'. *Science*, 313(5786), 523–6.

Keeter, B. (2017). NASA Office to Coordinate Asteroid Detection, Hazard Mitigation. https://www.nasa.gov/feature/nasa-office-to-coordinate-asteroid-detection-hazard-mitigation

Kellogg, E. A. (2013). 'C4 photosynthesis'. *Current Biology*, 23(14), R594–9.

Kelly, J. (2006). *The Great Mortality: An Intimate History of the Black Death, the Most Devastating Plague of All Time*. HarperCollins.

Kennedy, J. F. (1961). JFK Address at U.N. General Assembly, 25 September 1961. JFK Library Foundation.

—(1962). 'Message from the President John F. Kennedy to the Bulletin of the Atomic Scientists'. *Bulletin of the Atomic Scientists*, 18(10), 2.

—(1963). American University Address, 10 June 1963. Washington, D.C. John F. Kennedy Library.

King, D. et al. (2015). *Climate Change: A Risk Assessment*. Centre for Science and Policy.

Knight, F. H. (1921). *Risk, Uncertainty and Profit*. Houghton Mifflin.

Koch, A., Brierley, C., Maslin, M. M., and Lewis, S. L. (2019). 'Earth System Impacts of the European Arrival and Great Dying in the Americas after 1492'. *Quaternary Science Reviews*, 207, 13–36.

Kocić, J., Jovičić, N., and Drndarević, V. (2019). 'An End-to-End Deep Neural Network for Autonomous Driving Designed for Embedded Automotive Platforms'. *Sensors*, 19(9), 2,064.

Kolbert, E. (2014). *The Sixth Extinction: An Unnatural History*. Henry Holt and Company.

Kondratyev, K. Y., Krapivin, V. F., and Varotsos, C. A. (2003). *Global Carbon Cycle and Climate Change*. Springer.

Konopinski, E. J., Marvin, C., and Teller, E. (1946). Ignition of the Atmosphere with Nuclear Bombs [Report LA-602]. Los Alamos National Laboratory.

Krasovsky, V., and Shklovsky, I. (1957). 'Supernova Explosions and their Possible Effect on the Evolution of Life on the Earth'. *Proceedings of the USSR Academy of Sciences*, 116, 197–9.

Kristensen, H. M., and Korda, M. (2018). 'Indian Nuclear Forces, 2018'. *Bulletin of the Atomic Scientists*, 74(6), 361–6.

—(2019a). 'Chinese Nuclear Forces, 2019'. *Bulletin of the Atomic Scientists*, 75(4), 171–8.

—(2019b). 'French Nuclear Forces, 2019'. *Bulletin of the Atomic Scientists*, 75(1), 51–5.

—(2019c). 'Russian Nuclear Forces, 2019'. *Bulletin of the Atomic Scientists*, 75(2), 73–84.

—(2019d). Status of World Nuclear Forces. https://fas.org/issues/nuclear-weapons/status-world-nuclear-forces/

—(2019e). 'United States Nuclear Forces, 2019'. *Bulletin of the Atomic Scientists*, 75(3), 122–34.

Kristensen, H. M., and Norris, R. S. (2018). 'North Korean Nuclear Capabilities, 2018'. *Bulletin of the Atomic Scientists*, 74(1), 41–51.

Kristensen, H. M., Norris, R. S., and Diamond, J. (2018). 'Pakistani Nuclear Forces, 2018.' *Bulletin of the Atomic Scientists*, 74(5), 348–58.

Kühn, U., and Péczeli, A. (2017). 'Russia, NATO, and the INF Treaty'. *Strategic Studies Quarterly*, 11(1), 66–99.

Labelle, F. (2017). Elo Win Probability Calculator. https://wismuth.com/elo/calculator.html#system=goratings

Le Quéré, C., et al. (2018). 'Global Carbon Budget 2018'. *Earth System Science Data*, 10(4), 2,141–94.

Lebedev, A. (21 May 2004). 'The Man Who Saved the World Finally Recognized'. MosNews.

Leconte, J., Forget, F., Charnay, B., Wordsworth, R., and Pottier, A. (2013). 'Increased Insolation Threshold for Runaway Greenhouse Processes on Earth-Like Planets'. *Nature*, 504(7479), 268–71.

Lederberg, J. (1969). 'Biological Warfare and the Extinction of Man'. Stanford M.D., 8(4), 15–17.

Lee, C. (2009). 'Who Were the Mongols (1100–1400 CE)? An Examination of their Population History', in J. Bemmann, H. Parzinger, E. Pohl, and D. Tseveendorzh (eds), *Current Archaeological Research in Mongolia* (pp. 579–92). Rheinische Friedrich-Wilhelms-Universität Bonn.

Legg, S., and Kruel, A. (2011). Q & A with Shane Legg on Risks from AI. https://www.lesswrong.com/posts/No5JpRCHzBrWA4jmS/q-and-a-with-shane-legg-on-risks-from-ai

Leitenberg, M. (2001). 'Biological Weapons in the Twentieth Century: A Review and Analysis'. *Critical Reviews in Microbiology*, 27(4), 267–320.

Lepore, J. (30 January 2017). 'The Atomic Origins of Climate Science'. *The New Yorker*.

Leroy, E. M., et al. (2005). 'Fruit Bats as Reservoirs of Ebola Virus'. *Nature*, 438(7068), 575–6.

Leslie, J. (1996). *The End of the World: The Science and Ethics of Human Extinction.* Routledge.

Lewis, C. S. (1943). *The Abolition of Man.* Oxford University Press.

Lewis, G. (2018). Horsepox Synthesis: A Case of the Unilateralist's Curse? https://thebulletin.org/2018/02/horsepox-synthesis-a-case-of-the-unilateralists-curse/

Lewis, G., Millett, P., Sandberg, A., Snyder-Beattie, A., and Gronvall, G. (2019), 'Information Hazards in Biotechnology'. *Risk Analysis*, 39(5), 975–81.

Lightbown, S. (2017). VC Investment in Biotech Blasts through $10B Barrier in 2017. https://pitchbook.com/news/articles/vc-investment-in-biotech-blasts-through-10b-barrier-in-2017

Lindsey, R. (2018). Climate Change: Atmospheric Carbon Dioxide. https://www.climate.gov/news-features/understanding-climate/climate-change-atmospheric-carbon-dioxide

Lingam, M. (2019). 'Revisiting the Biological Ramifications of Variations in Earth's Magnetic Field'. *The Astrophysical Journal*, 874(2), L28.

Liu, M.-Y., and Tuzel, O. (2016). Coupled Generative Adversarial Networks. ArXiv, https://arxiv.org/pdf/1606.07536.

Livi-Bacci, M. (2017). *A Concise History of World Population* (6th ed.). John Wiley and Sons.

Longrich, N. R., Scriberas, J., and Wills, M. A. (2016). 'Severe Extinction and Rapid Recovery of Mammals across the Cretaceous-Palaeogene Boundary, and the Effects of Rarity on Patterns of Extinction and Recovery'. *Journal of Evolutionary Biology*, 29(8), 1,495–512.

Lordkipanidze, D., et al. (2006). 'A Fourth Hominin Skull from Dmanisi, Georgia'. *The Anatomical Record* Part A: Discoveries in Molecular, Cellular, and Evolutionary Biology, 288A(11), 1146–57.

Lovelock, J. (2019). *Novacene: The Coming Age of Hyperintelligence.* Penguin.

LSA (2014). Global sea level time series. Laboratory for Satellite Altimetry, NOAA/NESDIS/STAR.

Ma, W., Kahn, R. E., and Richt, J. A. (2008). 'The Pig as a Mixing Vessel for Influenza Viruses: Human and Veterinary Implications'. *Journal of Molecular and Genetic Medicine: An International Journal of Biomedical Research*, 3(1), 158–66.

MacAskill, W., Bykvist, K., and Ord, T. (n.d.). *Moral Uncertainty* (in press). Oxford University Press.

MacAskill, W. (2014). 'Normative Uncertainty' [PhD Thesis]. Faculty of Philosophy, University of Oxford.

—(2015). *Doing Good Better: Effective Altruism and a Radical New Way to Make a Difference*. Guardian Faber Publishing.

MacAskill, W., and Ord, T. (2018). 'Why Maximize Expected Choice-Worthiness?' *Noûs*, 1–27.

Macaulay, T. B. (1900). *The Complete Works of Thomas Babington Macaulay*, vol. 6. Houghton Mifflin.

Maddison, A. (2010). Historical Statistics of the World Economy: 1–2008 AD. https://datasource.kapsarc.org/explore/dataset/historical-statistics-of-the-world-economy-1-2008-ad/

Mainzer, A., et al. (2011). 'NEOWISE Observations of Near-earth Objects: Preliminary Results'. *The Astrophysical Journal*, 743(2), 156.

Mangus, S., and Larsen, W. (2004). 'Lunar Receiving Laboratory Project History' [Report S-924]. NASA.

Mann, C. C. (2018, January). 'The Book that Incited a Worldwide Fear of Overpopulation'. *Smithsonian Magazine*.

Marin, F., and Beluffi, C. (2018). Computing the Minimal Crew for a Multi-Generational Space Travel towards Proxima Centauri b. ArXiv, http://arxiv.org/abs/1806.03856.

Mason, B., Pyle, D., and Oppenheimer, C. (2004). 'The Size and Frequency of the Largest Explosive Eruptions on Earth'. *Bulletin of Volcanology*, 66(8), 735–48.

Masson-Delmotte, V., et al. (2013). 'Information from Paleoclimate Archives', in T. F. Stocker, et al. (eds), *Climate Change 2013: The Physical Science Basis. Contribution of Working Group I to the Fifth Assessment Report of the Intergovernmental Panel on Climate Change* (pp. 383–464). Cambridge University Press.

Mastrandrea, M., et al. (2010). 'Guidance Note for Lead Authors of the IPCC Fifth Assessment Report on Consistent Treatment of Uncertainties'. IPCC.

May, R. M. (1997). 'The Dimensions of Life on Earth', in P. H. Raven (ed.), *Nature and Human Society: The Quest for a Sustainable World*. National Academies Press.

McCarthy, J., Minsky, M. L., Rochester, N., and Shannon, C. E. (1955). 'A Proposal for the Dartmouth Summer Research Project on Artificial Intelligence'. Unpublished.

McDonald's Corporation (2018). Form 10-K, 'Annual Report Pursuant to Section 13 or 15(D) of the Securities Exchange Act of 1934 for the Fiscal Year ended December 31, 2017' (McDonald's Corporation 2017 Annual Report). McDonald's Corporation.

McEvedy, C., and Jones, R. (1978). *Atlas of World Population History*. Penguin.

McGuire, B. (1965). 'Eve of Destruction' [Lyrics]. Dunhill.

McInerney, F. A., and Wing, S. L. (2011). 'The Paleocene-Eocene Thermal Maximum: A Perturbation of Carbon Cycle, Climate, and Biosphere with Implications for the Future'. *Annual Review of Earth and Planetary Sciences*, 39(1), 489–516.

McKinnon, C. (2017). 'Endangering Humanity: An International Crime?' *Canadian Journal of Philosophy*, 47(2–3), 395–415.

McNamara, R. S. (14 October 1992). 'One Minute to Doomsday'. *The New York Times*.

Mecklin, J. (2018). 'It is 5 Minutes to Midnight'. *Bulletin of the Atomic Scientists*, 63(1), 66–71.

Medwin, T. (1824). *Conversations of Lord Byron*. H. Colburn.

Mellor, D. H. (1995). 'Cambridge Philosophers I: F. P. Ramsey'. *Philosophy*, 70(272), 243–62.

Melott, A. L., et al. (2004). 'Did a Gamma-Ray Burst Initiate the Late Ordovician Mass Extinction?' *International Journal of Astrobiology*, 3(1), 55–61.

Melott, A. L., and Thomas, B. C. (2011). 'Astrophysical Ionizing Radiation and Earth: A Brief Review and Census of Intermittent Intense Sources'. *Astrobiology*, 11(4), 343–61.

Menzel, P. T. (2011). 'Should the Value of Future Health Benefits Be Time-Discounted?' in *Prevention vs. Treatment: What's the Right Balance?* (pp. 246–73). Oxford University Press.

Metz, C. (9 June 2018). 'Mark Zuckerberg, Elon Musk and the Feud Over Killer Robots'. *The New York Times*.

Milanovic, B. (2016). *Global Inequality: A New Approach for the Age of Globalization*. Harvard University Press.

Minsky, M. (1984). Afterword, in *True Names*. Bluejay Books.

Mnih, V., et al. (2015). 'Human-Level Control through Deep Reinforcement Learning'. *Nature*, 518(7540), 529–33.

Montgomery, P. (13 June 1982). 'Throngs Fill Manhattan to Protest Nuclear Weapons'. *The New York Times*.

Moore, G. E. (1903). *Principia Ethica*. Cambridge University Press.

Moravec, H. (1988). *Mind Children: The Future of Robot and Human Intelligence*. Harvard University Press.

Morris, E. (2003). *The Fog of War.* Sony.

Muehlhauser, L. (2017). How Big a Deal Was the Industrial Revolution? http://lukemuehlhauser.com/industrial-revolution/

Mummert, A., Esche, E., Robinson, J., and Armelagos, G. J. (2011). 'Stature and Robusticity during the Agricultural Transition: Evidence from the Bioarchaeological Record'. *Economics and Human Biology*, 9(3), 284–301.

Musk, E. (2018). Q & A at South by Southwest 2018 Conference [Video]. https://youtu.be/kzlUyrccbos?t=2458

Mutze, G., Cooke, B., and Alexander, P. (1998). 'The Initial Impact of Rabbit Hemorrhagic Disease on European Rabbit Populations in South Australia'. *Journal of Wildlife Diseases*, 34(2), 221–7.

Naeye, R. (2008). A Stellar Explosion You Could See on Earth! https://www.nasa.gov/mission_pages/swift/bursts/brightest_grb.html

Narveson, J. (1973). 'Moral Problems of Population'. *Monist*, 57(1), 62–86.

NASA (2011). NASA Space Telescope Finds Fewer Asteroids Near Earth (NASA Content Administrator, ed.). https://www.nasa.gov/mission_pages/WISE/news/wise20110929.html%0A

National Research Council (2002). 'Appendix B: A History of the Lunar Receiving Laboratory', in *The Quarantine and Certification of Martian Samples.* National Academies Press.

—(2010). *Defending Planet Earth: Near-Earth-Object Surveys and Hazard Mitigation Strategies.* Washington, D.C.: National Academies Press.

Nesbit, M., and Illés, A. (2015). 'Establishing an EU "Guardian for Future Generations"' – Report and Recommendations for the World Future Council, Institute for European Environmental Policy.

Newton, I., and McGuire, J. E. (1970). 'Newton's "Principles of Philosophy": An Intended Preface for the 1704 "Opticks" and a Related Draft Fragment'. *The British Journal for the History of Science*, 5(2), 178–86.

Ng, Y.-K. (1989). 'What Should We Do About Future Generations?: Impossibility of Parfit's Theory X'. *Economics and Philosophy*, 5(2), 235–53.

—(2005), 'Intergenerational Impartiality: Replacing Discounting by Probability Weighting'. *Journal of Agricultural and Environmental Ethics*, 18(3), 237–57.

—(2016). 'The Importance of Global Extinction in Climate Change Policy'. *Global Policy*, 7(3), 315–22.

NHGRI (2018). Human Genome Project FAQ. https://www.genome.gov/human-genome-project/Completion-FAQ

NOAA (2019). Global Monthly Mean CO2. https://www.esrl.noaa.gov/gmd/ccgg/trends/

Norris, R. S., and Kristensen, H. M. (2012). 'The Cuban Missile Crisis: A Nuclear Order of Battle, October and November 1962'. *Bulletin of the Atomic Scientists*, 68(6), 85–91.

Nunn, N., and Qian, N. (2010). 'The Columbian Exchange: A History of Disease, Food, and Ideas'. *Journal of Economic Perspectives*, 24(2), 163–88.

O'Toole, G. (2013). If the Bee Disappeared Off the Face of the Earth, Man Would Only Have Four Years Left to Live. https://quoteinvestigator.com/2013/08/27/einstein-bees/

Obama, B. (2016). Remarks by President Obama and Prime Minister Abe of Japan at Hiroshima Peace Memorial. Obama White House.

Office for Technology Assessment (1979). *The Effects of Nuclear War.*

Ogburn, W. F. (1937). *Technological Trends and National Policy, including the Social Implications of New Inventions.* HathiTrust.

Oman, L., and Shulman, C. (2012). Nuclear Winter and Human Extinction: Q & A with Luke Oman. http://www.overcomingbias.com/2012/11/nuclear-winter-and-human-extinction-qa-with-luke-oman.html

Omohundro, S. M. (2008). 'The Basic AI Drives'. *Proceedings of the 2008 Conference on Artificial General Intelligence*, 483–92. IOS Press.

OPCW (2017). 'The Structure of the OPCW' [Fact Sheet]. Organisation for the Prohibition of Chemical Weapons.

—(2018). Decision – Programme and Budget of the OPCW for 2019. https://www.opcw.org/sites/default/files/documents/2018/11/c23dec10%28e%29.pdf

Ord, T. (2013). *The Moral Imperative toward Cost-Effectiveness in Global Health.* The Center for Global Development.

—(2015). 'Moral Trade'. *Ethics*, 126(1), 118–38.

Ord, T., and Beckstead, N. (2014). Chapter 10, in M. Walport and C. Craig (eds), *Innovation: Managing Risk, Not Avoiding It.* The Government Chief Scientific Adviser's annual report.

Ord, T., Hillerbrand, R., and Sandberg, A. (2010). 'Probing the Improbable: Methodological Challenges for Risks with Low Probabilities and High Stakes'. *Journal of Risk Research*, 13(2), 191–205.

Orseau, L., and Armstrong, S. (2016). 'Safely Interruptible Agents'. *Proceedings of the Thirty-Second Conference on Uncertainty in Artificial Intelligence*, 557–66. AUAI Press.

Orwell, G. (1949). *Nineteen Eighty-Four*. Secker and Warburg.

—(2013). 'To Noel Willmett/18 May 1944', in P. Davison (ed.), *George Orwell: A Life in Letters*. Liveright Publishing.

OSTP (2016). 'Request for Information on the Future of Artificial Intelligence: Public Responses'. White House Office of Science and Technology Policy.

Overbye, D. (12 April 2016). 'Reaching for the Stars, Across 4.37 Light-Years'. *The New York Times*.

Oxnard Press-Courier (12 March 1958). 'Accidents Stir Concern Here and in Britain'. *Oxnard Press-Courier*.

Parfit, D. (1984). *Reasons and Persons*. Oxford University Press.

—(2017a). 'Future People, the Non-Identity Problem, and Person-Affecting Principles'. *Philosophy and Public Affairs*, 45(2), 118–57.

—(2017b). *On What Matters*, vol. 3. Oxford University Press.

Pauling, L. (1962). 'Linus Pauling Nobel Lecture: Science and Peace'. The Nobel Peace Prize 1962. Nobel Media.

Peabody Energy. (2018) '2018 Annual Report'. Peabody Energy Corp.

Pearl, J. (2000). *Causality: Models, Reasoning and Inference*. Cambridge University Press.

Pearson, G. (1999). *The UNSCOM Saga: Chemical and Biological Weapons Non-Proliferation*. St Martin's Press.

Philips, A. F. (1998). 20 Mishaps That Might Have Started Accidental Nuclear War. http://nuclearfiles.org/menu/key-issues/nuclear-weapons/issues/accidents/20-mishaps-maybe-caused-nuclear-war.htm

Phillips, P. J., et al. (2011). 'Distinguishing Identical Twins by Face Recognition'. *Face and Gesture*, 185–92. IEEE.

Phoenix, C., and Drexler, E. (2004). 'Safe Exponential Manufacturing'. *Nanotechnology*, 15(8), 869–72.

Pierrehumbert, R. T. (2013). 'Hot Climates, High Sensitivity'. *Proceedings of the National Academy of Sciences*, 110(35), 14,118–19.

Pigou, A. C. (1920). *The Economics of Welfare* (1st ed.). Macmillan and Co.

Pimm, S. L., Russell, G. J., Gittleman, J. L., and Brooks, T. M. (1995). 'The Future of Biodiversity'. *Science*, 269(5222), 347–50.

Pinker, S. (2012). *The Better Angels of Our Nature: Why Violence has Declined*. Penguin.

—(2018). *Enlightenment Now: The Case for Reason, Science, Humanism, and Progress*. Penguin.

Piran, T., and Jimenez, R. (2014). 'Possible Role of Gamma Ray Bursts on Life Extinction in the Universe'. *Physical Review Letters*, 113(23), 231102-1–231102-6.

Pope, K. O., Baines, K. H., Ocampo, A. C., and Ivanov, B. A. (1997). 'Energy, Volatile Production, and Climatic Effects of the Chicxulub Cretaceous/Tertiary Impact'. *Journal of Geophysical Research: Planets*, 102(E9), 21,645–64.

Popp, M., Schmidt, H., and Marotzke, J. (2016). 'Transition to a Moist Greenhouse with CO2 and Solar Forcing'. *Nature Communications*, 7(1), 10,627.

Putin, V. (2018). Presidential Address to the Federal Assembly. http://en.kremlin.ru/events/president/news/56957

Quigley, J., and Revie, M. (2011). 'Estimating the Probability of Rare Events: Addressing Zero Failure Data'. *Risk Analysis*, 31(7), 1,120–32.

Radford, A., Metz, L., and Chintala, S. (2015). Unsupervised Representation Learning with Deep Convolutional Generative Adversarial Networks. ArXiv, https://arxiv.org/pdf/1511.06434.

Raible, C. C., et al. (2016). 'Tambora 1815 as a Test Case for High Impact Volcanic Eruptions: Earth System Effects'. *Wiley Interdisciplinary Reviews: Climate Change*, 7(4), 569–89.

Rampino, M. R., and Self, S. (1992). 'Volcanic Winter and Accelerated Glaciation Following the Toba Super-Eruption'. *Nature*, 359(6390), 50–2.

Ramsey, F. P. (1928). 'A Mathematical Theory of Saving'. *The Economic Journal*, 38(152), 543.

RAND (n.d.). RAND Database of Worldwide Terrorism Incidents. https://www.rand.org/nsrd/projects/terrorism-incidents.html

Ranjan, R., et al. (2018). 'Deep Learning for Understanding Faces: Machines May Be Just as Good, or Better, than Humans'. *IEEE Signal Processing Magazine*, 35(1), 66–83.

Rawls, J. (1971). *A Theory of Justice*. Belknap.

Reagan, R., and Weinraub, B. (12 February 1985). 'Transcript of Interview with President Reagan on a Range of Issues'. *The New York Times.*

Rees, M. (2003). *Our Final Century.* Random House.

Reisner, J., et al. (2018). 'Climate Impact of a Regional Nuclear Weapons Exchange: An Improved Assessment Based on Detailed Source Calculations'. *Journal of Geophysical Research: Atmospheres,* 123(5), 2,752–72.

Rhodes, C. J., and Anderson, R. M. (1996). 'Power Laws Governing Epidemics in Isolated Populations'. *Nature,* 381(6583), 600–2.

Rhodes, R. (1986). *The Making of the Atomic Bomb.* Simon and Schuster.

—(1995). *Dark Sun: The Making of the Hydrogen Bomb.* Simon and Schuster.

Riess, A. G., et al. (1998). 'Observational Evidence from Supernovae for an Accelerating Universe and a Cosmological Constant'. *The Astronomical Journal,* 116(3), 1,009–38.

Risø (1970). *Project Crested Ice. A Joint Danish-American Report on the Crash near Thule Air Base on 21 January 1968 of a B-52 Bomber Carrying Nuclear Weapons* (Report No. 213). Forskningscenter Risø, Atomenergikommissionen.

Ritchie, H., and Roser, M. (2019). CO_2 and Other Greenhouse Gas Emissions. Our World in Data. https://ourworldindata.org/co2-and-other-greenhouse-gas-emissions.

Roberts, Paul. (2004). *The End of Oil.* Bloomsbury.

Roberts, Priscilla. (2012). *Cuban Missile Crisis: The Essential Reference Guide.* Abc-clio.

Robock, A., et al. (2009). 'Did the Toba Volcanic Eruption of ~74 ka B.P. Produce Widespread Glaciation?' *Journal of Geophysical Research: Atmospheres,* 114(D10).

Robock, A., Oman, L., and Stenchikov, G. L. (2007). 'Nuclear Winter Revisited with a Modern Climate Model and Current Nuclear Arsenals: Still Catastrophic Consequences'. *Journal of Geophysical Research: Atmospheres,* 112(D13).

Robock, A., et al. (2007). 'Climatic Consequences of Regional Nuclear Conflicts'. *Atmospheric Chemistry and Physics,* 7(8), 2,003–12.

Rogelj, J., et al. (2016). 'Differences between Carbon Budget Estimates Unravelled'. *Nature Climate Change,* 6(3), 245–52.

Roser, M. (2015). The Short History of Global Living Conditions and Why it Matters that we Know it. https://ourworldindata. org/a-history-of-global-living-conditions-in-5-charts

Roser, M., and Ortiz-Ospina, E. (2019a). Global Extreme Poverty: Our World in Data. https://ourworldindata.org/extreme-poverty.

Roser, M., and Ortiz-Ospina, E. (2019b). Literacy: Our World in Data. https://ourworldindata.org/literacy.

—(2019). World Population Growth: Our World in Data. https:// ourworldindata.org/world-population-growth.

Rougier, J., Sparks, R. S. J., Cashman, K. V., and Brown, S. K. (2018). 'The Global Magnitude–Frequency Relationship for Large Explosive Volcanic Eruptions'. *Earth and Planetary Science Letters*, 482, 621–9.

Rowe, T., and Beard, S. (2018). 'Probabilities, Methodologies and the Evidence Base in Existential Risk Assessments' [working paper].

Rushby, A. J., et al. (2018). 'Long-Term Planetary Habitability and the Carbonate-Silicate Cycle'. *Astrobiology*, 18(5), 469–80.

Russell, B. (18 August 1945). 'The Bomb and Civilisation'. *Forward*.

—(March 1951). 'The Future of Man'. *The Atlantic*.

—(2002). '1955 address to the world's press assembled in London: The Russell-Einstein Manifesto', in K. Coates, J. Rotblat, and N. Chomsky (eds), *The Russell-Einstein Manifesto: Fifty Years On* (Albert Einstein, Bertrand Russell, Manifesto 50). Spokesman Books.

—(2009). *Autobiography*. Taylor and Francis.

—(2012). 'Letter to Einstein, 11 February 1955', in K. Coates, J. Rotblat, and N. Chomsky (eds), *The Russell-Einstein Manifesto: Fifty Years On* (Albert Einstein, Bertrand Russell, Manifesto 50) (pp. 29–30). Spokesman Books.

Russell, S. (2014). Of Myths And Moonshine. https://www.edge.org/ conversation/the-myth-of-ai#26015

—(2015). Will They Make Us Better People? https://www.edge.org/ response-detail/26157

—(2019). *Human Compatible: AI and the Problem of Control*. Allen Lane.

Russell, S., Dewey, D., and Tegmark, M. (2015). 'Research Priorities for Robust and Beneficial Artificial Intelligence'. *AI Magazine*, 36(4).

SAC (1969). 'Project CRESTED ICE: The Thule Nuclear Accident (U)', SAC Historical Study #113. Strategic Air Command.

Sagan, C. (1980). *Cosmos* (1st ed.). Random House.

—(1983). 'Nuclear War and Climatic Catastrophe: Some Policy Implications'. *Foreign Affairs*, 62(2).

—(1994). *Pale Blue Dot: A Vision of the Human Future in Space.* Random House.

Sagan, C., et al. (14 December 1980). *Cosmos: A Personal Voyage - Episode 12: Encyclopaedia Galactica* [TV Series]. PBS.

Sagan, C., and Ostro, S. J. (1994). 'Dangers of Asteroid Deflection'. *Nature*, 368(6471), 501.

Sandberg, A. (n.d.). Dyson Sphere FAQ. https://www.aleph.se/Nada/dysonFAQ.html

Sandberg, A., Drexler, E., and Ord, T. (2018). 'Dissolving the Fermi Paradox'. *ArXiv*, http://arxiv.org/abs/1806.02404.

Schaefer, K., et al. (2014). 'The Impact of the Permafrost Carbon Feedback on Global Climate'. *Environmental Research Letters*, 9(8), 085003, pp. 1–9.

Scheffler, S. (2009). 'Immigration and the Significance of Culture', in N. Holtug, K. Lippert-Rasmussen, and S. Lægaard (eds), *Nationalism and Multiculturalism in a World of Immigration* (pp. 119–50). Palgrave Macmillan UK.

—(2018). *Why Worry About Future Generations?* in Uehiro Series in Practical Ethics. Oxford University Press.

Schell, J. (1982). *The Fate of the Earth*. Avon.

—(14 June 2007). 'The Spirit of June 12'. *The Nation*.

Schindewolf, O. H. (1954). 'Über die möglichen Ursachen der grossen erdgeschichtlichen Faunenschnitte'. *Neues Jahrbuch für Geologie und Paläontologie, Monatshefte*, 1954, 457–65.

Schirrmeister, B. E., Antonelli, A., and Bagheri, H. C. (2011). 'The Origin of Multicellularity in Cyanobacteria'. *BMC Evolutionary Biology*, 11(1), 45.

Schlosser, E. (2013). *Command and Control: Nuclear Weapons, the Damascus Accident, and the Illusion of Safety*. Penguin.

Schneider, E., and Sachde, D. (2013). 'The Cost of Recovering Uranium from Seawater by a Braided Polymer Adsorbent System'. *Science and Global Security*, 21(2), 134–63.

Schneider von Deimling, T., Ganopolski, A., Held, H., and Rahmstorf, S. (2006). 'How Cold Was the Last Glacial Maximum?' *Geophysical Research Letters*, 33(14).

Schröder, K.-P., and Connon Smith, R. (2008). 'Distant Future of the Sun and Earth Revisited'. *Monthly Notices of the Royal Astronomical Society*, 386(1), 155–63.

Schulte, P., et al. (2010). 'The Chicxulub Asteroid Impact and Mass Extinction at the Cretaceous-Paleogene Boundary'. *Science*, 327(5970), 1,214–8.

Seneca, L. A. (1972). *Natural Questions*, vol. II (trans. T. H. Corcoran). Harvard University Press.

Serber, R. (1992). *The Los Alamos Primer: The First Lectures on How to Build an Atomic Bomb*. University of California Press.

Shapira, P., and Kwon, S. (2018). 'Synthetic Biology Research and Innovation Profile 2018: Publications and Patents'. *BioRxiv*, 485805.

Shelley, M. W. (1826). *The Last Man* (1st ed.). Henry Colburn.

—(2009). 'Introduction', in S. Curran (ed.), *Frankenstein* (vol. 1). University of Colorado, Boulder (original work published in 1831).

Sherwood, S. C., and Huber, M. (2010). 'An Adaptability Limit to Climate Change due to Heat Stress'. *Proceedings of the National Academy of Sciences*, 107(21), 9,552–5.

Shoham, D., and Wolfson, Z. (2004). 'The Russian Biological Weapons Program: Vanished or Disappeared?' *Critical Reviews in Microbiology*, 30(4), 241–61.

Shoham, Y., et al. (2018). 'The AI Index 2018 Annual Report'. AI Index Steering Committee, Human-Centered AI Initiative.

Shooter, R. A., et al. (1980). 'Report of the Investigation into the Cause of the 1978 Birmingham Smallpox Occurrence'. Her Majesty's Stationery Office.

Shu, D.-G., et al. (1999). 'Lower Cambrian Vertebrates from South China'. *Nature*, 402(6757), 42–6.

Siddiqi, A. A. (2010). *The Red Rockets' Glare: Spaceflight and the Russian Imagination, 1857–1957*. Cambridge University Press.

Sidgwick, H. (1907). Book III, Chapter IX, in *The Methods of Ethics* (2nd ed., pp. 327–31). Macmillan (original work published 1874).

Silver, D., et al. (2018). 'A General Reinforcement Learning Algorithm that Masters Chess, Shogi, and Go through Self-Play'. *Science*, 362(6419), 1,140 LP – 1,144.

Sims, L. D., et al. (2005). 'Origin and Evolution of Highly Pathogenic H5N1 Avian Influenza in Asia'. *Veterinary Record*, 157(6), 159–64.

Sivin, N. (1982). 'Why the Scientific Revolution Did Not Take Place in China – or Didn't It?' *Chinese Science*, 5, 45–66.

Slovic, P. (2007). '"If I look at the mass I will never act": Psychic Numbing and Genocide', in *Judgment and Decision Making* (vol. 2).

Smart, J. J. C. (1973). 'An Outline of a System of Utilitarian Ethics', in J. J. C. Smart and B. Williams (eds), *Utilitarianism: For and Against* (pp. 1–74). Cambridge University Press.

—(1984). *Ethics, Persuasion, and Truth*. Routledge and Kegan Paul.

Smith, C. M. (2014). 'Estimation of a Genetically Viable Population for Multigenerational Interstellar Voyaging: Review and Data for Project Hyperion'. *Acta Astronautica*, 97, 16–29.

Snow, D. R., and Lanphear, K. M. (1988). 'European Contact and Indian Depopulation in the Northeast: The Timing of the First Epidemics'. *Ethnohistory*, 35(1), 15.

Snyder-Beattie, A. E., Ord, T., and Bonsall, M. B. (2019). 'An Upper Bound for the Background Rate of Human Extinction'. *Scientific Reports*, 9(1), 11,054.

Sorenson, T. C. (1965). *Kennedy*. Harper and Row.

Sosin, D. M. (2015). 'Review of Department of Defense Anthrax Shipments'. House Energy and Commerce Subcommittee on Oversight and Investigations.

Speer, A. (1970), *Inside the Third Reich*. Simon and Schuster.

Spratt, B. G. (2007), 'Independent Review of the Safety of UK Facilities Handling Foot-and-Mouth Disease Virus'. UK Department for Environment, Food and Rural Affairs Archives.

Stathakopoulos, D. C. (2004). *Famine and Pestilence in the Late Roman and Early Byzantine Empire* (1st ed.). Routledge.

—(2008). 'Population, Demography, and Disease', in R. Cormack, J. F. Haldon, and E. Jeffreys (eds), *The Oxford Handbook of Byzantine Studies*. Oxford University Press.

Stern, D. I. (2004). 'The Rise and Fall of the Environmental Kuznets Curve'. *World Development*, 32(8), 1,419–39.

Stern, N. H. (2006). *The Economics of Climate Change: The Stern Review*. Cambridge University Press.

Stevens, B., and Bony, S. (2013). 'What Are Climate Models Missing?' *Science*, 340(6136), 1,053–4.

Stokes, G. H., et al. (2017). 'Update to Determine the Feasibility of Enhancing the Search and Characterization of NEOs'. Near-Earth Object Science Definition Team.

Strogatz, S. (26 December 2018). 'One Giant Step for a Chess-Playing Machine'. *The New York Times*.

Subramanian, M. (2019). 'Anthropocene Now: Influential Panel Votes to Recognize Earth's New Epoch'. *Nature*.

Sutton, R. (2015). 'Creating Human-level AI: How and When?' [Slides]. Future of Life Institute.

Sverdrup, H. U., and Olafsdottir, A. H. (2019). 'Assessing the Long-Term Global Sustainability of the Production and Supply for Stainless Steel'. *BioPhysical Economics and Resource Quality*, 4(2), 8.

Szilard, G. W., and Winsor, K. R. (1968). 'Reminiscences', by Leo Szilard, in *Perspectives in American History* (vol. 2). Charles Warren Center for Studies in American History.

Szilard, L., and Feld, B. T. (1972). *The Collected Works of Leo Szilard: Scientific Papers*. MIT Press.

Taagholt, J., Hansen, J., and Lufkin, D. (2001). 'Greenland: Security Perspectives' (trans. D. Lufkin). Arctic Research Consortium of the United States.

Tai, A. P. K., Martin, M. V., and Heald, C. L. (2014). 'Threat to Future Global Food Security from Climate Change and Ozone Air Pollution'. *Nature Climate Change*, 4(9), 817–21.

Tarnocai, C., et al. (2009). 'Soil Organic Carbon Pools in the Northern Circumpolar Permafrost Region'. *Global Biogeochemical Cycles*, 23(2), 1–11.

Tate, J. (2017). Number of Undiscovered Near-Earth Asteroids Revised Downward. https://spaceguardcentre.com/number-of-undiscovered-near-earth-asteroids-revised-downward/

Taubenberger, J. K., and Morens, D. M. (2006). '1918 Influenza: The Mother of all Pandemics'. *Emerging Infectious Diseases*, 12(1), 15–22.

Tegmark, M. (2014). *Our Mathematical Universe: My Quest for the Ultimate Nature of Reality*. Knopf Doubleday Publishing Group.

Tegmark, Max, and Bostrom, N. (2005). 'Is a Doomsday Catastrophe Likely?' *Nature*, 438(7069), 754.

Tertrais, B. (2017). '"On The Brink" – Really? Revisiting Nuclear Close Calls Since 1945'. *The Washington Quarterly*, 40(2), 51–66.

The, L.-S., et al. (2006). 'Are 44Ti-producing Supernovae Exceptional?' *Astronomy and Astrophysics*, 450(3), 1,037–50.

The Wall Street Journal. (6 January 2017). 'Humans Mourn Loss After Google Is Unmasked As China's Go Master'. *The Wall Street Journal*.

Thomas, J. M. (2001). 'Predictions'. *IUBMB Life* (International Union of Biochemistry and Molecular Biology: Life), 51(3), 135–8.

Tokarska, K. B., et al. (2016). 'The Climate Response to Five Trillion Tonnes of Carbon'. *Nature Climate Change*, 6(9), 851–5.

Toon, O. B., et al. (2007). 'Atmospheric Effects and Societal Consequences of Regional Scale Nuclear Conflicts and Acts of Individual Nuclear Terrorism'. *Atmospheric Chemistry and Physics*, 7, 1,973–2,002.

Tovish, A. (2015). The Okinawa missiles of October. https://thebulletin.org/2015/10/the-okinawa-missiles-of-october/

Toynbee, A. (1963). 'Man and Hunger: The Perspectives of History', in FAO (ed.), *Report of the World Food Congress, Washington, D.C., 4 to 18 June 1963*, vol. 2: *Major Addresses and Speeches*. Her Majesty's Stationery Office.

Trevisanato, S. I. (2007). 'The "Hittite Plague", an Epidemic of Tularemia and the First Record of Biological Warfare'. *Medical Hypotheses*, 69(6), 1,371–4.

Tritten, T. J. (23 December 2015). 'Cold War Missileers Refute Okinawa Near-Launch'. *Stars and Stripes*.

Tsao, J., Lewis, N., and Crabtree, G. (2006). 'Solar FAQs' [Working Draft Version 2006 Apr 20]. Sandia National Laboratories.

Tucker, J. B. (1999). 'Historical Trends Related to Bioterrorism: An Empirical Analysis. Emerging Infectious Diseases', 5(4), 498–504.

—(2001). Biological Weapons Convention (BWC) Compliance Protocol. https://www.nti.org/analysis/articles/biological-weapons-convention-bwc/

Turing, A. (1951) 'Intelligent Machinery, A Heretical Theory'. Lecture given to '51 Society' in Manchester.

U.S. Department of Homeland Security (2008). 'Appendix B: A Review of Biocontainment Lapses and Laboratory-Acquired Infections', in *NBAF Final Environmental Impact Statement*. United States Department of Homeland Security.

U.S. Department of State (n.d.). Treaty Between The United States Of America And The Union of Soviet Socialist Republics on The Elimination of Their Intermediate-Range and Shorter-Range Missiles (INF Treaty). https://2009–2017.state.gov/t/avc/trty/102360.htm

—(1963). 'State-Defense Meeting on Group I, II and IV Papers [Extract]. Gr. 59', Department of State, PM Dep. Ass. Sec. Records, 1961–1963, Box 2, Memoranda. National Archives.

U.S. House of Representatives (2013). *Threats from Space: A Review of U.S. Government Efforts to Track and Mitigate Asteroids and Meteors (Part I and Part II)* [Hearing]. U.S. Goverment Printing Office.

UCS (n.d.). What is Hair-Trigger Alert? https://www.ucsusa.org/nuclear-weapons/hair-trigger-alert

UN (n.d.). International Day for the Preservation of the Ozone Layer, 16 September. https://www.un.org/en/events/ozoneday/background.shtml

UN DESA (2019). World Population Prospects 2019, Online Edition.

UNESCO (1997). Declaration on the Responsibilities of the Present Generations Towards Future Generations. http://portal.unesco.org/en/ev.php-URL_ID=13178andURL_DO=DO_TOPICandURL_SECTION=201.html

UNOOSA (2018). *Near-Earth Objects and Planetary Defence* (Brochure ST/SPACE/73). United Nations Office of Outer Space Affairs.

Van Valen, L. (1973). 'Body Size and Numbers of Plants and Animals'. *Evolution*, 27(1), 27–35.

Van Zanden, J. L., et al. (2014). 'Global Well-Being since 1820', in *How Was Life?* (pp. 23–36). OECD.

Vellekoop, J., et al. (2014). 'Rapid Short-Term Cooling Following the Chicxulub Impact at the Cretaceous-Paleogene Boundary'. *Proceedings of the National Academy of Sciences*, 111(21), 7,537–41.

Vonnegut, K. (1963). *Cat's Cradle*. Holt, Rinehart and Winston.

Wallace, M. D., Crissey, B. L., and Sennott, L. I. (1986). 'Accidental Nuclear War: A Risk Assessment'. *Journal of Peace Research*, 23(1), 9–27.

Watson, C., Watson, M., Gastfriend, D., and Sell, T. K. (2018). 'Federal Funding for Health Security in FY2019'. *Health Security*, 16(5), 281–303.

Watson, F. G. (1941). *Between the Planets*. The Blakiston Company.

Weaver, T. A., and Wood, L. (1979). 'Necessary Conditions for the Initiation and Propagation of Nuclear-Detonation Waves in Plane Atmospheres'. *Physical Review A*, 20(1), 316–28.

Weeks, J. R. (2015). 'History and Future of Population Growth', in *Population: An Introduction to Concepts and Issues* (12th ed.). Wadsworth Publishing.

Weitzman, M. L. (1998). 'Why the Far-Distant Future Should Be Discounted at its Lowest Possible Rate'. *Journal of Environmental Economics and Management*, 36(3), 201–8.

—(2009). 'On Modeling and Interpreting the Economics of Catastrophic Climate Change'. *Review of Economics and Statistics*, 91(1), 1–19.

Wellerstein, A. (2014). Szilard's Chain Reaction: Visionary or Crank? http://blog.nuclearsecrecy.com/2014/05/16/szilards-chain-reaction/

Wells, H. G. (1894). *The Extinction of Man*. Pall Mall Budget.

—(1897). *The Star*. The Graphic.

—(1913). *The Discovery of the Future*. B. W. Huebsch.

—(1940). *The New World Order*. Secker and Warburg.

Wetterstrand, K. A. (2019). DNA Sequencing Costs: Data from the NHGRI Genome Sequencing Program (GSP). www.genome.gov/sequencingcostsdata

Wever, P. C., and van Bergen, L. (2014). 'Death from 1918 Pandemic Influenza during the First World War: A Perspective from Personal and Anecdotal Evidence'. *Influenza and Other Respiratory Viruses*, 8(5), 538–46.

Wheelis, M. (2002). 'Biological Warfare at the 1346 Siege of Caffa'. *Emerging Infectious Diseases*, 8(9), 971–5.

White, S., Gowlett, J. A. J., and Grove, M. (2014). 'The Place of the Neanderthals in Hominin Phylogeny'. *Journal of Anthropological Archaeology*, 35, 32–50.

WHO (2016). *World Health Statistics 2016: Monitoring Health for the SDGs, Sustainable Development Goals*. World Health Organisation.

Wiblin, R. (2017). How to Compare Different Global Problems in Terms of Impact. https://80000hours.org/articles/problem-framework/

Wiener, J. B. (2016). 'The Tragedy of the Uncommons: On the Politics of Apocalypse'. *Global Policy*, 7, 67–80.

Wiener, N. (1960). 'Some Moral and Technical Consequences of Automation'. *Science*, 131(3410), 1,355–8.

Wilcox, B. H., Mitchell, K. L., Schwandner, F. M., and Lopes, R. M. (2017). 'Defending Human Civilization from Supervolcanic Eruptions'. Jet Propulsion Laboratory/NASA.

Wilcox, B., and Wilcox, S. (1996). *EZ-GO: Oriental Strategy in a Nutshell*. Ki Press.

Williams, E. G. (2015). 'The Possibility of an Ongoing Moral Catastrophe'. *Ethical Theory and Moral Practice*, 18(5), 971–82.

Williams, M. (2012). 'Did the 73 ka Toba Super-Eruption have an Enduring Effect? Insights from Genetics, Prehistoric Archaeology, Pollen Analysis, Stable Isotope Geochemistry, Geomorphology, Ice Cores, and Climate Models'. *Quaternary International*, 269, 87–93.

Willis, K. J., Bennett, K. D., Bhagwat, S. A., and Birks, H. J. B. (2010). '4 °C and Beyond: What Did This Mean for Biodiversity in the Past?' *Systematics and Biodiversity*, 8(1), 3–9.

Willis, K. J., and MacDonald, G. M. (2011). 'Long-Term Ecological Records and Their Relevance to Climate Change Predictions for a Warmer World'. *Annual Review of Ecology, Evolution, and Systematics*, 42(1), 267–87.

Wilman, R., and Newman, C. (eds) (2018). *Frontiers of Space Risk: Natural Cosmic Hazards and Societal Challenges.* Taylor and Francis.

Wilson, I. (2001). *Past Lives: Unlocking the Secrets of our Ancestors.* Cassell.

Wilson, J. M., Brediger, W., Albright, T. P., and Smith-Gagen, J. (2016). 'Reanalysis of the Anthrax Epidemic in Rhodesia, 1978–1984'. *PeerJ*, 4, e2686.

Wise, J. (9 January 2013). 'About That Overpopulation Problem'. *Slate*.

Wolf, E. T., and Toon, O. B. (2015). 'The Evolution of Habitable Climates under the Brightening Sun'. *Journal of Geophysical Research: Atmospheres*, 120(12), 5,775–94.

Woodward, B., Shurkin, J. N., and Gordon, D. L. (2009). *Scientists Greater Than Einstein: The Biggest Lifesavers of the Twentieth Century*. Quill Driver Books.

World Bank (1993). *World Bank Report: Investing in Health*. Oxford University Press.

—(2019a). GDP (current US$). https://data.worldbank.org/indicator/ny.gdp.mktp.cd

—(2019b). Government Expenditure on Education, Total (%of GDP). https://data.worldbank.org/indicator/SE.XPD.TOTL.GD.ZS

Wright, L. (16 September 2002). 'The Man Behind Bin Laden'. *The New Yorker*.

Wright, S. (2001). 'Legitimating Genetic Engineering'. *Perspectives in Biology and Medicine*, 44(2), 235–247.

Xia, L., Robock, A., Mills, M., Stenke, A., and Helfand, I. (2015). 'Decadal Reduction of Chinese Agriculture after a Regional Nuclear War'. *Earth's Future*, 3(2), 37–48.

Yokoyama, Y., Falguères, C., Sémah, F., Jacob, T., and Grün, R. (2008). 'Gamma-Ray Spectrometric Dating of Late Homo Erectus Skulls from Ngandong and Sambungmacan, Central Java, Indonesia'. *Journal of Human Evolution*, 55(2), 274–7.

Yost, C. L., Jackson, L. J., Stone, J. R., and Cohen, A. S. (2018). 'Subdecadal Phytolith and Charcoal Records from Lake Malawi, East Africa, Imply Minimal Effects on Human Evolution from the ~74 ka Toba Supereruption'. *Journal of Human Evolution*, 116, 75–94.

Zalasiewicz, J., et al. (2017). 'The Working Group on the Anthropocene: Summary of Evidence and Interim Recommendations'. *Anthropocene*, 19, 55–60.

Zelicoff, A. P., and Bellomo, M. (2005). *Microbe: Are We Ready for the Next Plague?* AMACOM.

Zhang, B., and Dafoe, A. (2019). *Artificial Intelligence: American Attitudes and Trends*. Center for the Governance of AI, Future of Humanity Institute, University of Oxford.

Zhang, T., Heginbottom, J. A., Barry, R. G., and Brown, J. (2000). 'Further Statistics on the Distribution of Permafrost and Ground Ice in the Northern Hemisphere'. *Polar Geography*, 24(2), 126–31.

Zhu, M., et al. (2012). 'Earliest Known Coelacanth Skull Extends the Range of Anatomically Modern Coelacanths to the Early Devonian'. *Nature Communications*, 3(1), 772.

Ziegler, P. (1969). *The Black Death*. Collins.

Zilinskas, R. A. (1983). 'Anthrax in Sverdlovsk?' *Bulletin of the Atomic Scientists*, 39(6), 24–7.

Zonination (2017). Perceptions of Probability and Numbers. https://github.com/zonination/perceptions

INDEX

NOTE ON THE AUTHOR

Toby Ord is a Senior Research Fellow in Philosophy at Oxford University. His work focuses on the big-picture questions facing humanity. What are the most important issues of our time? How can we best address them?

His earlier work explored the ethics of global poverty, leading him to make a lifelong pledge to donate 10% of his income to the most effective charities helping improve the world. He created a society, Giving What We Can, for people to join this mission, and together its members have pledged over £1 billion. He then broadened these ideas by co-founding the effective altruism movement in which thousands of people are using reason and evidence to help the lives of others as much as possible.

His current research is on risks that threaten human extinction or the permanent collapse of civilization, and on how to safeguard humanity through these dangers, which he considers to be among the most pressing and neglected issues we face. Toby has advised the World Health Organization, the World Bank, the World Economic Forum, the US National Intelligence Council, and the UK Prime Minister's Office.

The Precipice is a landmark book that provides a new way of thinking about our time.

We live during the most important era of human history. In the twentieth century, we developed the means to destroy ourselves – without developing the moral framework to ensure we won't. This is the Precipice, and how we respond to it will be the most crucial decision of our time.

Oxford moral philosopher Toby Ord explores the risks to humanity's future, from the familiar man-made threats of climate change and nuclear war, to the potentially greater, more unfamiliar threats from engineered pandemics and advanced artificial intelligence. With clear and rigorous thinking, Ord calculates the various risk levels, and shows how our own time fits within the larger story of human history. Can we protect the legacy of the hundred billion who have come before us, and secure a future for the trillions that could follow? What can we do, in our present moment, to face the risks head on?

A major work that brings together the disciplines of physics, biology, earth and computer science, history, anthropology, statistics, international relations, political science and moral philosophy, *The Precipice* is a call for a new understanding of our age: a major reorientation in the way we see the world, our history, and the role we play in it.